OUVRAGE à RENDRE LE :

AGRICULTURE ISSUES AND POLITICS

DESALINATION: METHODS, COSTS AND TECHNOLOGY

AGRICULTURE ISSUES AND POLITICS

Additional books in this series can be found on Nova's website under the Series tab.

Additional E-books in this series can be found on Nova's website under the E-books tab.

AGRICULTURE ISSUES AND POLITICS

DESALINATION: METHODS, COSTS AND TECHNOLOGY

IRENA A. URBONIENĖ
EDITOR

Nova
Nova Science Publishers, Inc.
New York

For permission to use material from this book please contact us:
Telephone 631-231-7269; Fax 631-231-8175
Web Site: http://www.novapublishers.com

LIBRARY OF CONGRESS CATALOGING-IN-PUBLICATION DATA
Desalination : methods, costs and technology / editor: Irena A. Urboniene.
p. cm.
Includes index.
ISBN 978-1-61668-909-4 (hardcover)
1. Saline water conversion plants. I. Urboniene, Irena A.
TD479.4.D47 2010
628.1'67--dc22

Published by Nova Science Publishers, Inc. + New York

CONTENTS

PREFACE

Desalination refers to any of several processes that remove excess salt and other minerals from water. Water is desalinated in order to convert salt water to fresh water so it is suitable for human consumption or irrigation. Most of the modern interest in desalination is focused on developing cost-effective ways of providing fresh water for human use in regions where the availability of fresh water is limited. This book reviews research in the field of desalination including the growth of scientific and commercial interest to cogenerative and hybrid technologies for desalination and power generation; Membrane Distillaion (MD) process as a new membrane technology used for desalination and recycling; two alternative systems for conventional reverse osmosis desalination plants by incorporating the use of gravitational potential energy and others.

Chapter 1 - Hybridization of membrane processes with conventional thermally-based co-generative technologies was considered in this study that provides an analysis of integrated and hybrid schemes including the following systems: (1) gas turbine and unfired heat recovery steam generator, (2) auxiliary boiler, (3) multistage flush desalination, (4) reverse osmosis, and (5) ASR technology. It contains a comparison of the integrated system with the reference scheme. A comparison of the alternatives is based on the present values (PV) of expenses over the economic life of capital and "levelized" cost of water and specific carbon emissions allocated to the unit of the product. The analysis is based on the following assumptions: (1) depreciation period is equal to period of bank credit; (2) zero year is the base year of the project; (3) discount factor is dependent upon nominal interest rate and retail price index; (4) cost of primary fuel ranges from 1 to 8 \$/GJ; (5) assumed carbon tax varies from 0 to 50 \$ per tonne of emitted carbon. Thermal desalination (MSF) being more capital and resource consuming is characterised by higher PV of expenses. It is more vulnerable to growth of the following factors: (1) nominal interest rate, (2) cost of primary fuel, and (3) rate of carbon tax. Hybridisation of thermally- and electrically-driven processes (MSF and RO) will have the following positive impact: (1) decrease specific capital and energy consumption; (2) decrease the level of carbon tax (due to decrease of allocated CO_2 emissions); and (3) decrease the "levelized" cost of water. In addition, it will decrease the sensitivity and, in turn, commercial risk of the system to fluctuation of the following factors: (1) nominal interest rate, (2) cost of primary fuel, and (3) rate of imposed carbon tax. It was concluded that the combination of conventional co-generative systems with RO and ASR allows utilizing seasonally unused electricity while the ASR accumulates excess desalted water. The hybrid including RO & ASR technology is characterized by the following technological, economic and strategic advantages: (1) increased range of variation of power-to-water ratio; (2) possibility to use seasonal surplus of unutilized power for production of additional quantity of water; (3) decreased specific CO_2 emissions; (4) improved capital utilization factor and other economic indicators; (5) creation of strategic water reserve for the region, etc. Proposed method can be used as submodel in algorithm for synthesis of optimum configuration of the system. It can be of interest in evaluation of alternatives for feasibility studies as well.

Chapter 2 - In this chapter, a steady state mathematical model for multistage flash (MSF) desalination plants is developed. The model is based on basic principles of physics and chemistry that describe the physical phenomenon occurring in the process. The input plant parameters that are known to affect the performance of the MSF desalination plant are taken into account in the construction of the model. These parameters include make-up flow, brine recycle flow, seawater flow, seawater temperature, seawater concentration, top brine temperature, steam temperature and the plant load. For each stage, the developed model predicts the temperatures of the brine, distillate and cooling brine, and the flow rates of brine outlet and distillate production. For the purpose of this study, one of the MSF plant is selected for investigation. The developed model is evaluated with the MSF plant vendor simulation results and its actual operating data. Model predictions matched well with the vendor simulation results and the plant operating data. Thus, the developed model is sufficiently accurate and model predictions will be reliable over complete operating plant range. The developed model is thus used to carry out the optimization of the MSF plant performance with an objective to maximize its performance ratio or maximize its distillate production rate. It was found that the main variables that affect the performance of the plant were the top brine temperature and the brine recycle flow rate.

This model can be used for determining the optimum controller's set points of a running MSF desalination plant at different loads to maximize the water production or minimize the energy consumption or any other required objective function. This model can be integrated with the plant control software, and the optimum set points for controllers can be evaluated off line whenever plant loads are changed or management requirements change.

Chapter 3 - In today's life, environmental pollution has far reaching negative consequences on the lives of humans. As is known, because of their stability, many pesticides after their application may be leached and thus contaminate surface and waste waters. The majority of herbicides are resistant to chemical and/or photochemical degradation under typical environmental conditions. Recent investigations in the area of chemical treatment of both waste waters and drinking water are directed toward improving efficiency of degradation of organic compounds dissolved or dispersed in water by applying catalytic and photochemical methods. Since their early applications in the 1970's, advanced oxidation processes (AOPs) have shown great potential in degrading organic pollutants. Heterogeneous TiO_2 photocatalysis, ozonation and photo-Fenton's reaction are the most intensively investigated technologies. The aim of this review is to address the fundamentals of one kind of AOP, namely, photocatalytic processes and their application for herbicide degradation in aqueous suspensions using TiO_2 as catalyst. Since AOPs rely on the generation and subsequent reaction of highly reactive oxygen radicals with herbicides, there are many factors that can affect the efficiency of this process. Hence, this review will attempt to summarize and highlight the effect of a variety of conditions on TiO_2-photocatalyzed removal of herbicides, such as the amount of catalyst, reaction pH, light intensity, concentration of herbicide, and the presence of some additives. The review will also summarize the pathways of herbicides degradation, including some of the intermediates that are generated during the degradation process.

Chapter 4 - The authors present in this chapter an overview and a state of the art on the Membrane Distillation (MD) process. Such a process is a new, rapidly increasing membrane technology, which can be used for desalination and also for recycling. It is a separation process that involves transport of vapor through porous hydrophobic membranes. The principle, the main applications and the benefits of this separation technique are outlined. Approaches to improve the process performance are also discussed.

A detailed theoretical model for the transport phenomena in an Air Gap Membrane Distillation used for desalination was developed. This model is based on the conservation equations for the mass, momentum, energy and species within the feed water solution as well as on the mass and energy balances on the membrane sides.

In order to enhance the MD process performance, a composite hydrophilic-hydrophobic membrane is proposed in an Air Gap Membrane Distillation (AGMD) unit. The configuration considered here is a vertical tubular one with two concentric tubes. Hot saline water flows inside the inner tube which has a porous hydrophilic-hydrophobic wall. The external tube maintained at a constant and low temperature is the condenser.

The results analyzed the effect of the operating parameters such as the velocity and the temperature of the hot feed water at the inlet of the unit, the properties of the membranes, the cooling plate temperature and air gap thickness on the production rate of the vapor and the efficiency of the process. These results show that this desalination process is very efficient. More than 90% of the total heat flux coming from the hot side of the unit (feed water side) is converted into latent heat for the generation of the permeate (water vapor). The process performance is very sensitive to several parameters in particular to the porosity of the hydrophobic membrane. An increase of this parameter induces a fast growth of the mass and temperature boundary layers of the feed solution. A significant improvement of the permeate flux and the thermal efficiency are obtained by the incorporation of the composite hydrophilic-hydrophobic membrane.

Chapter 5 - 'Necessity drives invention' that is the basis of Desalination Technology. In this chapter, principles of different membrane desalination techniques (viz. Electrodialysis, Reverse osmosis, membrane distillation) are discussed. The emphasis of the chapter is on reverse osmosis technology. The membrane-based technology is dealt with special attention on technology and economics; also membrane preparation methods and materials got brief attention. Present scenario of the technology is brightened by comparing other technologies with it. The different energy sources are also coupled with the technology. The economics of the technology is discussed. Different factors influencing the cost of production are discussed. RO cost prediction techniques are also sketched.

Chapter 6 - Supported liquid membranes have being recognized as an effective method for the selective separation of compounds in a wide range of processes. Nevertheless, their industrial application is still limited, mainly due to concerns about supported liquid membrane stability and long-term performance. The use of ionic liquids as a liquid membrane phase could overcome these inconveniences due to their negligible vapour pressure, their greater viscosity and the possibility of minimizing their solubility in the surrounding phases by adequate selection of the cation and anion. The possibility of design suitable ionic liquids for specific separation problems has also opened up new fields of application of supported ionic liquid membranes. In this review an overview is given of recent advances in supported membranes based on ionic liquids, including issues such as methods of preparation, fields of application, stability and process intensification using supported ionic liquid membranes.

Chapter 7 - This chapter proposes two alternative systems for conventional reverse osmosis (RO) desalination plants by incorporating the use of gravitational potential energy. In the proposals, this kind of energy is obtained by storing water to be desalinated (sea or brackish -water).

The primary source of energy of both systems is wind. According to the chosen device (windmill or wind generator) a percentage of the available wind energy is used to pump water to a reservoir located at a certain height (less than 20 m). The stored water has the potential energy that will make a special plunger pump work. The piston of this special plunger pump is designed so that high pressure (over 55 bar) can be achieved in a different way from the conventional one. Conventionally, RO plants use a high-pressure pump, powered by electricity or fossil fuel. The function of the pump is to send a flux of saline water to a group of semi-permeable membrane modules, capable of "filtering" the dissolved salts.

In the alternative systems, here proposed, to lift water to the reservoir, the first one uses multi-bladed windmills and the second one uses wind generators that power an electric pump. To pump water to the RO membranes, the special pistons go downwards due to their own weight, and are lifted, through a system of pulleys, with a counterweight filled with water obtained from the reservoir. The technical viability of the alternatives was theoretically proven by deductions based on physics and mathematics and with a prototype.

Case studies are presented and, along them, the two systems are compared based on some chosen performance parameters such as specific consumption of energy, occupied area, availability etc.

Chapter 8 - There are few reports in the open literature describing the vapor side corrosion (VSC) of condenser tubes in multistage flash desalination (MSF) plants. This chapter describes findings of a laboratory investigation on the vapor side corrosion of copper-nickel base alloys, which are extensively used in desalination plant. The materials under investigation were Cu-10%Ni-base alloy and Cu-30%Ni-base alloy. The experiments comprised study of the behavior of these two copper-nickel alloys in condensing-vapour conditions under real operating conditions similar to those in MSF plants. Some of the experiments were conducted to test pure water vapour whilst others were done to examine the effects of injection of carbon dioxide and air into the water vapour. The findings were discussed in terms of the weight-loss measurements and microscopically examination of specimens. The experimental tests in pure water vapour indicated that the Cu-30%Ni-base alloy was more corrosion resistance than Cu-10%Ni-base alloy. When carbon dioxide was injected into the impinging water vapour, the severity of the corrosion of both Cu-10%Ni and Cu-30%Ni alloys was observed to be increased. Moreover, the corrosion of Cu-30%Ni specimen appeared to be of a similar magnitude to that of the Cu-30%Ni specimen. Hence, the superiority of Cu-30%Ni-base alloy over Cu-10%Ni-base alloy, which was seen with pure water vapour, appeared to be diminished. The microscopically examinations of specimens demonstrated a good agreement with the weight loss measurements of the two alloys. Also, the results indicated similar features on the corroded Cu-10%Ni tube removed from the MSF plant and the laboratory Cu-10%Ni specimens. The usefulness of this work comes from the importance of the MSF desalination plants since they are considered as one of the main sources of drink water in the Gulf countries.

Chapter 9 - Direct per capita availability of freshwater resources decreases as world population continues its growth and human activities further intensify. This fact threatens the well-being and ultimately the survival of humanity as a whole. Among the scientific community and dominant political establishment, the belief that knowledge and human ingenuity will ultimately solve water scarcity problems through the efficiency improvement of freshwater production is widely held. In this vision, seawater desalination seems very promising and especially reverse osmosis technologies are increasingly portrayed as a panacea for much of the world's water woes. However, the cost of such technology in terms of energy and its environmental impacts must be taken into account, together with the different purposes served by desalinated water.

The primary aim of this chapter is to question the long-term sustainability of the desalination technology as a solution in tackling the global freshwater crisis. A North-South approach is used to raise certain questions on the significance of *scarcity*. The issue of water services for tourism is far removed from water scarcities for the Poor: water satisfying luxury or maintaining livelihood? An approach based on the "basic needs" scenario is relevant to address these issues. The city of Laâyoune (Moroccan Sahara), the Canary Island of Lanzarote (Spain), the Gaza Strip (Occupied Palestinian Territories) and the urban slums of Chennai (South India) are taken as case studies to illustrate the different purposes served by the desalination technology: alleviation of scarcity situations where the dominant component are either physical, socially-constructed, politically-constructed or structural. Beyond technological *change*, this chapter focuses on technological *equity* as a cornerstone for debate.

Chapter 10 - The need for desalting seawater is becoming more and more pressing in many parts of the world. In the Mediterranean, water needs are definitely increasing. An increase of around 30% of the water demand by the year of 2010 has been estimated for southern and eastern countries.

Seawater is practically an unlimited resource of water and seawater desalination has a long time been a major source of water in many regions of the Mediterranean.

The matching of Renewable Energy Sources (RES) with seawater desalination is a slowly growing market and thus mainly due to the high cost of the produced water.

Parameters like the size of the unit, the design, the energy consumption and availability and other factors that should be considered for the final decision of technologies selection will be examined in this work.

Several RES desalination units have been installed around the world. Several combinations such as photovoltaic units with reverse osmosis plants, wind energy with reverse osmosis and solar thermal distillation plants become in operation, most of them within research programmes. Several scientific publications on the matching of the two technologies exist and a lot of work has been done in particular from research centers. However it seems that we are still at the beginning.

There is no doubt that world's water needs can be covered and satisfied if also non-conventional resources are utilized. There is no doubt that world's energy needs can be covered and satisfied if also Renewable Energy sources are utilized. The matching of the two technologies under certain circumstances is an important prospect to the present and future potable water needs.

In: Desalination: Methods, Cost and Technology
Editor: Irena A. Urbonienė pp.1-53
ISBN: 978-1-61668-909-4

Chapter 1

HYBRIDIZATION OF MEMBRANE PROCESSES WITH CONVENTIONAL THERMALLY-BASED CO-GENERATIVE TECHNOLOGIES (SUSTAINABILITY AND TECHNO- ECONOMIC ADVANTAGES)

Sergey P. Agashichev
National Energy & Water Research Center (ADWEA),
Abu Dhabi, United Arab Emirates

ABSTRACT

Hybridization of membrane processes with conventional thermally-based co-generative technologies was considered in this study that provides an analysis of integrated and hybrid schemes including the following systems: (1) gas turbine and unfired heat recovery steam generator, (2) auxiliary boiler, (3) multistage flush desalination, (4) reverse osmosis, and (5) ASR technology. It contains a comparison of the integrated system with the reference scheme. A comparison of the alternatives is based on the present values (PV) of expenses over the economic life of capital and "levelized" cost of water and specific carbon emissions allocated to the unit of the product. The analysis is based on the following assumptions: (1) depreciation period is equal to period of bank credit; (2) zero year is the base year of the project; (3) discount factor is dependent upon nominal interest rate and retail price index; (4) cost of primary fuel ranges from 1 to 8 $/GJ; (5) assumed carbon tax varies from 0 to 50 $ per tonne of emitted carbon. Thermal desalination (MSF) being more capital and resource consuming is characterised by higher PV of expenses. It is more vulnerable to growth of the following factors: (1) nominal interest rate, (2) cost of primary fuel, and (3) rate of carbon tax. Hybridisation of thermally- and electrically-driven processes (MSF and RO) will have the following positive impact: (1) decrease specific capital and energy consumption; (2) decrease the level of carbon tax (due to decrease of allocated CO_2 emissions); and (3) decrease the "levelized" cost of water. In addition, it will decrease the sensitivity and, in turn, commercial risk of the system to fluctuation of the following factors: (1) nominal interest rate, (2) cost of primary fuel, and (3) rate of imposed carbon tax. It was concluded that the combination of conventional co-generative systems with

RO and ASR allows utilizing seasonally unused electricity while the ASR accumulates excess desalted water. The hybrid including RO & ASR technology is characterized by the following technological, economic and strategic advantages: (1) increased range of variation of power-to-water ratio; (2) possibility to use seasonal surplus of unutilized power for production of additional quantity of water; (3) decreased specific CO_2 emissions; (4) improved capital utilization factor and other economic indicators; (5) creation of strategic water reserve for the region, etc. Proposed method can be used as submodel in algorithm for synthesis of optimum configuration of the system. It can be of interest in evaluation of alternatives for feasibility studies as well.

Keywords: cogeneration, hybrid desalination, membrane technology, eco-taxation, sustainable development

1. INTRODUCTION

Recent publications indicate the growth of scientific and commercial interest to co-generative and hybrid technologies for desalination and power generation. The concept of integrated water- resource management through innovative combination and hybridization of different technologies has been scrutinized intensively, [1-7]. Conventional technologies, referred to as co-generative technologies, are based on integration of the power plant with thermal desalination systems such as multistage flush (MSF) of multieffect distillation (MED). These types of co generative technologies are characterized by inflexible limits of the power-to-water production and therefore can't meet existing changes of the pattern of power and water demand in many regions (e.g., Middle East). Along with that, they are characterized by some techno and environmental disadvantages such as high energy consumption and green house gas emissions [8-9]. Implementation of a policy of emission trading and reconfiguration of the system of eco-taxation enhance the shift towards more sustainable technological options. Hybridization of thermally and electrically driven desalination processes is an example of a new generation of co-generative technology. It can provide essential economic, environmental and operational advantages over conventional dual and mono-purpose stand-alone plants in particular: (1) increased technological flexibility that is characterized by ability to diversify range of the power-to-water ratio; (2) possibility to use seasonal surplus of idle power and unused water; (3) decrease of specific fuel consumption while RO is less energy consuming than thermal desalination; (4) growth of utilization of installed capital and decreases in capital recovery cost per unit of production; (5) decrease of "carbon footprints" in terms of specific CO_2 emissions and finally decrease cost of produced water .

Driven by these factors one can see the development of integrated or hybrid schemes where application of membrane-based operations such as nano-filtration (NF), ultra-filtration (UF), and micro-filtration (MF) within different technological segments takes place. There are different levels and types of hybridization, e.g. ,(A) incorporation of reverse osmosis (RO) into existing co-generative schemes; (B) application of membrane-based operations such as NF, UF, and MF as a pretreatment before RO; (C) low pressure RO as a post treatment for boron elimination; (D) NF softening before MSF and so on. Evolution from pure theoretical concept to diversified spectrum of different technological combinations with various levels of integration occurs [13-26]. There is a wide spectrum of technological combinations from using common intake for MSF- RO; NF pretreatment for MSF; membrane-based pretreatment for RO to mutual conjugation of technological regimes between power generating and desalination plants. The term "hybrid" used in literature has rather wide connotation; it covers the cases from just using common intake and blending permeate and distillate to integration of existing co-generative plants with RO and ASR.

The development of a new generation of hybrid technologies requires estimation of their sustainability, which can be quantified by the group of multi-parameter indicators. The following groups of sustainability indicators were proposed: (1) indicators for estimation of water quality; (2) estimation of resource consumption; (3) estimation of environment, and group of indicators for estimation of (4) social aspects and (5) efficiency indicators [27, 28]. They can be used for comprehensive evaluation of sustainability. Along with it requires development of methodology for techno-economic analysis, accounting for carbon emissions. In addition to technological and economic characteristics traditionally used, there is a wide spectrum of specific indicators such as flexibility, resource and energy consumption, utilization factor of installed capital, carbon emission, and so on. Those are becoming absolutely essential for analysis of hybrid and integrated technologies. Among R&D trends, one should focus on development of methodology for quantitative allocation of carbon emission in cost-benefit analysis. Different aspects of integrated environmental and economic accounting were considered in [29-36]. Integrated and hybrid technologies being characterized by multiple inputs and outputs (technologies producing water along with energy with different potential) require a systemic approach for design and analysis. In this regard the proposed study focuses on the following aspects: (1) feasibility of hybridization of membrane processes with conventional thermally-based co generative technologies; (2) different generations and configuration of hybrid and integrated schemes; (3) analysis of integrated scheme; (4) exergy-based characteristics in accounting for CO_2 emissions in economic analysis of integrated or multi-purpose co-generative systems; (5) sensitivity of water cost to imposed carbon tax in comparative evaluation of carbon risk; (6) sustainability and techno-economic advantages of integrated and hybrid systems.

2. Existing Co-Generative Technologies and Water/Power Demand Pattern

Analysis of published statistics on water demand for Abu Dhabi Emirate [37] reveals a high rate of growth of seasonal variation of water demand over the last years and increase of cumulative gross annual water demand over the same period. According to this data, seasonal difference between the highest and lowest water demand increases from 1,452 MG/month (in 1998) to 2,629 MG/month in 2003, which represents up to 80% growth. Cumulative annual water demand over the same period ranges from 63064 MG/year (in 1998) to 113484 MG/year in 2003. Relying upon the ADWEC statistics [37], the prevalent trends in development of demand pattern for Abu Dhabi region can be summarized as follows: (1) growth of range of seasonal variation of demand; (2) growth of water and electricity demand is characterized by different functional behavior; and (3) increase of cumulative annual gross power and water demand.

Existing trend in variation of power to water demand cannot be provided by conventional co-generative systems currently used while existing technologies are characterized by inflexible limits of variations of the "power-to-water ratio". (The criteria referred to as "power-to-water ratio", is an essential indicator specifying the performance of dual purpose co-generative plant. This criterion is equal to the ratio of produced power to water output). Conventional systems are preferred for countries where demand pattern is characterized by high value of the power-to-water ratio. (Unlike the majority of countries, United Arab Emirates with high level of water demand is characterized by low value of this criterion).

There is a wide spectrum of possible technological combinations between power-generating and desalination systems refereed to as conventional co-generative systems. In particular MSF, MED or RO, as a desalination plant, can be conjugated with different types of power plants. There are different configuration of power plants currently used, they are: (A) gas turbine, heat recovery steam generator (HRSG) connected with back pressure steam turbine (BP-ST); (B) gas turbine, fired heat recovery steam generator (HRSG) connected with extraction-condensing steam turbine (EC-ST) and so on.

Example of the conventional co-generative plants shown in Figure 1 consists of the power and desalination plant. This technological example includes gas turbine (GT), unfired heat recovery steam generator (HRSG); auxiliary boiler (AB) and thermal desalination system, e.g. MSF. This technological combination is one of the most environmentally and technologically unfavorable examples.

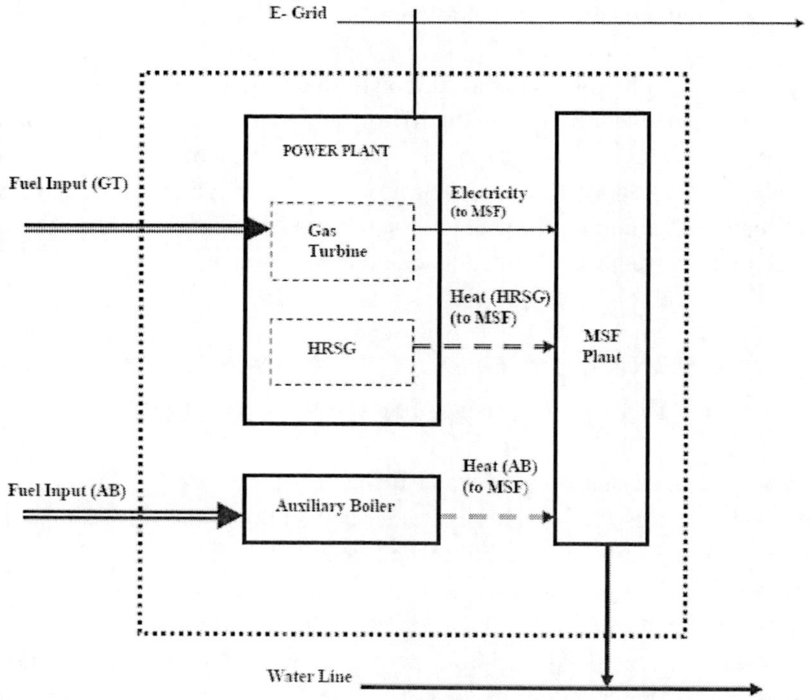

Figure 1. Schematic diagram of conventional co-generative plant (Gas turbine, unfired heat recovery steam generator and auxiliary boiler are connected to multistage flush desalination)

The most severe mismatch problem has been experienced in Abu Dhabi, Kuwait and parts of Saudi Arabia. Some desalination plants operate for as much as 27 % of time without benefits of cogeneration using mainly boiler- raised steam. Annual utilization of power generating capacity is 43 % [9]. The growth of annual demand and range of seasonal fluctuation of the demand pattern require reorganization of management, in particular it dictates implementation of the concept of integrated water- resource management through innovative combination and hybridization of different technologies, [38] that would make the system more flexible and in turn, allows optimization of water resource utilization. Unfavorable disproportions between water and power demand can be reduced by implementation of additional desalination capacities and creation of water reserve. Many authors intensively scrutinized theoretical and practical aspects of this issue.

3. Carbon Taxation for Accounting of "Carbon Footprints" in Cost- Benefit Analysis. Exergy–Based Characteristics for Reallocation of "Carbon Footprints" between Material and Energy Fluxes in Economic Analysis of Multi-Purpose Co-Generative Systems

Carbon "footprints" is an essential indicator of sustainability. Carbon reduction management as a part of the strategy is becoming essential for creation of low-carbon economy. Implementation of emission trading and policy of carbon taxation are considered to be important tools for carbon reduction. Recent documents and guidelines, such as Kyoto protocol, Treaties of Maastricht and Amsterdam, Rio and Oslo Accords put forward the foundation of environmentally sound technological policy. Conferences on sustainable development held in Oslo recommended shifting tax burden from labor to use of resource and damage of environment. Within the this context the growth of scientific, engineering and commercial interest to sustainable technologies characterized by low level of fuel consumption and decreased CO_2 emission takes place. Accounting environmental damage in cost-benefit analysis is becoming essential for implementation of ecological taxes and policy of emission trading. An importance of the issue was stressed by different studies, [39- 42] Greenhouse gas emission can became one of the world's largest traded commodities in the future and accounting of different aspects of carbon risk is getting absolutely essential. Carbon risk management is getting inextricably linked with engineering and technological consulting nowadays. Different aspects and advantages of policy of emission trading and carbon taxation are widely scrutinized by different groups of experts. Comparison between two leading approaches, such as carbon markets and carbon taxes is given by Ronald Baile, [43]. A *Wall Street Journal Survey* found that 54 percent of economists favor a carbon tax over all other approaches among them Paul Volcker, Alan Greenspan and others, [43]. Carbon tax offers less opportunity for corruption because it does not create artificial scarcities and monopolies. Carbon allowances can be analogized to quotas in international trade while carbon taxes to tariffs. It's been a lot easier to manage tariffs than quotas. "The great political advantage of carbon taxes is that they raise large revenues which governments can use to reduce other unpopular and more distorting taxes, or finance popular spending programs," says Robert Shapiro. The level of recommended CO_2 taxes being one of key economic assumption in forecast is still disputable issue. Published data on level of carbon tax are still randomized owing to lack of coherent policy. According to, [44], "level $25 to $85 per tonne of carbon, the level many experts think may be needed if industrialized countries are serious about the emission targets agreed on in 1997 at Kyoto". Energy taxation according to its carbon content at the level of EUR 40 per tone CO_2 equivalent was recommended by Deroubaix and Leveque in [45]. Tax level can be adjusted over time to achieve whatever limits to balance the costs of climate change with the benefits of economic progress. Nordhaus suggests that the optimal carbon tax trajectory, balancing costs and benefits, would start with a tax of about $17 per ton, rising to $84 in 2050 and $270 in 2100. According to [43], carbon tax is proposed to start with a $5-per-ton tax on carbon and raising it by $5 per ton every other year. The first year would raise $9 billion in revenues for the Treasury, rising to $25 billion by 2010 and $75 billion by 2020. A $25-per-ton carbon tax translates into a 5 percent increase in average electricity rates and a boost of about 6 cents per gallon of gasoline. The 2007 IPCC report contains an assessment of the literature on the SCC: Peer-reviewed estimates of the SCC for 2005 have an average value of USD 43 per tone of emitted carbon with a standard deviation at the level USD 83 per tone of emitted carbon. The wide range of estimates is explained mostly by underlying uncertainties in the science of climate change (e.g., the climate sensitivity), different choices of discount rate, different valuations of economic and non-economic impacts, treatment of equity, and how potential catastrophic impacts are estimated. According to *The Daily Yomiuri*, [46] Japan will study the introduction of a tax on fossil fuels in 2010 to curb emissions of carbon dioxide and help make up revenue from the abolition of a provisional tax on gasoline.

Recent tendencies in theory of macroeconomic analysis are characterized by implication of exergy - based characteristics, being exogenous to economics, into economic methodology, in particular, R. Ayres in [41], consider exergy as a factor of production along with the labor and capital. Exergy being proportional to potential future entropy production (or measure of "distance" from thermodynamic equilibrium) is used as an exogenous input into economic model. In proposed study the exergy-based characteristics were used for reallocation of CO_2 emissions between consumed fluxes of thermal and electrical energy. It is based on the cost of exergy embodied into corresponding flux. Embodied exergy ecological footprint (instead of conventional "footprint") was proposed to be considered as an indicator of sustainability by Chen in [47] as well.

4. Hybrid Scheme as an Object for Modeling and Analysis

According to published studies hybridization of RO desalination with existing co-generative plants is accompanied by technological, environmental and economic benefits and can be considered as a sustainable technological solution. Hybrid scheme for power generation and desalination including GT, Unfired HRSG and AB (power generation plant), MSF or MED (thermal desalination plant) and RO (membrane desalination) is shown in Figure 2. Power generating plant is characterized by influx of primary fuel; output of low grade heat raised by HRSG and AB; electricity generated by GT along with CO_2 emissions. Low grade heat in turn is used for thermal desalination through MED of MSF. Electricity consumed by greed along with thermal and membrane desalination (see Figure2) Similar case was considered in [10].

Hybrid scheme can be represented as an array of subsystems of different hierarchy levels, the hybrid including RO, MSF and power generation plant can be subdivided into three groups of models underlying the system, and they are: (A) models describing the power-generating technologies; (B) models for thermal MSF desalination and (B) model for pressure-driven RO desalination plant. Any group of individual models (A, B or C), in turn, contains set of sub-models of different hierarchical level specifying behavior of different variables such as technological, resource consumption, ecological and economic ones.

These sub-models *generate data*: on electricity production, low- grade heat output, CO_2 emission, efficiency of the primary fuel utilization at different load of generator. *The technological submodel* is focused on calculation of technological characteristics at different operating load of the power generating systems; *The fuel or energy submodel* covers influx of primary energy and output of electricity and low- grade heat with different exergetic values depending upon operating regimes. The *ecological submodel* is based on correlation between exergy and carbon fluxes through the system. It covers redistribution of carbon per unit of product proportional to consumed exergy. The *economic submodel* quantifies economic indicators such as (a) cost of water, (b) cost of electricity and low-grade energy, and (c) cost of allocated CO_2 emissions through imposed carbon tax. Cost of energy raised by GT, HRSG and AB is based on load-dependent manufacturing expenses. The submodel covers calculation of economic indicators for base year and long term projections. Schematic structure of the model is shown in Figure 2. Technological prototype (GT, Unfired HRSG and AB) underlying this model is the most environmentally unfavorable technological example.

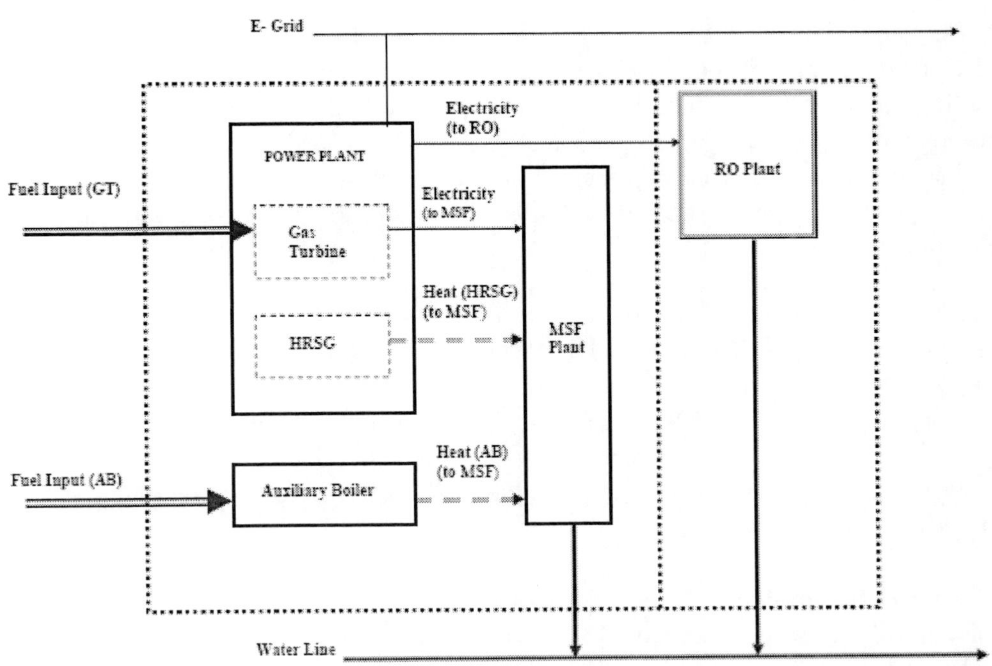

Figure 2. Hybrid scheme for power generation and desalination (Gas turbine, unfired heat recovery steam generator and auxiliary boiler are connected to multistage flush and reverse osmosis desalination)

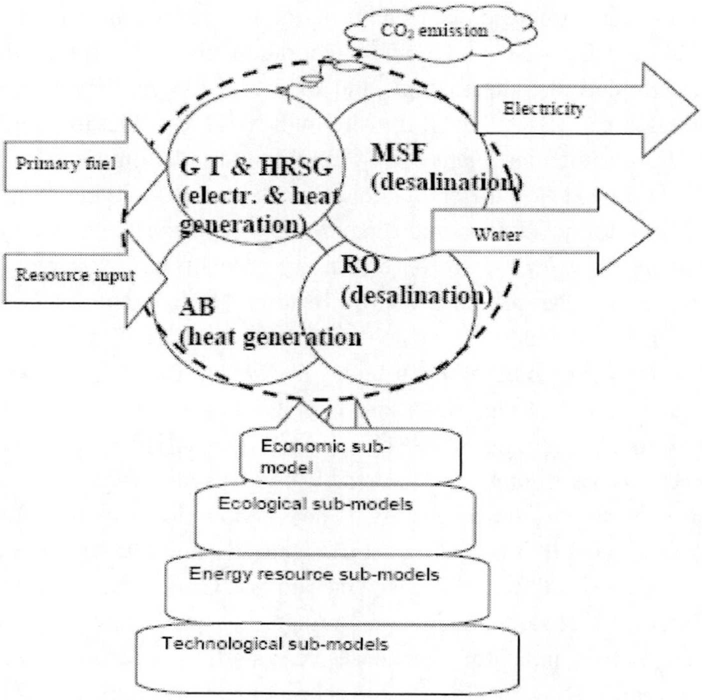

Figure 3 Schematic structure of the model

The model allows analysis of behavior of technological, economic and ecological indicators at various economic assumptions and technological parameters such as: (1) load, specific fuel consumption and efficiency of energy generating system, (2) specific energy consumption for desalination, (3) specific emissions of CO_2, and (4) taxes on CO_2 emissions. Analysis of similar systems was presented in [10-11] where (A) evaluation of indicators for the base year of the project and (B) long- term projections of the present value of expenses were estimated.

5. Techno-Economic Assumptions

Current analysis is based on the following premises and assumptions:

Base year for financial transaction is the year when the production supposed to start (zero year of the project).

Depreciation period being equivalent to **economic life of capital** is assumed to be equal to period of bank credit, in proposed study it is 10 yrs. Money borrowed from the bank are equal to sum of direct and indirect capital (DC+IC) including start up expenses. (Start up expenses and interest during construction assumed to be capitalised and should be recovered through depreciation).

Strait-line depreciation with **no salvage value** is assumed in evaluation.

I_N- **Nominal Rate of Interest** on borrowed capital. External source of capital is assumed and interest is considered as manufacturing cost. Interest is charged on the total capital investment (DC+IC) at nominal interest rate, [48- 50]. In proposed study the nominal rate of interest was assumed to be from 1 to 10% yearly.

RPI- Retail Price Index (Rate of inflation) The RPI being measure of inflation is expressed as characterising increase of average price of good and services. Real value of money over N-years can be evaluated as follows:

$$\text{Real value} = S/(1+RPI)^n \tag{1}$$

Where, S, is current value of money.

R- Real Discount Rate (real rate of interest) represents the combined effect of the nominal rate of interest I_N, and the overall rate of inflation (RPI). It is a measure of real cost of money. It can be expressed from the following expression:

$$(1+I_N) = (1+R)(1+RPI) \tag{2}$$

Being solved for R, it gives:

$$R = (1+I_N)/(1+RPI) - 1 \tag{3}$$

Discount factor, $f(R,n)$, being reciprocal to discount rate, R, is expressed as follows:

$$f(R,n) = 1/(1+R)^n \tag{4}$$

Structure of annual expenses includes the following groups:

(A) Expenses for recovery of borrowed principal and interest, they assumed to be equal to depreciated assets; (B) O&M expenses; (C) Energy expenses; (D) Carbon tax.

Recovery of borrowed principal, P, plus interest, I, is expressed in form of **annuities**, [49]

$$X = \frac{P\,(1+I_N)^n I_N}{(1+I_N)^n - 1} \tag{5}$$

Where: P- Borrowed principal; I_N- Nominal interest rate per period, n- number of periods

Expenditures in any year suppose to be converted to its *present value*, PV, by multiplying by discount factor, $f(R,n)$, see Eq. (4). Where, n, number of years after the base year in which transaction occurred, **n** varies from n=1 to n= n_{max}, being equal to useful life of asset.

Calorific Value and Carbon Allocated to the Primary Fuel (Assumptions for Ecological Model)

Calorific value of primary fuel = 45580 kJ / kg (primary fuel); Specific carbon content of the primary fuel =1.7552 E-05 kg C/ kJ (primary fuel); 0.8 kg C/ kg (primary fuel); Potential carbon emission allocated per kJoule of primary fuel 6.436E-05 kg CO2 / kJ (primary fuel); (where the carbon fraction in CO_2, = 0.27 kg carbon/ kg CO_2;)

Cost of primary fuel is assumed to vary from 1 to 12 USD per GJ, for calorific value 45580, kJ / kg of primary fuel. Evolving fuel price 3% per year is assumed in this study

Specific consumption of primary fuel, $f(L)$, depends upon GT-load. For the technological option selected in this study, namely GT+ Unfired HRSG+ AB, the specific fuel consumption of GT was approximated by the following correlations.

$$f_{GT}(L_{GT}) = \left[A_{GT} L_{GT}{}^2 + B_{GT} L_{GT} + C_{GT}\right] f_{L=1} \tag{6}$$

Where A_{GT}=**4.17**, B_{GT}=**-8.34**, C_{GT}= **7.70** and $f_{L=1}$ =3.7 J(fuel)/ J(electr)

Load dependent fuel consumption of auxiliary boiler, $f_{AB}(L_{GT})$, can be expressed in terms of L_{GT} as follows:

$$f_{AB}(L_{GT}) = \left[A_{AB}\frac{(1-L_{GT})}{(1-L_{\min(GT)})} + B_{AB}\right] f_{AB(L=L\min(GT))} \tag{7}$$

Where A_{AB}= -0.07, B_{AB}= 1.07, $f_{AB(L=L\min(GT))}$, specific fuel consumption of AB at its rated output that corresponds to minimum load of GT, $f_{AB(L=L\min(GT))}$=1.17 J(fuel)/J(therm).

For the treatment to be simplified the load of auxiliary boiler, L_{AB}, is expressed in terms of L_{GT} as follows: $L_{AB} = (1-L_{GT})/(1-L_{\min(GT)})$. (See the term in parenthesis, Eq. (7) where L_{AB} varies from L_{AB} =0 to L_{AB} =1 while L_{GT} drops from L_{GT}=1 to $L_{GT}=L_{GT(MIN)}$.

Rate of carbon tax in current study is assumed to vary from 0 to 50 USD per tone- of emitted carbon (in CO_2 form); [44- 47]. The value of carbon tax is based on rate of emitted carbon that, in turn, is influenced by specific fuel consumption that depends upon the GT load.

Specific CO_2 emissions, $c(L)$, depends upon GT- load. Technique for reallocation of emissions between heat and electricity fluxes based on their exergetic values is considered in [10-11]. For evaluation of emitted carbon the following linear correlation was used.

$$c(L) = AL + B \tag{8}$$

Where A & B are coefficients for evaluation of allocated CO_2 emissions:

A =-0.4148 B= 0.9522 kg CO_2/kWh(el) (For CO_2 allocate to produced electricity)

A =-0.0939 ...B= 0.2254 kgCO_2/kWh(heat-HRSG) (For CO_2 allocate to produced heat)

Future value, FV, of present amount, PV, can be estimated as:

$$FV = PV\left(1+R\right)^n \tag{9}$$

Where: PV- present value of expenses (evaluated at base year), FV- future value of expenses, R- Real Discount Rate;

Present value, PV, of future amount, C_Σ, can be estimated as:

$$PV\left(R,n\right) = C_\Sigma\left(n\right)/\left(1+R\right)^n \tag{10}$$

Where, R, real discount rate (see Eq. 3). When determining PV, the taxes are included as expenses. The profit on operations, income tax, federal and local taxes are outside the scope of the study. Only projected level of carbon taxes is considered in this study.

Cumulative present value, CPV, is equal to sum of annual present values, PV(R, n), over economic life of capital. It can be expressed as:

$$CPV\left(R,n\right) = \sum_{n=0}^{n=\max} PV\left(R,n\right) = \sum_{n=0}^{n=\max} \frac{C_\Sigma\left(n\right)}{\left(1+R\right)^n} \tag{11}$$

Main assumptions for specific cost of installed capital, energy consumption and O&M expenses were adopted from [51-55]

6. Technological Model for Energy Generation [System: Gas Turbine, Unfired Heat Recovery Steam Generator and Auxiliary Boiler]

Technological model is based on energy- and material balances. Simplified energy balance is shown below.

$$E_\Sigma(l) = E_{GT\_EL}(l) + E_{GT\_TH}(l) + E_{AB\_TH}(l) + E_{loss} \tag{12}$$

The term on the left hand side corresponds to total influx of primary fuel into the system (GT and AB), the first term on the right hand side represents electricity generated by GT: the second and third terms represents low- grade heat produced by HRSG and by AB respectively; the last one is measure of irreversible dissipation. All the terms depend upon the load of power generating system.

$$E_{GT\_EL}(l) = A_{GT\_EL}\, l + B_{GT\_EL} \tag{13}$$

$$E_{GT\_TH}(l) = A_{GT\_TH}\, l + B_{GT\_TH} \tag{14}$$

$$E_{AB\_TH}(l) = A_{AB\_TH}\, l + B_{AB\_TH} \tag{15}$$

Where: $E_{GT\_EL}(l)$ and $E_{GT\_TH}(l)$, electrical and thermal production vs. load of power generating system; $E_{AB\_TH}(l)$, thermal output of AB vs. load of power generating system, Fuel sub-model is based on the correlation between fuel consumption $F_i(L)$ vs. load, L, of the power generating system (GT).

$$F_{GT}(l) = \left[A_{F/GT}\, l^2 + B_{F/GT}\, l + C_{F/GT}\right] El_{GT}(l) \tag{16}$$

Where, $F_{GT}(l)$, primary fuel influx; A, B& C coefficients describing specific fuel consumption vs. the GT- load. Term in brackets (eq. 16) represents specific fuel consumption at current GT-load. Similar equation was used for auxiliary boiler (AB).

$$F_{AB}(l) = A_{F/ABi}\, l + B_{F/AB} \tag{17}$$

Where, $F_{AB}(l)$, fuel influx

The model describing fuel consumption gives an input data for estimation of specific CO_2 emissions. The system is characterized by influx of primary energy and out-coming fluxes of electricity and heat accompanied by unused losses. Any energy flux leaving the system is accompanied by definite amount of CO_2 emissions.

7. Accounting CO_2 Emissions (Correlation between Emitted Carbon and Exergy Flux)

Ecological sub-model allocates CO_2 emissions among the fluxes of produced heat and electricity. Allocation of CO_2 emission is based on exergy balance and accounting of carbon fluxes through the system. Exergy- based characteristics, being exogenous to economics, is becoming widely used in economic methodology, in particular, Ayres [41], considered exergy as a factor of production along with the labor and capital. Exergy being proportional to potential future entropy production (or measure of "distance" from thermodynamic equilibrium) is used as an exogenous input into economic model. In proposed study the exergy-based characteristics were used for reallocation of CO_2 emissions between thermal or electrical fluxes and between water produced through thermal or membrane desalination. Carbon accounting was based on the cost of exergy embodied into corresponding flux. In proposed study carbon taxes were applied depending upon consumed carbon of primary fuel and exergetic value of generated flux. (It's essential in case when low-grade heat raised by HRSG and AB are characterised by different potential and load- dependent specific fuel consumption of generator). Implying the fact we assume that the overall flux of CO_2 emitted by the system can be subdivided into three subfluxes being proportional to its exergy content, namely subflux of energy raised by HRSG, by AB and GT.

Carbon balance is based on the following assumptions: Calorific value of primary fuel = 45580, kJ / kg (primary fuel); Specific carbon content of the primary fuel =1.7552 E-05 kg C/ kJ (primary fuel); 0.8 kg C/ kg (primary fuel); Potential carbon emission allocated per kJoule of primary fuel 6.436E-05 kg CO2 / kJ (primary fuel); (where the carbon fraction in CO_2, = 0.27 kg carbon/ kg CO_2)

Accounting of CO2 emission in energy generation Estimation of share of emitted CO_2 allocated to definite flux of generated energy is proportional to the cost of exergy allocated to that flux. The following assumptions were used: (1) Cost of exergy embodied in losses is assumed to be zero; (2) efficiency of reference cycle is assumed to be 0.24; (3) dependence of CO_2 emissions upon the GT load takes place.

Energy fluxes generated by GT, HRSG and AB are characterized by different exergetic values.

$$\varepsilon_\Sigma(l) = \varepsilon_{GT\_EL}(l) + \varepsilon_{GT\_TH}(l) + \varepsilon_{AB\_TH}(l) + \varepsilon_{loss}(l) \tag{18}$$

Where the first term on the right hand side represents exergy allocated to electricity generated by GT; the second and third terms represent exergy allocated to low- grade heat produced by HRSG and by AB respectively. All the terms depend upon the load of power generating system and can be rewritten in terms of energy and efficiency of reference cycle.

$$\varepsilon_\Sigma(l) = E_{GT\_EL}(l)\ \xi_{GT} + E_{GT\_TH}(l)\ \xi_{HRSG} + E_{AB\_TH}(l)\ \xi_{AB} + \varepsilon_{loss}(l) \tag{19}$$

Where ξ – efficiency of the reference cycle (it's a measure of maximal possible conversion of heat to work. It's proportional to exergy content of the energy generated by AB, GT or HRSG respectively; that depends in turn on the steam temperature and temperature level of the surroundings)

$$\xi_i = 1 - T_i / T_0 \tag{20}$$

Where T_0 -temperature level of the surroundings (T_0= 298). Distribution of carbon proportional to exergy allocated to corresponding flux $\varepsilon_i(l)$ can be expressed as

$$C_i(l) = \frac{\varepsilon_i(l)}{\varepsilon_\Sigma(l)}\ C_\Sigma(l) \tag{21}$$

Total emitted carbon $C_{\Sigma}(l)$, suppose to be distributed between energy output from GT, HRSG and AB, therefore

$$C_{\Sigma}(l) = C_{GT\_EL}(l) + C_{GT\_TH}(l) + C_{AB\_TH}(l) \tag{22}$$

The first term on the right hand side (eq. 22) represents carbon allocated to electricity generated by GT; the second and third terms represents carbon allocated to low- grade heat produced by HRSG and by AB respectively. All the terms depend upon the load of power generating system. Terms of the right hand side can be approximated as:

$$C_{GT\_EL}(l) = A_{C/GT(EL)}\, l + B_{C/GT(EL)} \tag{23}$$

$$C_{GT\_TH}(l) = A_{C/GT(TH)}\, l + B_{C/GT(TH)} \tag{24}$$

$$C_{AB-TH}(l) = A_{C/AB(TH)}\, l + B_{C/AB(TH)} \tag{25}$$

Where $C_{GT-EL}(l)$, $C_{GT-TH}(l)$ and $C_{AB-TH}(l)$ are specific CO_2 emission by GT with respect to electricity and heat raised by GT, HRSG and AB respectively, kg CO_2/kWh (electricity) and kg CO_2/ kWh (therm). Behavior of these functions is shown in Figure 4.

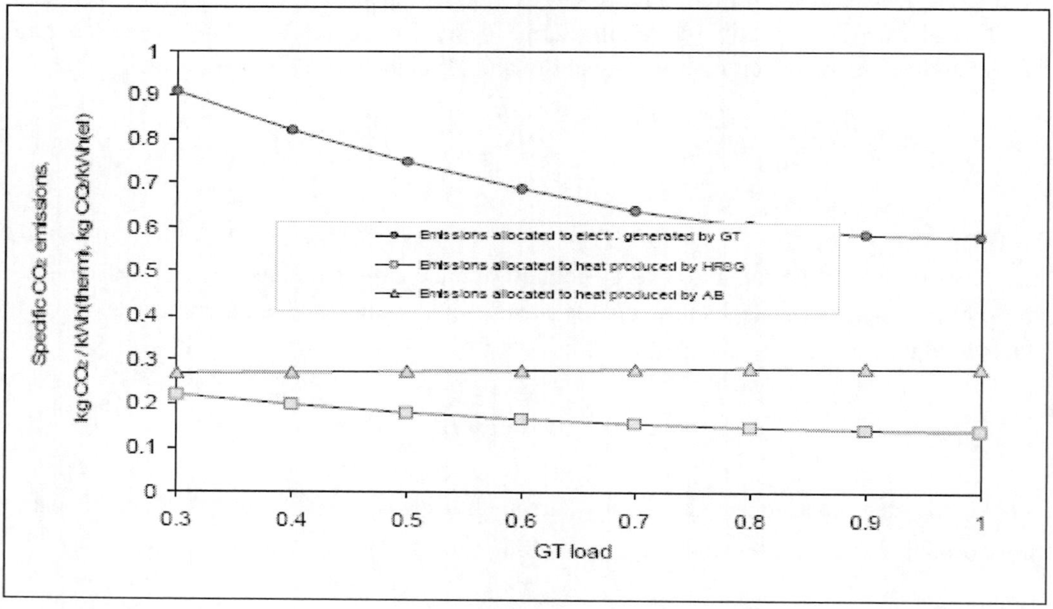

Figure 4. Specific CO_2 emission allocated to thermal or electrical output by GT+ Unfired HRSG +AB

Further reallocation of carbon on the stage of desalination is based on combination of carbon, exergy and energy equations for power generation with material and energy balance of desalination.

Accounting of CO2 emission in desalination Energy required for thermal MSF desalination includes thermal and electrical terms. Data on required energy consumption can be found in [51]. Total energy required for thermal desalination is

$$E_{MSF} = E_{GT\_EL} + E_{GT\_TH}(l) + E_{AB\_TH}(l) \tag{26}$$

The first term on the right hand side E_{GT_EL}, represents auxiliary electrical power requirement (to be consumed for pumping), the second one – thermal energy generated by HRSG , the third one- thermal output raised by AB. Thermal energy generated by HRSG and AB is used in distiller. For constant distiller production $E_{GT\_TH}(l) + E_{AB\_TH}(l) = const$ and $E_{GT\_EL} = const$

As soon as the terms on the right hand side of Eq. (26), E_{GT_EL}; $E_{GT\_TH}(l)$; $E_{AB\_TH}(l)$, are depends upon GT load and characterized by different exegetic value and therefore different specific carbon emissions that makes essential accounting carbon allocated to any energy influx.

$$C_{\Sigma\_MSF} = E_{GT\_EL} * c_{GT\_EL}(l) + E_{GT\_TH}(l) * c_{GT\_TH}(l) + E_{AB\_TH}(l) * c_{AB\_TH}(l) \tag{27}$$

Water production of distiller is based on the balance

$$W_{MSF} = \lfloor E_{GT\_TH}(l) + E_{AB\_TH}(l) \rfloor PR/\lambda \tag{28}$$

Where λ – latent heat, PR- performance ratio, the numerator represents thermal energy (generated by HRSG and AB) to be consumed by distiller. Calculated profile of specific CO_2 emissions per cubic meter of MSF water (based on Eqs. 27 and 28) is shown in Figure 5.

[Input data: RO specific energy consumption is 5 kWh (el)/m^3(permeate), MSF-specific thermal consumption is 60 kWh (thermal)/ m^3(distillate) at PR= 10, MSF-auxiliary electrical energy consumption is 3 kWh (el)/m^3 (distillate)]

Energy required for pressure driven RO desalination can be estimated as

$$E_{RO} = W_{RO}\ r \tag{29}$$

Where r – specific energy consumption that in turn depends on type of energy recovery equipment, degree of permeate recovery, water characteristics and so on.

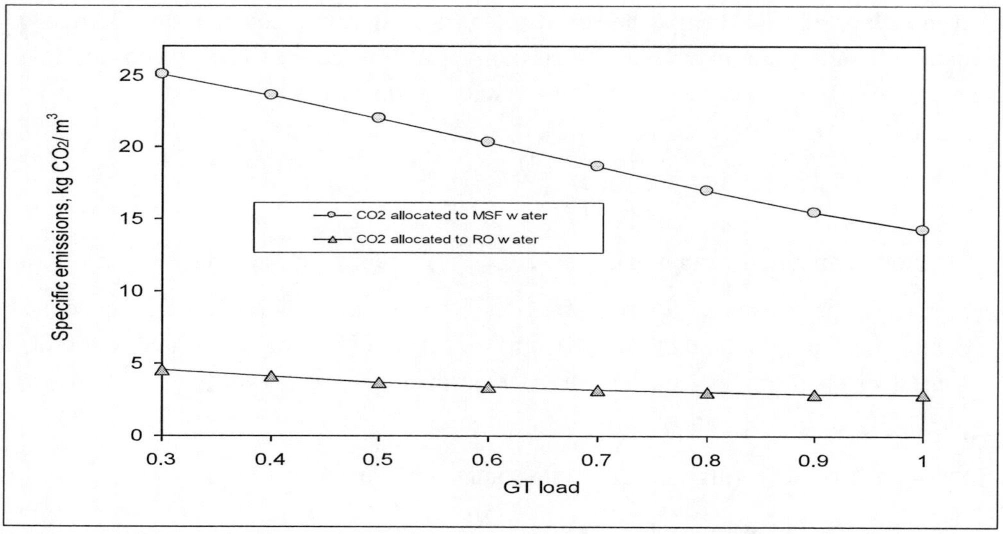

Figure 5. Specific CO_2 emission per cubic meter of water vs. load of the GT

$$E_{RO} = E_{GT\_EL}(l) \tag{30}$$

$$C_{\Sigma\_RO} = E_{GT\_EL}(l) * c_{GT\_EL}(l) \tag{31}$$

As soon as the term on the right hand side of Eq. (31), $E_{GT\_EL}(l)$; depends upon the GT load that determines carbon emission as a function of load. Calculated profile of specific CO_2 emissions allocated to cubic meter of RO water (based on Eqs. 30 and 31) is shown in Figure 5.

8. Economic Model for Energy Generation (Gas Turbine, Unfired Heat Recovery Steam Generator)

This model generates there groups of characteristics firstly load-dependent behavior of manufacturing expenses for technological unit namely GT, unfired HRSG, AB; secondly calculation of itemized structure of manufacturing expenses for base year (at existing GT-load distribution over the year), thirdly calculation of long-term projection over economic life of the capital and "levelized" cost of product. The model is based on assumptions listed in table 1 using data generated by technological and environmental sub-models as an input.

Structure of manufacturing expenses consists of load-dependent and load-independent segments (1) load-dependent influx of primary fuel; (2) load-dependent outputs of electricity; (3) load-dependent output of heat, (4) load-dependent output of CO_2 emissions, (5) load-independent recovery of capital (interest plus principal) and (6) load-independent O&M costs. The first group of characteristics of the model estimate load- dependent behavior of cost of energy generated by GT and HRSG. Set of projections is shown in Figures 6-9

Table 1. Basic assumptions (GT and Unfired HRSG system)

	Basic Assumptions	
1	Size of GT	100 MW
2	Specific capital cost, [14]	500 USD /kW(electr);
3	Initial capital investment (DC+IC)	50 mln USD
4	Nominal Interest Rate	0.1
5	Retail price index (Inflation rate)	0.01
6	Real Interest Rate	0.09
7	Cost of primary fuel	2.0 USD/GJ
8	Carbon tax	20.0 USD/tone of emitted carbon
9	Annual O&M	5 mln USD /year

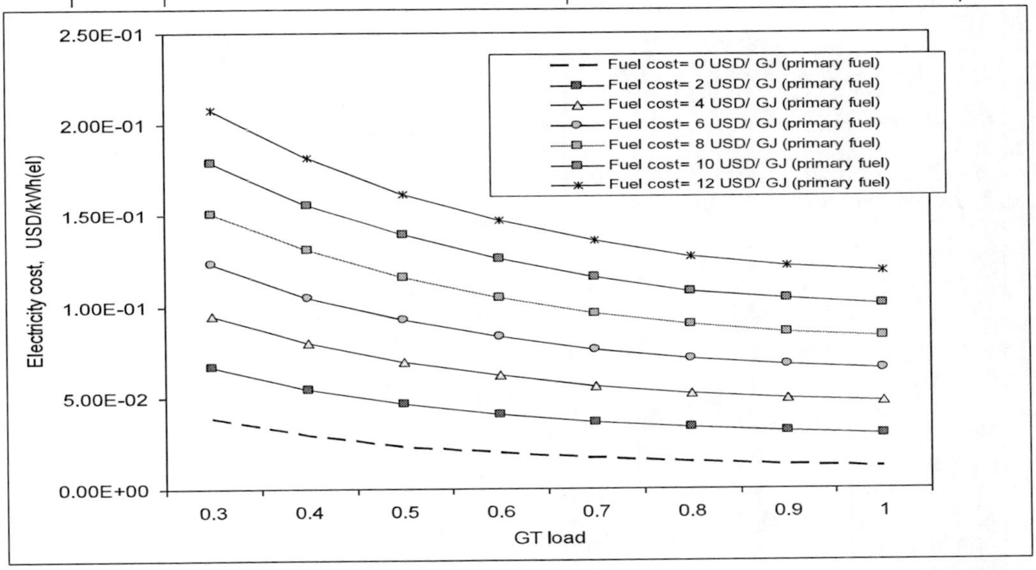

Figure 6. The cost of generate electricity vs. load of the power generating system at various level of the primary fuel cost (No carbon tax).

Annual expenses are subdivided into load-independent and load-dependent segments as well. They are: (1) load- independent charges for recovery of capital (principal and interest); (2) O&M expenses; (3) fuel expenses; (4) carbon tax. Future value of expenses is estimated using real discount rate (see Eq. 3). Annual cost of capital recovery is evaluated as annuity, (see Eq. 5). Accounting CO_2 emission is done through imposition of carbon tax, where consumed fuel carbon is used as base for evaluation of carbon tax, (see [10, 11]). Sum of annual expenses, $C_{\Sigma(GT-HRSG)}$, is expressed as follows:

$$C_{\Sigma(GT-HRSG)} = \left[C_{capital} + C_{O\&M}\right] + \left[C(l)_{fuel} + C(l)_{C-tax}\right] \tag{26}$$

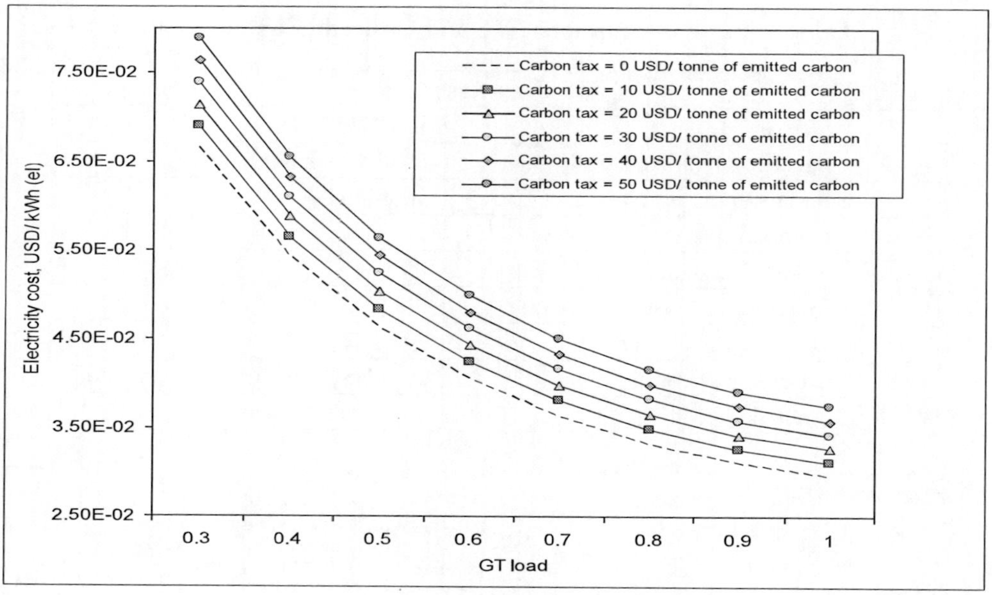

Figure 7. The cost of generated electricity vs. load of the power generating system at various level of carbon tax (Cost of primary fuel is 2 USD/GJ)

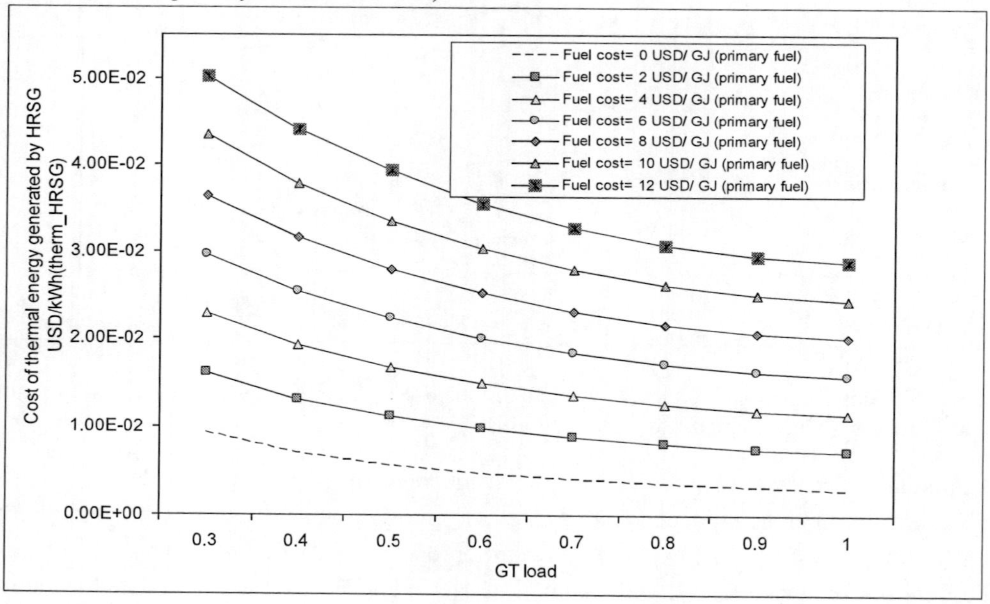

Figure 8. The cost of heat generated by HRSG vs. load of the power generating system at various level of the primary fuel cost (No carbon tax)

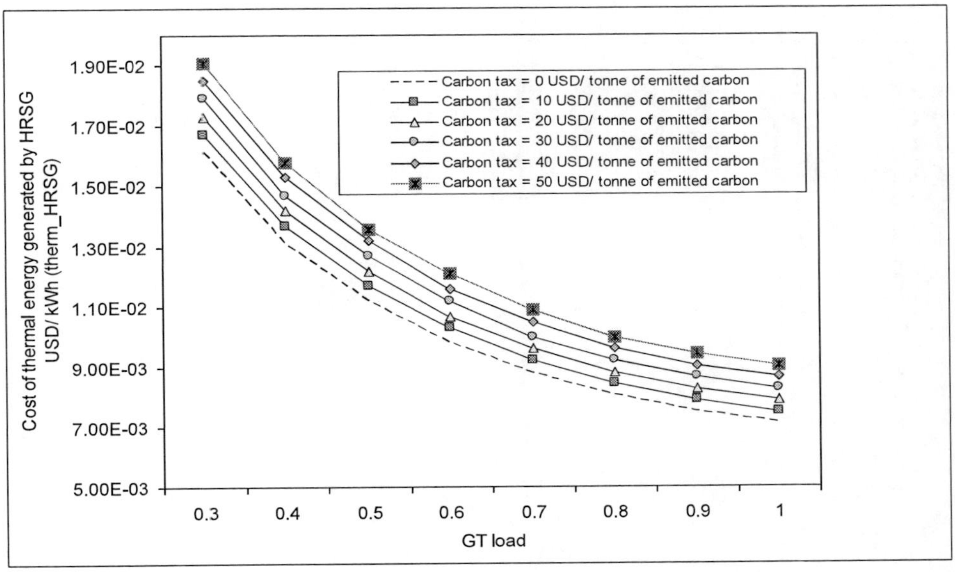

Figure 9. The cost of heat generated by HRSG vs. load of the power generating system at various level of carbon tax (Cost of primary fuel is 2 USD/GJ)

Long term projections over economic life of capital based on economic submodel are presented in Appendix **A.** The matrix contains itemised structure of annual expenses and gives indicators such as PV and cost of energy (in form of low-grade heat and electricity) produced by GT and unfired HRSG, respectively. Any row corresponds to definite group of expenses. Any column labeled "year of operation" contains the expenditures allocated to corresponding year, $C_{\Sigma(GT-HRSG)}$. Their value has been converted to its present value (PV) by multiplying the discount factor, (see Eq. 4). Cumulative PV of expenses and "levelized" cost of produced heat and electricity are key indicators for evaluation of the subsystem, where the "levelised" cost is expressed as the ratio of cumulative PV of expenses over economic life of capital to overall production over the same period. It is influenced by parameters such as GT- load, cost of primary fuel, nominal interest rate, retail price index (rate of inflation), and rate of carbon tax. Redistribution of expenses between heat and electricity is proportional to exergy embedded into corresponding energy outflux.

Table 2. Key performance characteristics (GT and Unfired HRSG system)

		Key Indicators	
		L_{GT}=1	L_{GT}=0.7
1	Fuel consumption	1.06E+07 GJ(primary fuel)/y	8.24 E+06 GJ(primary fuel)/y
2	Annual CO_2 emissicn	6.81 E+05 tone CO_2/y	5.27 E+05 tone CO_2)/y
3	Cumulative PV of expenses (10 years)	[3.56E+02]] mln USD	[3.00E+02] mln USD
4	Levelised cost of electricity	3.02E-02 USD/kWh(electr)	3.63E-02 USD/kWh(electr)
5	Levelised cost of heat	7.30E-03 USD/kWh(therm)	8.78E-03 USD/kWh(therm)

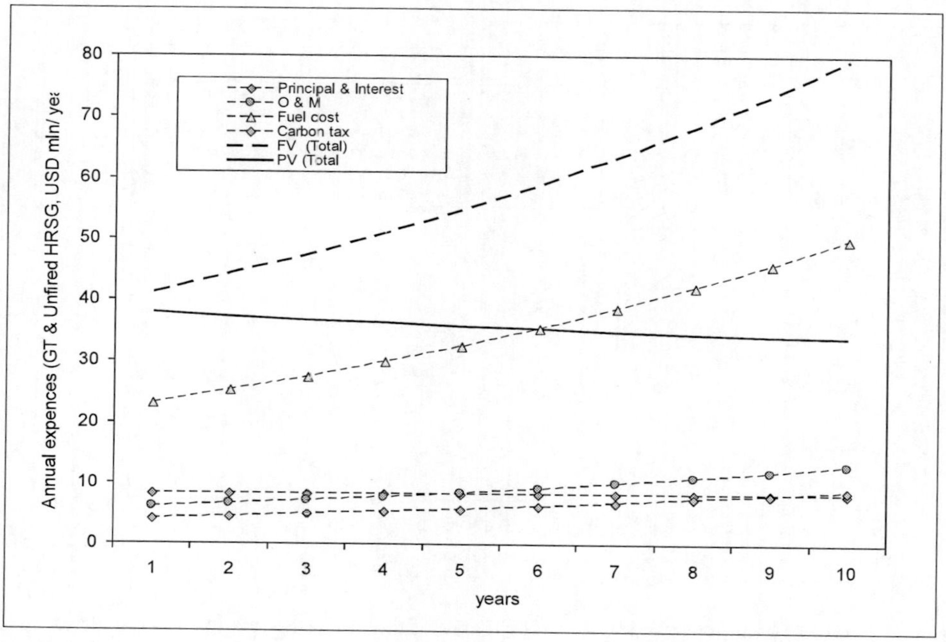

Figure 10. Calculated projections of manufacturing expenses (GT_ Unfired HRSG subsystem): (1) recovery of borrowed capital (principal and interest); (2) primary fuel cost; (3) O&M cost; (4) carbon tax; (5) FV of overall expenses; (6) PV of overall expenses. Basic techo- economic assumptions (for L_{GT}=1) are given in Table 1

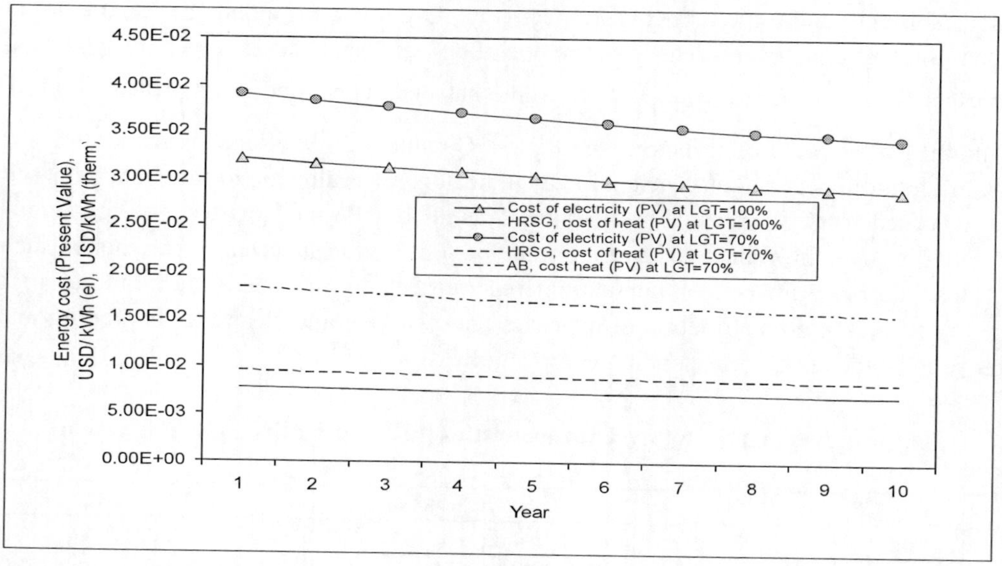

Figure 11. Lon- term projections of electricity and low- grade heat cost produced by HRSG and AB at different operating load of GT

9. ECONOMIC MODEL FOR ENERGY GENERATION BY AUXILIARY BOILER (AB)

The boiler is used to supply shortage of heat required to MSF, (in order to maintain constant production of MSF). Operating parameters of AB are dependent upon load of GT. For the analysis to simplify, the technological parameters of boiler such as fuel consumption and heat output are expressed against load of GT. Load of AB can be expressed in terms of load of GT as follows: $L_{AB} = (1 - L_{GT})/(1 - L_{\min(GT)})$. Specific fuel consumption of AB vs. load is approximated by Eq. (7). Estimation of required size of AB is based on shortage of heat energy when load of GT drops to 20%, that corresponds to required size of AB. In considered case the AB capacity is assumed to be 3.9 10+9 MJ(heat)/year (or ~ 124 MW(thermal)). According to outlined premises the thermal output of the boiler ranges from zero to 124 MW(thermal) when GT-load varies from 100 to 20%. (When the GT operates at 100% load the boiler is off).

This model generates there groups of characteristics, first-description of load-dependent behavior of manufacturing expenses for AB; second- calculation of itemized structure of manufacturing expenses for the base year (at existing GT-load distribution over the year), third- calculation of long-term projection over the economic life of the capital and "levelized" cost of heat raised by AB. The model is based on assumptions listed in table 1 using data generated by technological and environmental sub-models as an input.

Structure of manufacturing expenses consists of load-dependent and load-independent segments (1) load-dependent influx of primary fuel; (2) load-dependent output of heat, (3) load-dependent output of CO_2 emissions, (4) load-independent recovery of capital (interest plus principal) and (5) load-independent O&M costs. The first group of characteristics of the model estimate load- dependent behavior of cost of heat generated by AB. Set of projections is shown in Figures 12-13

Annual expenses are subdivided into load-independent and load-dependent segments as well. They are: (1) load- independent charges for recovery of capital (principal and interest); (2) O&M expenses; (3) fuel expenses; (4) carbon tax. Future value of expenses is estimated using real discount rate (see Eq. 3). Annual cost of capital recovery is evaluated as annuity, (see Eq. 5). Accounting of CO_2 emission is done through imposition of carbon tax, where consumed fuel carbon is used as base for evaluation of carbon tax, (see [10-11]). Sum of annual expenses, $C_{\Sigma AB}$, is expressed as follows:

$$C_{\Sigma(AB)} = \left[C_{capital} + C_{O\&M}\right] + \left[C(l)_{fuel} + C(l)_{C-tax}\right] \tag{27}$$

Calculated data on long term projections over economic life of the capital based on this submodel are attached in Appendix B. The matrix contains itemised structure of annual expenses and gives indicators such as PV and cost of low-grade heat produced by AB. Any row corresponds to definite group of expenses. Any column labelled "year of operation" contains the expenditures allocated to corresponding year, $C_{\Sigma(AB)}$ Their value has been converted to its present value (PV) by multiplying the discount factor, (see Eq. 4). Cumulative PV of expenses and "levelized" cost of produced heat and electricity are key indicators for evaluation of the subsystem, where the "levelized" cost is expressed as the ratio of cumulative PV of expenses over economic life of capital to overall production over the same period. It is influenced by regime parameters such as GT- load, cost of primary fuel, nominal interest rate, retail price index (rate of inflation), and rate of carbon tax.

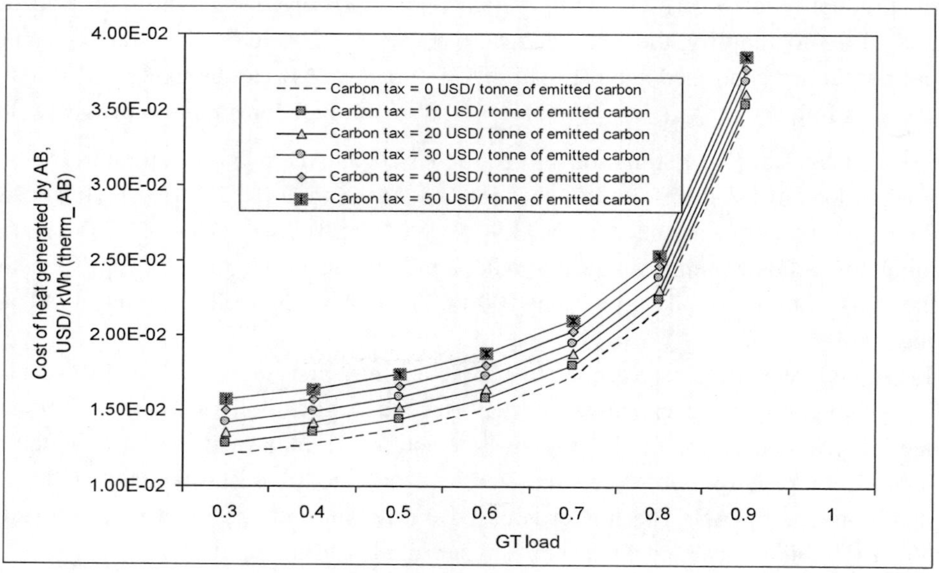

Figure 12. The cost of heat generated by AB vs. load of the power generating system at various level of carbon tax (Cost of primary fuel is 2 USD /GJ)

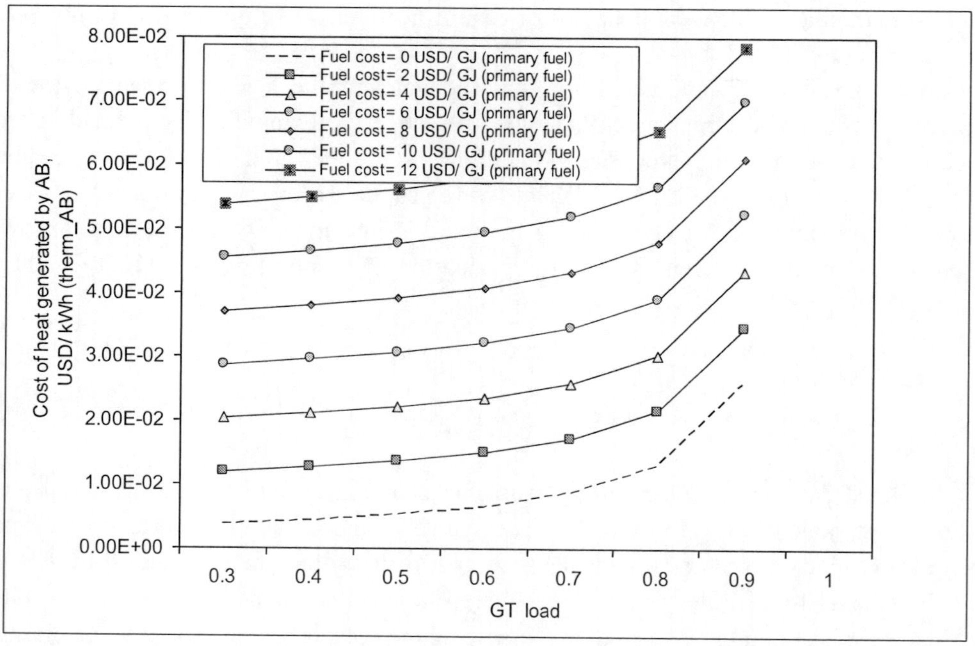

Figure 13 The cost of heat generated by AB vs. load of the power generating system at various level of the primary fuel cost (No carbon tax)

Table 3. Basic assumptions (AB system)

	Basic Assumptions	
1	Size of AB	124 MW(therm)
2	AB- load variation	from 0 to 100 % (equivalent thermal output: from 0 to 124 MW(therm)*
3	Specific capital cost	130 USD/kW(therm).
4	Initial capital investment (DC+IC)	15 mln USD
5	Nominal Interest Rate	0.1
6	Retail price index (Inflation rate)	0.01
7	Real Interest Rate	0.09
8	Cost of primary fuel	1.0 USD/GJ
9	Carbon tax	20.0 USD/tone of emitted carbon
9	O&M costs:	1.12 mln USD /year

Table 4. Key performance characteristics of AB

	Key Indicators		
		L_{GT}=1	L_{GT}=0.7
1	Fuel consumption	AB is switched off	1.73 E+06 GJ (primary fuel)/y
2	Annual CO_2 emission		1.1 E+05 tonne CO_2)/y
3	Cumulative PV of expenses (10 years)		[71.2 mln USD]
4	Levelised cost of heat		1.67E-02 USD /kWh(therm)

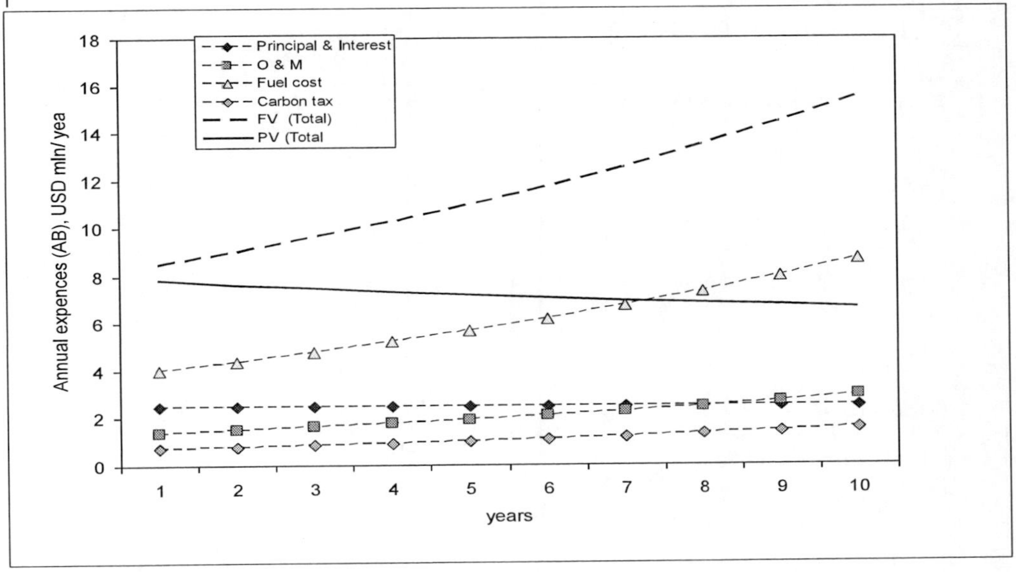

Figure 14 Calculated projections of manufacturing expenses for auxiliary boiler (AB): (1) recovery of borrowed capital (principal and interest); (2) primary fuel cost; (3) O&M cost; (4) carbon tax; (5) FV of overall expenses; (6) PV of overall expenses. Basic techo- economic assumptions (for LGT=0.7) are given in Table 3.

10. ECONOMIC MODEL FOR MSF DESALINATION

This model generates there groups of characteristics, firstly load-dependent behavior of manufacturing expenses for MSF system; second- calculation of itemized structure of manufacturing expenses for base year (at existing GT-load distribution over the base year), third- calculation of long-term projection over economic life of the capital and "levelized" cost of product. The model is based on assumptions listed in table 1 using data generated by technological and environmental sub-models as an input.

Table 5. Basic assumptions for MSF system

	Basic Assumptions	
1	Size of MSF subsystem:	60000 m^3/day
2	PR	10
3	Specific heat energy consumption	61 kWh(therm) /m^3(distillate);
4	Auxiliary electricity consumption [24]	3 kWh(el)/m^3(distillate)
5	Specific capital cost	1500 USD/ m^3/day
6	Initial capital cost (DC+IC)	90 mln USD,
7	Nominal interest rate	0.1
8	Retail price index (Inflation rate)	0.01
9	Real interest rate	0.09
10	Cost of primary fuel	2.0 USD/GJ
11	Carbon tax	20.0 USD/tone of emitted carbon
12	Annual O&M costs (MSF)	5.4 mln USD/year

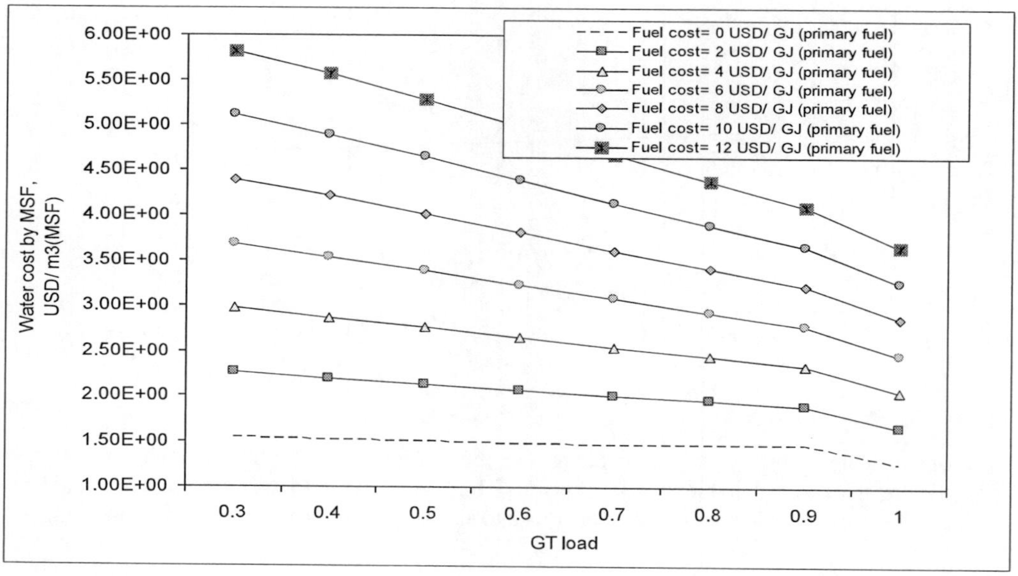

Figure 15. The cost of water produced by the MSF vs. load of the power generating system at various level of the primary fuel cost (No carbon tax)

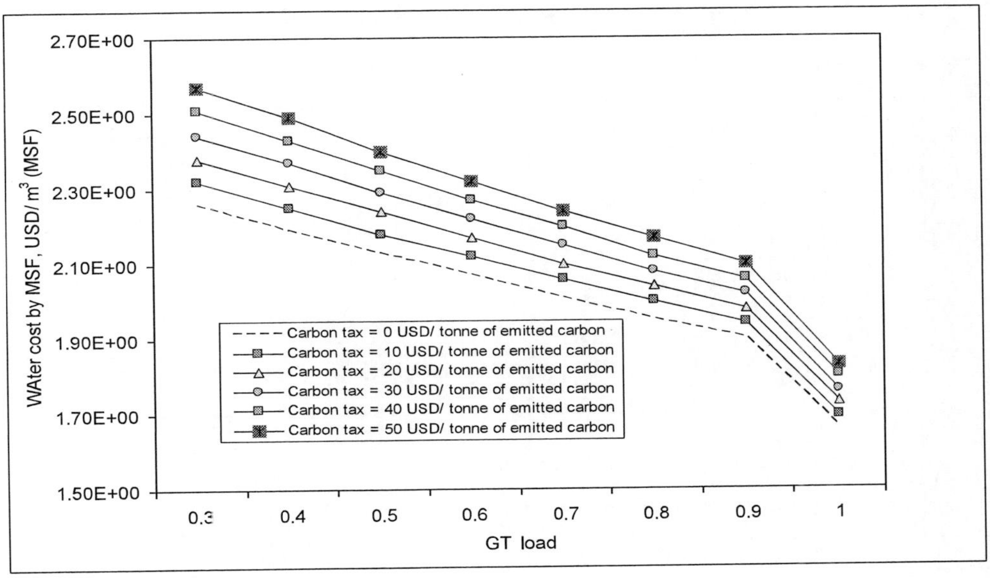

Figure 16. The cost of water produced by the MSF vs. load of the power generating system at various level of carbon tax (The cost of primary fuel is 2 USD/GJ)

Structure of manufacturing expenses consists of load-dependent and load-independent segments (1) load-dependent input of energy generated by GT, HRSG and AB; (2) load-dependent output of water; (3) load-independent recovery of capital (interest plus principal) and (4) load-independent O&M costs. The first group of characteristics of the model estimate load- dependent behavior of water produced by MSF. Set of projections is shown in Figures 15-16

Annual expenses are subdivided into load-independent and load-dependent segments as well.. They are: (1) load- independent charges for recovery of capital (principal and interest); (2) O&M expenses; (3) energy (thermal & electrical) expenses; Future value (FV) of expenses is estimated using real discount rate (see Eq. 3). Annual cost of capital recovery is evaluated as annuity, (see Eq. 5). Sum of annual expenses, $C_{\Sigma(MSF)}$, is expressed as follows:

$$C_{\Sigma(MSF)} = \left[C_{capital} + C_{O\&M}\right] + \left[C(l)_{heat-HRSG} + C(l)_{heat-AB} + C(l)_{electr}\right] \tag{28}$$

Calculated data on long-term projections over economic life of the capital based on economic submodel are attached in Appendix C. The matrix contains itemised structure of annual expenses and gives indicators such as PV and cost of water. . Any row corresponds to definite group of expenses. Any column labeled "year of operation" contains the expenditures allocated to corresponding year, $C_{\Sigma(MSF)}$. Their value has been converted to its present value (PV) by multiplying the discount factor, (see Eq.4). Cumulative PV of expenses and "levelized" cost of water is key indicator for evaluation of the subsystem, where the "levelised" cost is expressed as the ratio of cumulative PV of expenses over economic life of capital to overall production over the same period. It is influenced by regime parameters such as GT- load, cost of primary fuel, nominal interest rate, retail price index (rate of inflation), and rate of carbon tax.

Table 6. Key performance characteristics for MSF

	Key Indicators		
		L_{GT}=1	L_{GT}=0.7
1	Cumulative PV of expenses (10 years)	[2.94E+02] mln$	[3.34E+02]
2	Levelized cost of water	1.48 USD/m^3(MSF)	1.86 USD/m^3 (MSF)
3	Energy consumption		
	(1) Heat generated by HRSG,	1.46 E+09 kWh (therm-HRSG)/y	1.02 E+09 kWh (therm-HRSG)/y
	(2) Heat generated by AB	0 (AB is switched off)	4.07 E+08 kWh (therm-AB)/y
	(3) Electricity generated by GT	6.57 E+07 kWh (electr)/y	6.57 E+07 kWh (electr)/y
4	Allocated CO2 emissions		
	(1) Allocated to heat by HRSG	1.92 E+05 tone CO_2/y	1.63 E+05 tone CO_2/y
	(2) Allocated to heat by AB	0 (AB is switched off)	1.11 E+05 tone CO_2/y
	(3) Allocated to electr by GT	7.85 E+05 tone CO_2/y	9.46 E+05 tone CO_2/y
	Total CO_2 emissions	9.77 E+05 tone CO_2/y	12.2 E+05 tone CO_2/y

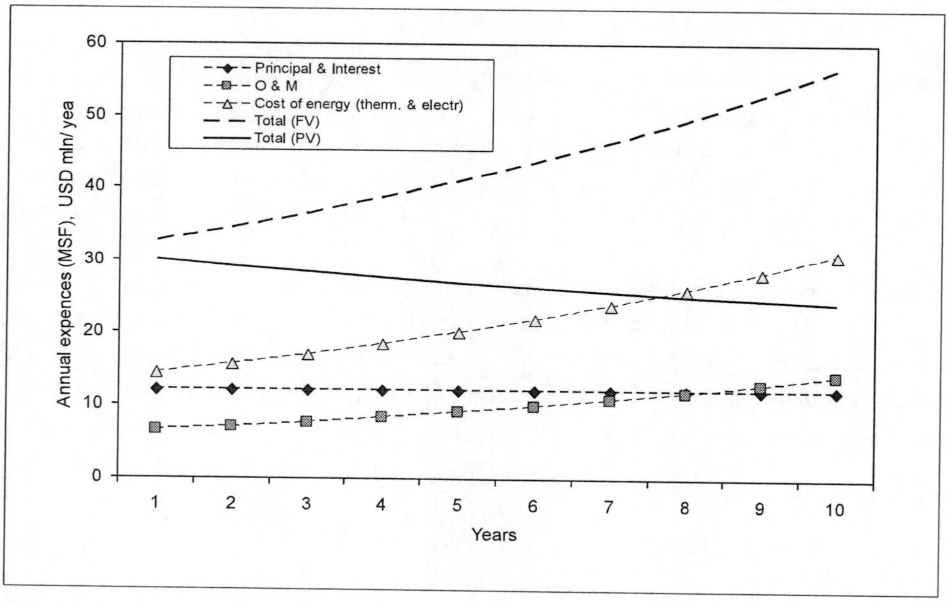

Figure 17. Calculated projections of manufacturing expenses (MSF system): (1) recovery of borrowed capital (principal and interest); (2) cost of energy (thermal & electrical); (3) O&M cost; (4) FV of overall expenses; (5) PV of overall expenses

11. Economic Model for RO Desalination

This model generates there groups of characteristics, firstly load-dependent behavior of manufacturing expenses of RO system; second- calculation of itemized structure of manufacturing expenses for base year (at existing GT-load distribution over the base year), third- calculation of long-term projection over economic life of the capital and "levelized" cost of product. The model is based on assumptions listed in table 1 using data generated by technological and environmental sub-models as an input.

Structure of manufacturing expenses consists of load-dependent and load-independent segments namely: (1) load-dependent influxes of energy generated by GT (2) load-dependent output of water; (3) load-independent recovery of capital (interest plus principal) and (4) load-independent O&M costs. The first group of characteristics of the model estimate load-dependent behavior of the water cost produced by RO. Set of projections is shown in Figures 18-19

Table 7. Basic assumptions (RO system)

	Basic Assumptions	
1	Size of RO subsystem	60000 m^3/day;
2	Specific energy consumption	5 kWh(el)/m^3(perm)
3	Specific capital cost	1000 USD /m3/day
4	Initial capital investment (DC+IC)	60 mln USD,
5	Nominal Interest Rate	0.1
6	Retail price index (Inflation rate)	0.01
7	Real Interest Rate	0.09
8	Cost of primary fuel	2.0 USD /GJ of primary fuel
9	Carbon tax	20.0 USD /tone of emitted carbon
10	Annual O&M costs (RO)	7.2 mln USD /year.

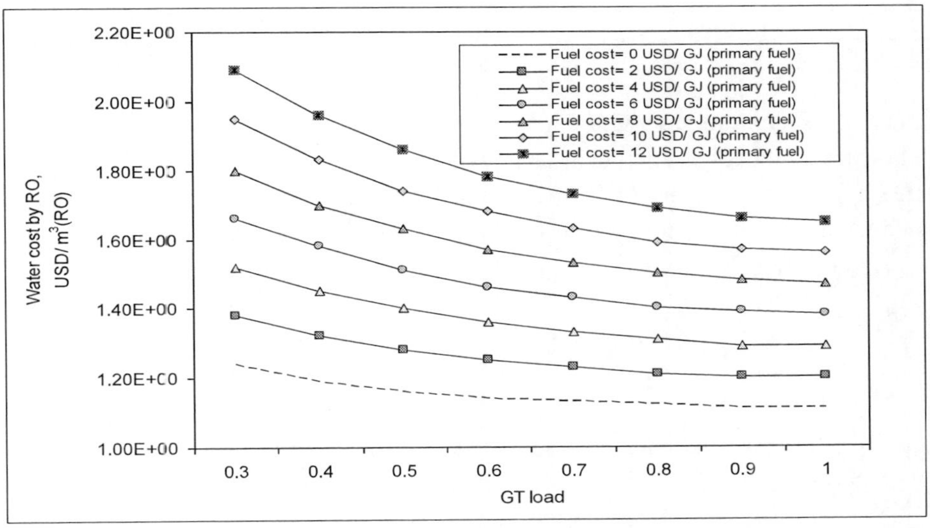

Figure 18. The cost of water produced by RO vs. load of the power generating system at various level of the primary fuel cost (No carbon tax)

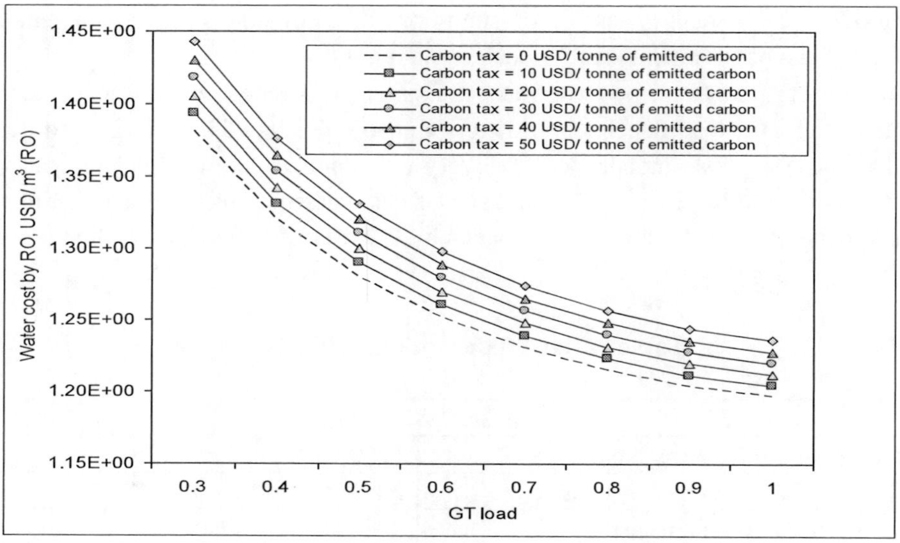

Figure 19. The cost of water produced by RO vs. load of power generating system at various level of carbon tax (Cost of primary fuel is 2 USD/GJ)

Table 8. Key performance characteristics for RO system

	Key Indicators		
		L_{GT}=1	L_{GT}=0.7
1	Cumulative PV of expenses (10 years)	[1.99E+02] mln USD	[2.06E+02] mln USD
2	Levelised cost of water	1.05 USD /m^3(RO)	1.09 USD /m^3(RO)
3	Energy consumption	1.10E+08 kWh (el)/y	1.10E+08 kWh (el)/y
4	Allocated CO_2 emissions	5.88E+04 tone CO_2/y	7.25E+04 tone CO_2/y

Annual expenses are subdivided into load-independent and load-dependent segments as well. They are: (1) load- independent charges for recovery of capital (principal and interest); (2) O&M expenses; (3) energy (thermal & electrical) expenses; Future value of expenses is estimated using real discount rate (see Eq. 3). Annual cost of capital recovery is evaluated as annuity, (see Eq. 5). Sum of annual expenses, $C_{\Sigma(RO)}$, is expressed as follows:

$$C_{\Sigma(RO)} = \left[C_{capital} + C_{O\&M}\right] + \left[C(l)_{electr}\right] \tag{29}$$

Long term projections over economic life of capital based on economic submodel are presented in Appendix D. The matrix contains itemised structure of annual expenses and gives indicators such as PV and cost of water. Any row corresponds to definite group of expenses. Any column labelled "year of operation" contains the expenditures per corresponding year, $C_{\Sigma(RO)}$. Their value has been converted to its present value (PV) by multiplying the discount factor, (see Eq.4). The cumulative PV of expenses and "levelized" cost of water is key indicator for evaluation of the subsystem, where the "levelised" cost is expressed as the ratio of cumulative PV of expenses over economic life of capital to overall production over the same period. It is influenced by parameters such as GT- load, cost of primary fuel, nominal interest rate, retail price index (rate of inflation), and rate of carbon tax.

12. Comparison of Long–Term Projections for RO and MSF (Sensitivity of PV of Expenses and "Levelized" Water Cost with Respect to Nominal Interest Rate, Cost of Primary Fuel and Carbon Tax Rate)

A ccomparison of alternative processes (MSF and RO) is based on the following indicators: (1) Cumulative PV of expenses over economic life of capital; (2) "Levelized" cost of water (expressed as ratio of cumulative PV of expenses to total output over the same period); (3) Sensitivity of water cost with respect to nominal interest rate, cost of primary fuel, carbon tax rate etc. Set of calculated profiles for RO and MSF at different cost of energy (expressed through different GT load) is shown below.

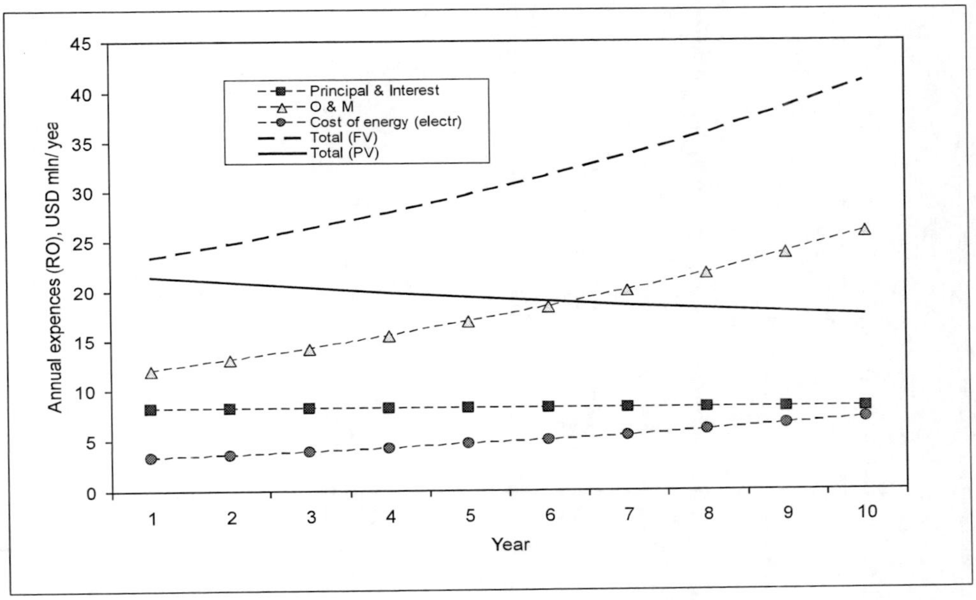

Figure 20 Calculated projections of PV and FV of manufacturing expenses for RO system: (1) recovery of borrowed capital (principal and interest); (2) cost of electrical energy; (3) O&M cost; (5) FV of overall expenses; (6) PV of overall expenses. Basic techo- economic assumptions (for L_{GT}=1) are given in Table 7

Table 9. Cumulative PV of expenses over economic life of capital vs. cost of primary fuel (MSF and RO systems). Techno- economic assumptions: Retail Price Index= 0.01; Nominal Interest Rate =0.1; Carbon tax= 0; GT-Load= 100%

	Cost of Primary Fuel, USD /GJ	0	2	4	6
1	Cumulative PV of expenses for MSF subsystem over 10 years, mln USD	1.82E+02	2.54E+02	3.26E+02	3.98E+02
2	Cumulative PV of expenses for RO subsystem over 10 years, mln USD	1.73E+02	1.89E+02	2.06E+02	2.22E+02

12.1. Influence of Cost of Primary Fuel

The calculated projections of "levelized" water cost vs. cost of primary fuel are shown in Fig. 21 where the water cost is expresses as ratio of cumulative PV of expenses over economic life of capital (10 years) to total output over the same period.

The sensitivity of water cost for price of primary fuel ranges from 3.92E-01 to 1.80E-01 and from 9.70E-02 to 7.38E-02 in case of MSF and RO respectively. (The sensitivity is expressed as a ratio of nondimensional growth of water cost to the increase of the cost of primary fuel as $[\Delta C / C]$ $[\Delta F / F]$ Where, DF -is the variation of the cost of primary fuel; DC, is the growth of the "levelized" water cost corresponding to ΔF

An analysis of calculated projections confirms that the MSF being more energy consuming is more vulnerable to the growth of fuel cost. The PV of expenses and cost of water produced by MSF are more dependent upon the fuel cost than RO.

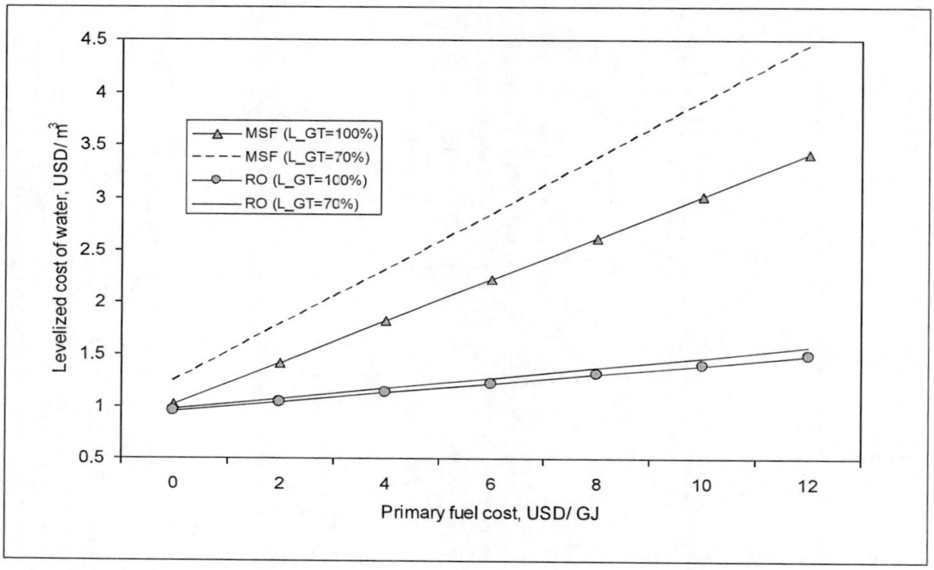

Figure 21. "Levelized" cost of water (over 10 years) vs. cost of primary fuel. Techno-economic assumptions: nominal interest rate = 0.1 retail price index = 0.01: carbon tax= 0: (L_{GT}= 100% and 70%) [Techno-economic assumptions are in Tables 5 and 7].

12.2. Influence of Carbon Tax Rate

The cumulative PV of expenses is influenced by the rate of the carbon tax. Calculated values for MSF and RO subsystems are shown in Table 10.

The calculated projections of "levelized" water cost vs. rate of carbon tax are shown in Figure 22 where the water cost is expresses as a ratio of cumulative PV of expenses over economic life of the capital to total output over the same period. The sensitivity of water cost with respect to the rate of carbon tax ranges from 2.11E-02 to 2.63E-02 and from 9.43 E-03 to 9.62E-03 in case of MSF and RO respectively. (The sensitivity is expressed as ratio of dimensionless growth of water cost to an increase of the rate of carbon tax as $[\Delta C / C]/[\Delta T / T]$ Where, ΔT - is the variation of the rate of carbon tax; ΔC , is the growth of "levelized" water cost corresponding to ΔT. An analysis of calculated data confirms that MSF being environmentally more unfavourable is more vulnerable to an increase of the carbon tax.

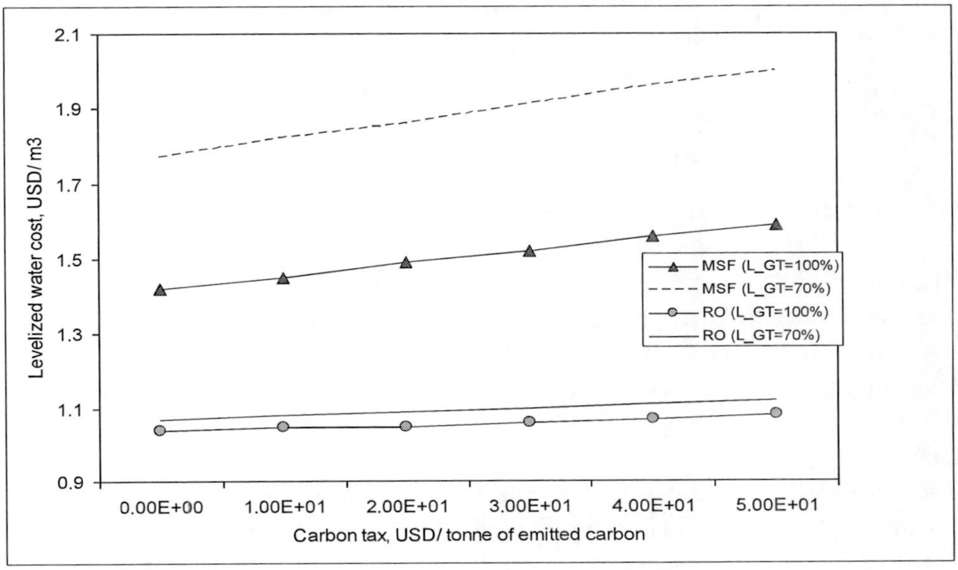

Figure 22. "Levelized "cost of water (over 10 years) vs. carbon tax. Techno-economic assumptions: nominal interest rate = 0.1 retail price index = 0.01: cost of primary fuel= 2 USD/GJ: (L_{GT}= 100% and 70%) [Techno-economic assumptions are in Table 5 & 7

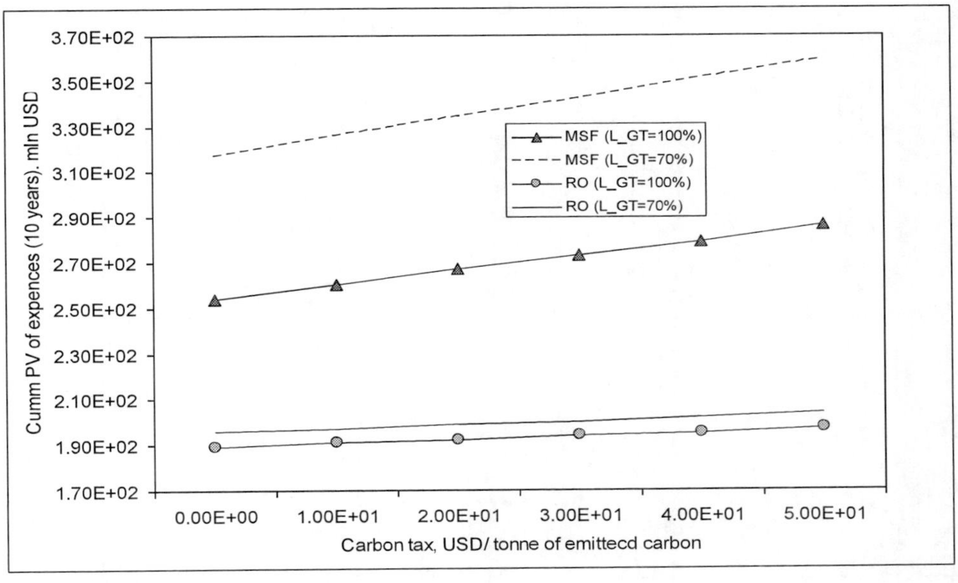

Figure 23 Cumulative PV of expenses (over 10 years) vs. carbon tax. Techno-economic assumptions: nominal interest rate = 0.1 retail price index = 0.01: cost of primary fuel= 2 USD/GJ: (L_{GT}= 100% and 70%) [Techno-economic assumptions are in Table 5 & 7.

Table 10. Cumulative PV of expenses vs. rate of carbon tax (MSF and RO). Techno-economic assumptions: Nominal Interest Rate =0.1; Retail Price Index= 0.01; Cost of primary fuel =2 USD /GJ; L_{GT}=1

	Carbon tax, USD/tonne of emitted carbon	0	10	20	30	40	50
1	Cumulative PV of expenses for MSF subsystem (10years), mln USD	2.54E+02	2.60E+02	2.67E+02	2.73E+02	2.79E+02	2.86E+02
2	Cumulative PV of expenses for RO subsystem (10 years), mln USD	1.89E+02	1.91E+02	1.92E+02	1.94E+02	1.95E+02	1.97E+02

12.3. Influence of Nominal Interest Rate

The nominal interest rate, a measure of a cost of capital, is an essential economic factor that the cumulative PV of expenses is influenced by. calculated projections of PV for the MSF and RO subsystems are given in Table 11

The calculated projections of "levelized" water cost vs. nominal interest rate are shown in Figure 24. (where the water cost is expresses as a ratio of cumulative PV of expenses over economic life of the capital to total output over the same period).

The sensitivity of water cost with respect to interest rate ranges from 7.75E-03 to 1.05E-02 and from 2.02E-03 to 2.41E-03 in case of MSF and RO respectively. The sensitivity is expressed as the ratio of dimensionless changes in water cost to growth of nominal interest rate as $[\Delta C / C]/[\Delta i / i]$ Where, Δi - is the variation of the nominal interest rate; ΔC , is the growth of "levelized" water cost corresponding to Δi .

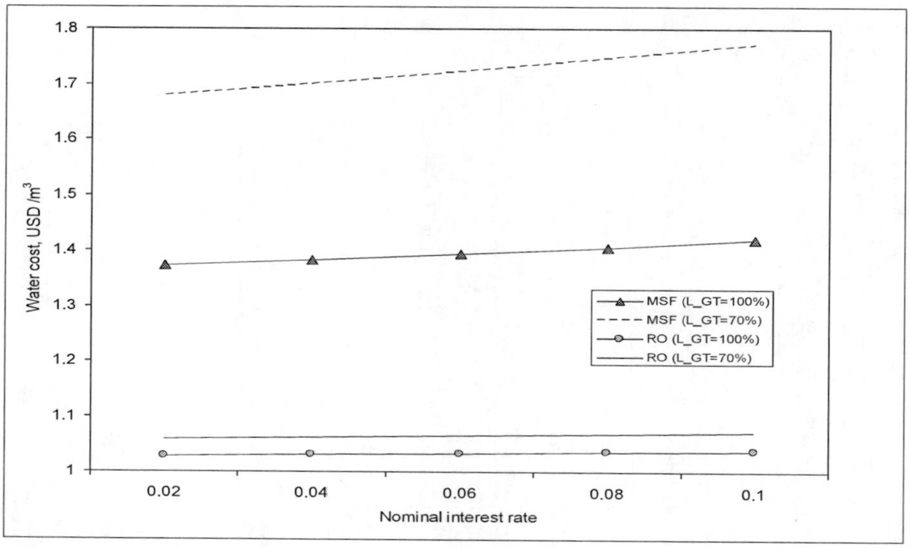

Figure 24 "Levelized "water cost (over 10 years) vs. nominal interest rate. Techno-conomic assumptions: cost of primary fuel= 2 USD/GJ; carbon tax =0: (L_{GT}= 100% and 70%) [Techno-economic assumptions are in Table 5 & 7.

Table 11. Cumulative PV of expenses over economic life of capital vs. nominal interest rate (MSF and RO subsystems) Techno- economic assumptions: Retail Price Index= 0.01; Cost of primary fuel= 2 USD /GJ; Carbon tax= 0; L_{GT} =1

	Nominal Interest Rate, dimensionless	0.02	0.04	0.06	0.08	0.1
1	Cumulative PV of expenses for MSF subsystem over 10 years, mln USD	2.468E+02	2.48E+02	2.49E+02	2.52E+02	2.540E+02
2	Cumulative PV of expenses for RO subsystem over 10 years, mln USD	1.88E+02	1.88E+02	1.88E+02	1.89E+02	1.89E+02

An analysis of calculated data confirms that MSF, which is more capital heavy, is more vulnerable to an increase of the nominal interest rate. The PV of expenses and cost of water produced by MSF are more rate-dependent than for RO.

13. Hybridization of Co-Generative Plant with Reverse Osmosis and Aquifer Storage (Enhancement of Flexibility of Hybrid System)

The growth and seasonal fluctuations of the power and water demand require reconfiguration and increase of flexibility of conventional co-generative systems though their integration with RO and the ASR technologies,[12, 56-57].

13.1. Aqua Storage and Recovery (ASR) Technology

The concept of ASR technology has been extensively scrutinized, [57-62]. Alkatheeri in [12, 57-58] proposed restructuring the management through development of hybrid systems including the ASR. According to the study, [12, 57, 58], the hybridization of existing co-generative plants with the ASR technology is characterized by technological and economic advantages. The ASR technology is an example of multipurpose technology; it can be used for (A) seasonal and water reuse water storage; (B) system for creation of strategic water reserve and it can be (C) integrated or hybridized with existing co-generative (or desalination) system.

Different functional, technical and economic aspects of the ASR technology are considered in [57-59]. Semi-analytical model for predicting the quality of water recovered by an ASR system is proposed by in [60]. There are many domains of potential applications of the ASR technology they are:

(A) **ASR for seasonal conservation and water reuse** The ASR is a well known technology for accumulating large volumes of water where natural underground aquifer is used as a storage. Along with seasonal storage the ASR technology can be applied as storage for reuse of reclaimed wastewater as well. Different aspects of the ASR technology were considered by Pyne in [59]. The ASR-based technologies have been implemented in Australia, England, Canada, South Africa, and India. The ASR technology has become attractive water conservative technology in Australia since

1980. The cases where the ASR technology was used as storage in water reuse were considered by Dillon in [61]. In order to preserve native groundwater and restoration of groundwater, the ASR technology can be applied for reuse of reclaimed wastewater by artificial recharge of reclaimed wastewater into aquifer.

(B) ASR as a strategic water storage facility According to published data [62] the ASR can be considered as a strategic water storage facility along with being a tool to reduce overall cost of water production. According to data [62] Many GCC Countries have insufficient strategic water reserve facility [62]. An average strategic reserve would be sufficient only for 2 days; in this regard the ASR technology can be considered as storage for strategic purpose. The total capacity of water reservoirs of Abu Dhabi Water and Electricity Authority (ADWEA) are equal to 400 MIG that can satisfy only two days of water demand [57]. According to [63] the UAE is investigating the possibility of storing desalinated water underground. At the moment, the Emirate Abu Dhabi only has two or three days' water storing capacity. Injecting surplus water into would protect the country against red tides and opens up the possibility of using off- peak power supply to produce water storage. It is an idea that could change the concept of water and electricity management drastically.

The ADWEA has tendered the first contract in its program to establish the emirate's first full-scale aquifer storage and recovery project. the aim is to develop sufficient capacity to serve Abu Dhabi city and the surrounding area for up to 40 MIGD (181,800 m3/d) [63]. The water would be reserved and used in case of emergency. Peak water demand in the emirate of Abu Dhabi currently stands at approximately 2.8 million m3/d, though the emirate's existing water storage capacity consists largely of reserves at water production facilities, and its sufficient only for 2- 3 days of supply. The potential vulnerability of the emirate's desalination plants to phenomena such as algal blooms was demonstrated in December 2008, when the operation of the Fujairah plant was severely affected by red tide. Aqua storage and recovery ASR pilot project was initiated by Abu Dhabi National Oil Company in 2002 [63]. The level of water withdrawal under this project was at the level 10,000 m3/day continuously available for approximately one month. Feasibility study and pilot project for water produced by Fujairah Water and Power Plant was completed in 2003 in Eastern Region of Abu Dhabi [63].

(C) Hybridization of ASR with existing co-generative plant Seasonal and long-term variation of pattern of power and water demand cannot be provided by conventional co-generative systems with inflexible limits of variations of the power-to-water ratio. Published statistics [37] reveals the growth of cumulative gross annual demand and increase of amplitude of seasonal difference between the highest and lowest water demand. In this regard hybridization of the ASR technology with existing cogeneration plants makes the system more flexible in order to meet seasonal demand that allows optimization of water and power resource utilization [12, 64]. The ASR allows a treated water to be injected underground during the period with low water demand and recovered to the water distribution system during the periods with elevated water demand. Thus the scheme integrated with the ASR can provide season-dependent water demand. Relying upon the published data one of the prevalent tendencies consist in development and reorganization of existing co-generative systems aimed at extension of the limits of variation of the power to water ratio. Integration and hybridization of thermal desalination with electrically driven processes such as RO and the ASR technologies allows utilizing idle power and unused water that makes broader the range of variation of the power-to-water ratio. Techno- economic analysis of hybrid systems given in [10-12] is essential for countries like the UAE.

Some published studies by Write, Maliva, Missimer, Almulla, Awerbuch [62- 65, 14] emphasize the fact that hybridization of the ASR with desalination system reduce overall cost of water production and the hybrid system can be considered as a cost –effective technological option. According to [62] the ASR is a tool to reduce overall cost of water production. Data on capital costs of the ASR system [64] ranges from 30 to 70 [USD/m3/day].

13.2. Hybridization of Existing Co-Generative Plant with Reverse Osmosis and Aquifer Storage (Comparison of Schemes and Calculated Projections)

Analysis of the hybrid including gas turbine (GT), heat recovery steam generator (HRSG), auxiliary boiler (AB), MSF and RO desalination was presented in [10-12]. The proposed study is an attempt of comparison of the conventional co-generative scheme as a reference case with the hybrid innovative scheme. Unlike the reference case, the hybrid scheme implies the concept where the seasonal surplus of available unutilized power-generating capacity is consumed by RO and seasonal excess of water accumulated within underground aquifer. Feasibility and techno-economic aspects of hybridization of the existing co-generative plant with reverse osmosis and aquifer storage is considered below.

Reference case represents conventional co-generation plant attached to grid. (See Figure 1) It includes power plant and system for thermal desalination. Power plant, in turn, includes gas turbine (GT), heat recovery steam generator (HRSG) and auxiliary boiler (AB). This scheme (reference case) represents one of the technological options widely used. The reference case considered in this paper is based on the following assumptions: (1) maximum capacity of the power plant is 180 MW; (2) available power gen. capacity ranges from 80 to 100 % of the maximum capacity; (3) Seasonal production of the power plant ranges from 40 to 75% (that corresponds to the range 70-135 MW) of the maximum capacity; (4) maximum production capacity of existing MSF desalination plant is equal to 38 MGD; (5) available MSF desalination capacity is assumed to be 100 % of its maximum capacity; (6) seasonal production of water by MSF ranges form 400 to 560 MG per month.

Calculated profiles of seasonal variation of power production, external grid demand, internal consumption for thermal desalination and available power gen. capacity are shown in Figure 25. Current production of power plant is adjusted to current demand and it's equal to sum of external grid demand and internal demand required for thermal desalination. Current production of the power plant is less than the available capacity.

The required water production is assumed to be equal to water demand as well. There is no enormous seasonal variation of water demand. Seasonal profile of total water production (being equal to demand) is shown in Figure 26. For analysis to be simplified, the available published statistics [37] were approximated by parabolic function.

2.2 Hybrid schemes. Along with technologies incorporated into the reference scheme the hybrid scheme includes electrically-driven (RO) desalination and aquifer storage and recovery (ASR) technology where RO suppose to utilize seasonally unused electricity while the ASR accumulates excess of desalted water. Conceptual flow- diagram of this scheme is shown in Figure 27.

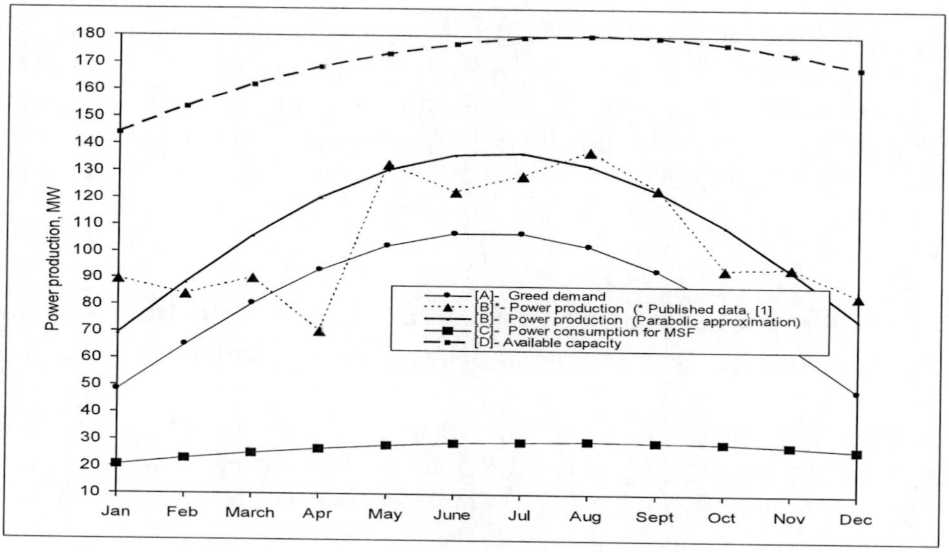

Figure 25. Calculated projections of seasonal variation of power demand (reference case)

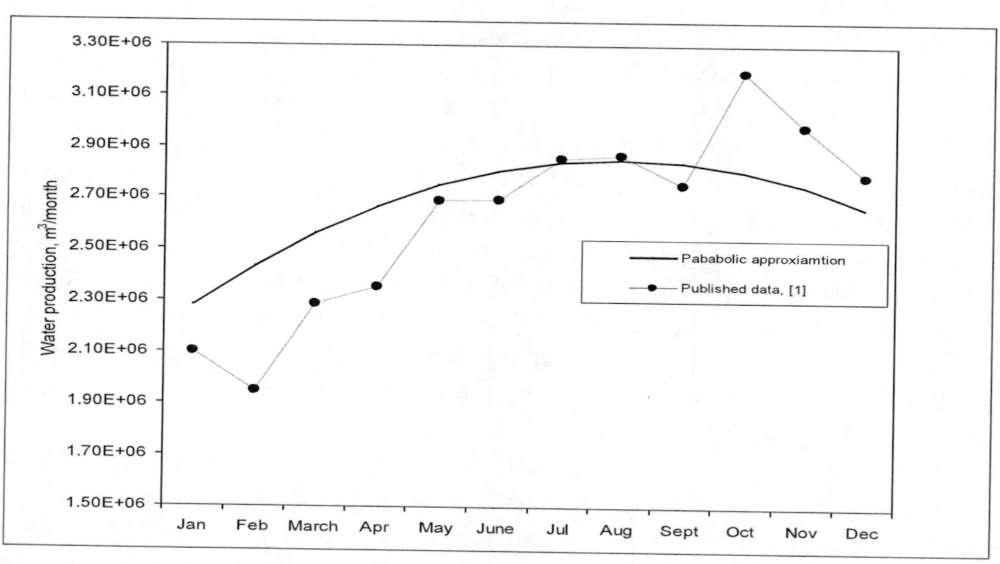

Figure 26. Calculated projections of seasonal variation of water demand

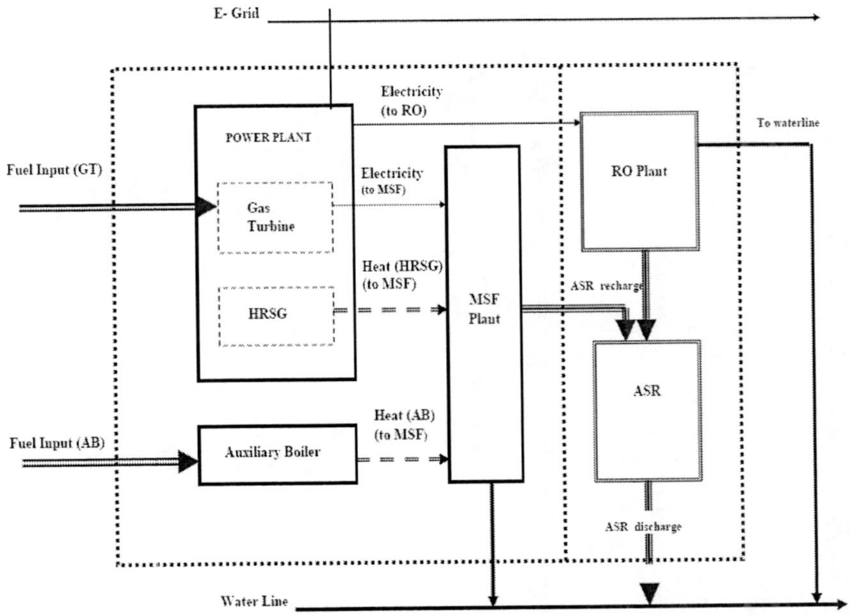

Figure 27. Flow-diagram of the hybrid scheme

Modeling of the hybrid scenario is based on the following assumptions: (1) Maximum capacity of the power plant is 180 MW; (2) Available power gen. capacity ranges from 80 to 100 % of the maximum output; (3) Current load of the power plant within the hybrid scheme is assumed to be equal to available power generating capacity (80 to 100 % of the maximum output); (4) Total production capacity of existing thermal desalination plant is equal to 38 MGD; (5) Seasonal variation of both water and power demand is considered in the model; (6) Specific energy consumption of projected RO is 6 kWh/m^3.

Seasonal variation of production of power plant (being equal available power gen. capacity), external grid demand, internal consumption for thermal desalination and unused power capacity (or surplus) is shown in Figure 28.

Current production of power plant being demand-independent is equal to the available capacity. This regime of operation being close to maximum generating capacity is characterized by elevated efficiency of fuel utilization, decreased specific CO_2 emissions and improved economic indicators. Within this regime the system produces extra power that can be stored or used for water desalination. Seasonal variation of (1) power production, (2) grid demand, (3) consumption for thermal desalination and (4) surplus of produced power is shown in Figure 28).

For the scenario based on co-generative system (180 MW- 38 MGD) operating at 80-100% of the maximum generating capacity, the unutilized capacity of power production ranges from 75 MW to 40 MW.

Seasonal variation of water production by MSF and RO, water demand and rate of ASR recharge is shown in Figure 29. Water production in this scenario is assumed to be demand-independent and based on available power generated by the power plant. The surplus of produced power can be used for RO-desalination. It's sufficient for production of 40- 80 MGD by RO. Unconsumed water can be to be accumulated within the underground aquifer. Calculated rate of seasonal variation of aquifer recharge ranges between 50 and 80 MGD.

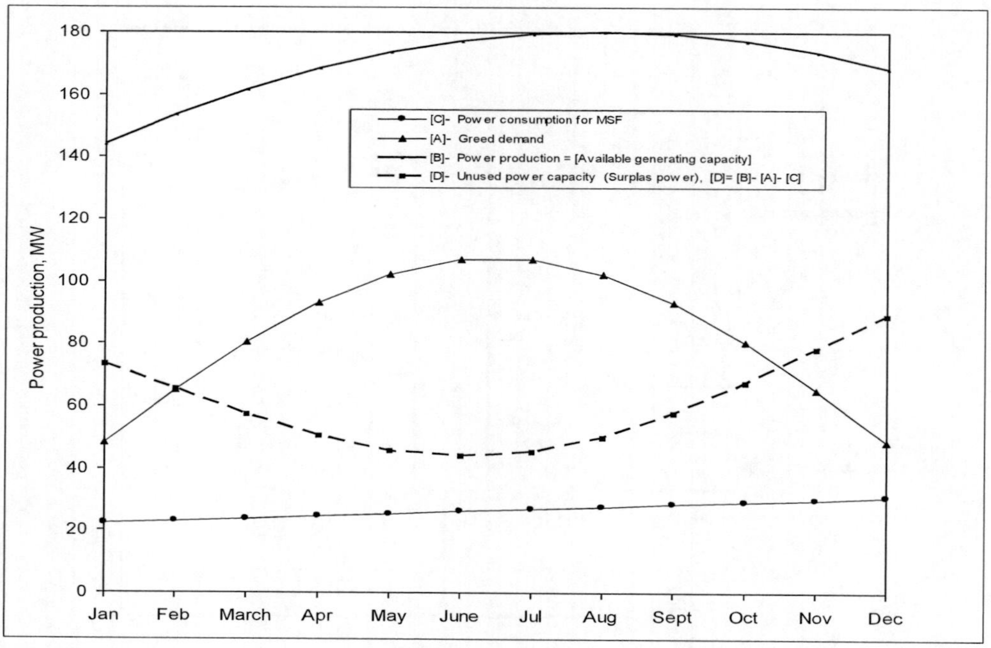

Figure 28 Calculated projections of seasonal variation of power generation and demand Hybrid scheme).

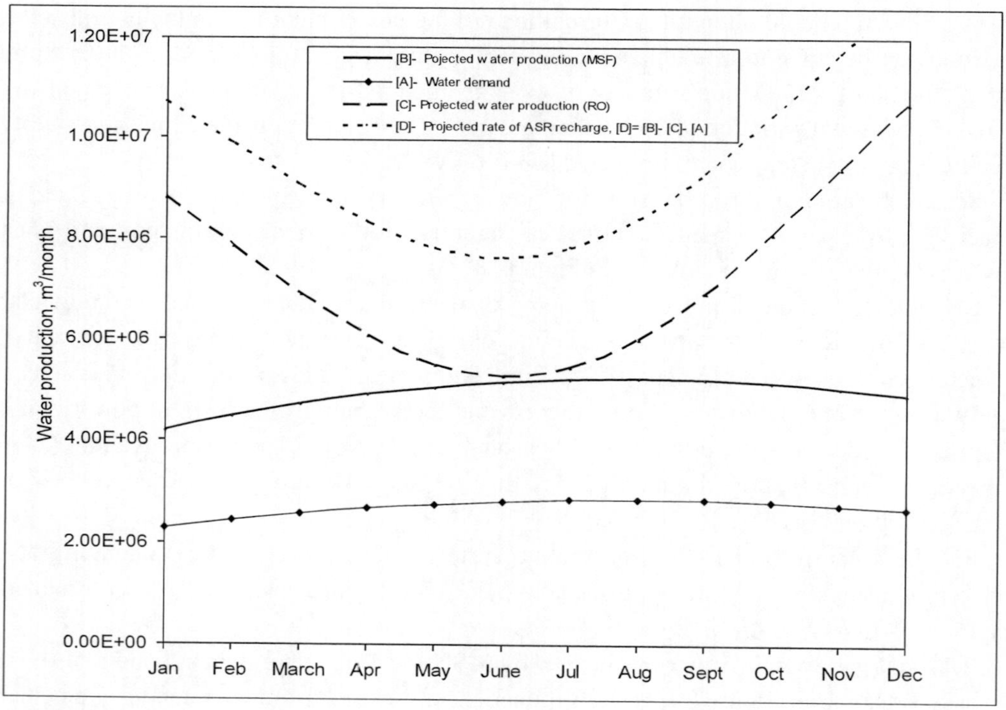

Figure 29 Calculated projections of seasonal variation of water demand and production (Hybrid scheme).

14. Conclusions

The current study confirms that the RO can successfully coexist with MSF rather than be a process that should replace it. An analysis of economic indicators such as PV of expenses and "levelized" cost of water favours the integrated RO-MSF system.

Thermal desalination (MSF) being more capital and resource consuming is characterised by higher PV of expenses. It is more vulnerable to growth of the following factors: (1) nominal interest rate, (2) cost of primary fuel, and (3) rate of carbon tax.

Hybridisation of thermally- and electrically-driven processes (MSF and RO) will have the following positive impact: (1) decrease specific capital and energy consumption; (2) decrease the level of carbon tax (due to decrease of allocated CO_2 emissions); and (3) decrease the "levelized" cost of water. In addition, it will decrease the sensitivity and, in turn, commercial risk of the system to fluctuation of the following factors: (1) nominal interest rate, (2) cost of primary fuel, and (3) rate of imposed carbon tax.

Based upon the current study and published data it was concluded that the combination of conventional co-generative systems with RO and ASR allows utilizing seasonally unused electricity while the ASR accumulates excess of desalted water. This system is characterized by the following technological, economic and strategic advantages: (1) increased range of variation of power-to-water ratio; (2) possibility to use seasonal surplus of unutilized power for production of additional quantity of water; (3) decreased specific CO_2 emissions; (4) improved capital utilization factor and other economic indicators; (5) creation of strategic water reserve for the region, etc.

The proposed method can be used as a submodel in an algorithm for synthesis of optimum configuration of the system. It can be of interest in evaluation of alternatives for feasibility studies as well.

Acknowledgments

The author is grateful to the National Energy and Water Research Center (ADWEA) for supporting this study and Eng. Entissar Elkatheeri for important comments and advices

15. Symbols

$C_{capital}$	Annual cost for recovery of capital (principal +interest), mln USD/y
$C_{O\&M}$	Annual O&M cost, mln USD/y;
$C(l)_{fuel}$	Annual load-dependent expenses for fuel, mln USD/y;
$C(l)_{C-tax}$	Carbon tax (is based on mass of consumed carbon), mln USD/y;
$C(l)_{heat-HRSG}$	Annual load-dependent cost of heat consumed by HRSG, mln USD/y;
$C(l)_{heat-AB}$	Annual load-dependent cost of heat consumed by AB, mln USD/y;
$C(l)_{electr}$	Annual load-dependent cost of electricity, mln USD/y;

$c(L)$	Specific CO_2 emissions, $kgCO_2/kWh$
E	Energy , J, kWh;
ε	Exergy , J, kWh;
ξ	Efficiency of the reference cycle, $\xi_i = 1 - T_i/T_0$, nondimensional
CPV	Cumulative present value, mln USD
FV	Future value of expenses, mln USD
$f(R,n)$	Discount factor, dimensionless
I_N	Nominal Rate of Interest, dimensionless
RPI	Retail Price Index (Rate of inflation), dimensionless
R	Real Discount Rate, dimensionless
X	Annuities, mln$
PV	Present value of expenses, mln USD
L	Load of operating system (LGT- for gas turbine, LAB- for auxiliary boiler)

Subscripts

TH	Thermal
EL	Electrical
AB	Auxiliary Boiler
GT	Gas Turbine
HRSG	Heat Recovery Steam Generator
MSF-	Multistage flush
RO	Reverse osmosis
O&M	Operating and maintenance

REFERENCES

[1] Desalination Market 2005- 2015 (A global assessment & forecast), *Media Analytics Ltd.*, Oxford, UK, 2004.

[2] The McILVAINE Company (www.mcilvainecompany.com).

[3] Strong growth forecast for membrane technology, *Membrane Technology (An International Newsletter)*, February 2007, 5-7.

[4] Membrane market set to reach $ 1 billion, *Membrane Technology (An International Newsletter)*, April 2007, 6.

[5] Daniel, E. Suk, Takeshi Matsuura, Membrane-based hybrid processes: a review, *Separation Science and Technology*, 2006, 41, 595-626.

[6] Matsuura, T. Progress in membrane science and technology for seawater desalination – a review, *Desalination*, 2001, 134 47-54.

[7] Fritzmann, C; Lowenberg, J; Wintgens, T; Melin, T. State-of-the art of reverse osmosis desalination, *Desalination*, 2007, 216, 1-76.

[8] Sherman May, Hybrid Desalination Systems, *MEDREC Series R& D Reports*, Project 97-AS-008b MEDRC, Muscat, Oman 2000.

[9] Hoffman, D; Zfati, A. *Hybrid desalination systems- Effective integration of membrane /thermal desalination and power technology*, MEDREC Series R& D Reports, Project 97-AS-008a MEDRC, Muscat, Oman, 2001.

[10] Agashichev, S; Ali, M. El-Nashar, Systemic approach for techno- economic evaluation of triple hybrid (RO, MSF and power generation) scheme including accounting of CO_2 emission, In*: Sustainable Development of Energy, Water and Environment Systems*, Ed. N.Lior, Swets & Zeitlinger Pbl, 2003.

[11] Agashichev, S. Analysis of integrated co-generative schemes including MSF, RO and power generating systems (Present value of expenses and "levelized" cost of water), *Desalination*, 2004, 164, 281-302.

[12] Entisar Al-Katheeri, Sergey, P. Agashichev, Feasibility of the concept of hybridization of existing co-generative plant with reverse osmosis and aquifer storage *Desalination*, 2008, 222, 87-95.

[13] Munasinghe, M. Water Supply and Environmental Management, (Developing world application), Westview Press, Inc. 1992.

[14] Awerbuch, L. Power-desalination and importance of hybrid ideas, In*: IDA World Congress on Desalination*, vol. 4, 184-192, 6- 9 Oct. 1997, Spain.

[15] Al-Marafie, MR. Prospects of hybrid RO/MSF desalting plants in Kuwait, *Desalination*, 1989, 72, 395-404.

[16] Al-Sofi, MAK; Hasan, AH; El-Sayed, EF. Optimum integrated power /MSF/RO plants. In: 1993 Research Activities and Studies, *Kingdom of Saudi Arabia*, Vol. 3, 1993, 61-83.

[17] El-Sayed, A; Ebrahim, S; Al-Saffar, A; Jawad, MA. Pilot study of MSF/RO hybrid systems, *Desalination*, 1998, 120, 121-128.

[18] Cogeneration– a New Generation at Middle East Electricity, In: *The Newsletter of the Middle East Desalination Research Centre*, Issue 13, August 2001.

[19] Mark Wilf, Klinko Kenneth, Optimization of seawater RO systems design, *Desalination*, 2001, 138, 299-306.

[20] Javier Uche, Luis Serra, Antonio Valero, Hybrid desalting systems for avoiding water shortage in Spain, *Desalination* , 2001, 138, 329-334.

[21] Al-Bahri, ZK; Hanburry, WT; Hodgkiess, T. Optimum feed temperatures for seawater reverse osmosis plant operation in MSF/RO hybrid plant, *Desalination*, 2001, 138, 335-339.

[22] Nashar, Ali M. The role of desalination in water management in gulf region In: Int. Conference. *Spanish Hydrological Plan and Sustainable Water Management*, (Environmental Aspects, Water Reuse and Desalination) Zaragossa, Spain, 13-14 June, 2001.

[23] Abdel-Jawad, M et al, Fifteen years of R&D program in seawater desalination at KISR. Part I. RO system performance, *Desalination*, 2001, 135, 155-167.

[24] Ebrahim, S; Abdel-Jawad, M; et al, Fifteen years of R&D program in seawater desalination at KISR. Part II, Pretreatment technologies for RO systems, *Desalination*, 2001, 135, 141-153.

[25] Kamaluddin, BA; Talavera, JL; Wangnik, K. Reverse osmosis seawater desalination systems in co generation plants for additional production of drinking water using off-peak electrical energy, In: *IDA World Congress on Desalination*, vol.1, 387-413, 6-9 Oct. 1997, Spain.

[26] Wangnik, K. A Global Overview of Water Desalination Technology and the Perspectives, In: Int. Conference. Spanish Hydrological Plan and Sustainable Water Management, (Environmental Aspects, Water Reuse and Desalination) *Zaragossa*, Spain, 2001, 13-14 June.

[27] Afgan, N; Carvalho, M. Multi-criteria assessment of new and renewable energy power plants, *Energy*, 2002, 27, 739-755.

[28] Naim, H; Afgan, Maria G. Carvalho, Darwish Al-Gobaisi, Indicators for Sustainability of Water Desalination System, *IDA World Congress on Desalination and Water Reuse*, Manama, Bahrain, March , 2002, 8-13.

[29] United Nations, Integrated Environmental and Economic Accounting, Series F, No 61, United Nations, New York, 1993.

[30] Andersen, S. The green tax reform in Denmark: shifting the focus of tax liability, *"Environmental Liability*, 1994, Vol 2, (No2), 29-41.

[31] Sterner, T. Environmental tax reform: the Swedish experience. Studies in Environmental and Development, Department of Economics, Goteborg University, 1994.

[32] European Parliament Fact Sheets 2001, (4.9.1. Environmental policy: General Principles).

[33] European Parliament Fact Sheets 2001, (3.4.7. The taxation of energy). 2001.

[34] Federal Plan for Sustainable Development 2000- 2004. (Adopted by the federal Government of Belgium on 20 July 2000, and laid down by Royal Decree of 19 Sept. 2000), Brussels, 2000.

[35] Agenda 21:Programme of Action for Sustainable Development, (Rio Declaration of Environment and Development), Rio de Janeiro, Brazil, 2-14 June 1992.

[36] Hanley, N; Spash, C. *Cost- Benefit Analysis and the Environment*, Edward Elgar Publ. Co., 1995.

[37] Statistical Report 1998-2004, Abu Dhabi Water & Electricity Company (ADWEC), Bin Desmal Publ, Abu Dhabi, 2005.

[38] Durham, B; Guendert, D; Rinch- Pfeifer, S. Integrated water resource management-through desalination, reuse and aquifer recharge, In: *IDA World Congress on Desalination and Water Reuse- Desalination: The Source o Sustainable Water Supplies*, Bahamas, USA, Sept 28- Oct 3, 2003.

[39] Costanza, R. *Ecological Economics: (The science and management of Sustainability)*, Columbia University Press, NY, 1991.

[40] Tim O'Riordan, *Ecotaxation*, Earthscan Pbl., UK, 1997.

[41] Ayres, R. U. Eco-thermodynamics and the second law, *Ecological Economics*, 26, 1998, 189-209.

[42] Duic, N; Juretic, F; Zeljko, M; Bogdan, Z. Croatia energy planning and Kyoto Protocol, *Energy policy*, 2005, 33, 1003-1010.

[43] Ronald Bailey, May 24, 2007, Carbon Taxes Versus Carbon Market (What's the best way to limit emissions? (http://reason.com/archives/2007.

[44] A renaissance may not come. *Economist,* May 19th, 2001, 29-31.

[45] J- Fr. Deroubaix, Fr. Leveque, The rise and fall of French Ecological Tax Reform: social acceptability versus political feasibility in the energy tax implementation process, *Energy policy*, 34, 2006, 940- 949.

[46] Alex Devine, Japan Studies Introducing Fossil-Fuels Tax in 2010, Yomiuri Says October 30, 2009, (www.bloomberg.com) Last Updated: October 30, 2009, 21, 01 EDT.

[47] Chen, B; Chen, G. *Modified ecological footprint accounting in analysis based on embodied exergy- a case study of the Chinese society* 1981-2001, *Ecological Economics*, 2007, 61, 355-376.

[48] Peters, MS; Timmerhaus, KD. *Plant Design and Economics for Chemical Engineers,* 4th Ed, McGraw-Hill, Inc, 1991.

[49] William, D. Baasel, *Preliminary Chemical Engineering Plant Design,* 2nd Ed., Van Nostrand Reinhold, NY, 1990.

[50] Szonyi, A; Fenton, R; White, J; Agee, M; Case, K. *Principles of engineering economic analysis*, Wall & Emerson, Inc., Toronto, Canada, 2003.

[51] Neil, M. Wade. Distillation plant development and cost update, *Desalination*, 136, 2001, 3-12.

[52] Master Plan for the Power and Water Requirements During the Period 1995 -2010 for WED. (Report HI Operation and User Manual for Optimization Model and Probabilistic model for Reliability), Bechtel Co. Ltd. 1996.

[53] 1999 Update to Master Plan for the Power and Water Requirements During the Period 1995 -2010 for ADWEA, Vol. 1. *Bechtel Co. Ltd*. 1999.

[54] Eastop, TD; Croft, DR. *Energy efficiency for engineers and technologists*, Longman Scientific & Technical, UK, 1990.

[55] Leitner, GF; Leitner, W. Life cycle and present worth concepts, applicable for large desalination plant? *Desalination*, 1994, 97, 291-300.

[56] Quteishat, K. Again to "Desalination and Integrated Management of Water", In: *The Newsletter of the Middle East Desalination Research Center*, Issue 31, July 2006.

[57] Entisar, S. Al-Katheeri, Towards an establishment of water resource management in Abu Dhabi emirate, *Water Resource Management*, 2006, Available from: http://dx.doi.org/10.1007/s11269- 006- 9151.

[58] Al-Katheeri, ES. Evaluation and policy alternatives for water resource development in Abu Dhabi, In: *2nd International Conference on Water Resources and Arid Development,* Riyadh, Saudi Arabia, 26- 29 Nov, 2006.

[59] Pyne, RD; Howard, JB. *Desalination/Aquifer Storage Recovery (DASR): a cost effective combination for Corpus Christi*, Texas, *Desalination*, 2004, 165, 363- 367.

[60] Ali Sedighi, Harald Klammler, Chris Brown and Kirk Hatfield, A semi-analytical model for predicting water quality from an aquifer storage and recovery system, *Journal of Hydrology, 2006,* Vol. 329, 3-4, 403-412.

[61] Dillon, P; Pavelic, P; Toze, S; Rink- Pfeiffer, S; Martin, R; Knapton, A; Pidsley, D. Role of aquifer storage in water reuse, *Desalination*, 2006, 188, 123-134.

[62] Almulla, A; Hamad, A; Gagalla, M. Aquifer storage and recovery (ASR): a strategic cost-effective facility to balance water production and demand fro Sharjah, *Desalination*, 2005, 174, 193-204.

[63] Maliva, R; Th. Missimer, M. Aquifer Storage and Recovery to Improve efficiency and Economics of water project, In: *IDA World Congress on Desalination and Water Reuse- Desalination: The Source o Sustainable Water Supplies*, Bahamas, Sept 28- Oct, 3, 2003.

[64] ADWEA mulls large-scale aquifer storage, *Global Water Intelligence,* Vol 10, (N10), Oct, 2009 (see also www.globalwaterintel.com)
[65] Write, RR; Th. Missimer, M. Power Optimization in Membrane Plants using Aquifer Storage and Recovery, In: *IDA World Congress on Desalination and Water Reuse*, 1999, Vol. 3, 361-376, San Diego, USA.
[66] Awerbuch, L. Integration of Desalination and Power The challenge in the Middle East, In: *Middle East Energy*, Dubai, UAE, 8-10 Oct, 2001.

APPENDIX A. CASH FLOW CHART FOR GT- HRSG SYSTEM

		Assumptions
1	Size of GT	100 MW
2	Specific capital cost, [14]	500 USD /kW(electr);
3	Initial capital investment (DC+IC)	50 mln USD
4	Nominal Interest Rate (	0.1
5	Retail price index (Inflation rate)	0.01
6	Real Interest Rate	0.09
7	Cost of primary fuel	2.0 USD /GJ
8	Carbon tax	20.0 USD /tone of emitted carbon
	Annual O&M	5 mln USD /year

	Year	1	2	3	4	5	6	7	8	9	10
	FV- discount factor	1.09	1.19	1.29	1.41	1.53	1.67	1.82	1.98	2.16	2.35
	PV-discount factor	0.92	0.84	0.77	0.71	0.65	0.60	0.55	0.51	0.46	0.43
		Load- independent segment of annual cost									
A1	Recovery of principal & Interest, mln USD/y	8.14	8.14	8.14	8.14	8.14	8.14	8.14	8.14	8.14	8.14
A2	O&M cost, mln USD/y	5.94	6.46	7.04	7.67	8.35	9.10	9.91	10.79	11.75	12.80
		Load- dependent segment of annual cost at L_{GT}=1									
A3	Fuel cost, mln USD/y	2.31E+01	2.51E+01	2.73E+01	2.98E+01	3.24E+01	3.53E+01	3.85E+01	4.19E+01	4.56E+01	4.97E+01
A4	Carbon-tax, mln USD/y	4.05E+00	4.41E+00	4.80E+00	5.23E+00	5.69E+00	6.20E+00	6.75E+00	7.35E+00	8.01E+00	8.72E+00
B1	Total cost (in FV), mln USD/y	4.12E+01	4.41E+01	4.73E+01	5.08E+01	5.46E+01	5.88E+01	6.33E+01	6.82E+01	7.35E+01	7.94E+01
B2	Total cost (in PV), mln USD/yr	3.78E+01	3.72E+01	3.66E+01	3.61E+01	3.56E+01	3.52E+01	3.48E+01	3.44E+01	3.41E+01	3.38E+01
C1	Cost of electricity (in FV), USD/kWh(electr)	3.50E-02	3.75E-02	4.02E-02	4.31E-02	4.64E-02	4.99E-02	5.37E-02	5.79E-02	6.24E-02	6.74E-02
C2	Cost of heat (in FV), USD/kWh(therm)	8.45E-03	9.06E-03	9.71E-03	1.04E-02	1.12E-02	1.21E-02	1.30E-02	1.40E-02	1.51E-02	1.63E-02
C3	Cost of electricity (in PV), USD/kWh(electr)	3.21E-02	3.16E-02	3.11E-02	3.07E-02	3.03E-02	2.99E-02	2.96E-02	2.92E-02	2.90E-02	2.87E-02

Appendix A. (Continued)

	Year	1	2	3	4	5	6	7	8	9	10
C4	Cost of heat (in PV), USD/kWh(therm)	7.76E-03	7.63E-03	7.52E-03	7.41E-03	7.32E-03	7.23E-03	7.14E-03	7.07E-03	7.00E-03	6.94E-03
C5	Cumulative PV of expenses at L_{GT}=1 (10 years)					[3.56E+02] mln USD over 10 years					
C6	Levelized cost of electricity at L_{GT}=1					3.02E-02 USD /kWh(electr)					
C7	Levelized cost of heat at L_{GT}=1					7.30E-03 USD/kWh(therm)					
		Load- dependent segment of annual cost at L_{GT} =0.7									
A3	Fuel cost, mln USD/y	1.79E+01	1.94E+01	2.12E+01	2.31E+01	2.51E+01	2.74E+01	2.98E+01	3.24E+01	3.53E+01	3.85E+01
A4	Carbon-tax, mln USD /y	3.13E+00	3.41E+00	3.72E+00	4.05E+00	4.41E+00	4.80E+00	5.23E+00	5.70E+00	6.20E+00	6.76E+00
B1	Total cost (in FV), mln USD/y	3.51E+01	3.75E+01	4.01E+01	4.29E+01	4.60E+01	4.94E+01	5.31E+01	5.71E+01	6.14E+01	6.62E+01
B2	Total cost (in PV), mln USD/y	3.22E+01	3.16E+01	3.10E+01	3.05E+01	3.00E+01	2.96E+01	2.92E+01	2.88E+01	2.85E+01	2.82E+01
C1	Cost of electricity (in FV), USD/kWh(electr)	4.25E-02	4.54E-02	4.86E-02	5.20E-02	5.58E-02	5.99E-02	6.43E-02	6.92E-02	7.44E-02	8.02E-02
C2	Cost of heat (in FV), USD/kWh(therm)	1.03E-02	1.10E-02	1.17E-02	1.26E-02	1.35E-02	1.45E-02	1.55E-02	1.67E-02	1.80E-02	1.94E-02
C3	Cost of electricity (in PV), USD/kWh(electr)	3.90E-02	3.83E-02	3.76E-02	3.70E-02	3.64E-02	3.59E-02	3.54E-02	3.49E-02	3.45E-02	3.42E-02
C4	Cost of heat (in PV), USD/kWh(therm)	9.43E-03	9.25E-03	9.09E-03	8.94E-03	8.80E-03	8.67E-03	8.55E-03	8.45E-03	8.35E-03	8.26E-03
C5	Cumulative PV of expenses at L_{GT}=0.7 (10 years)					[3.00E+02] mln USD over 10 years					
C6	Levelised cost of electricity at L_{GT}=0.7					3.63E-02 USD/kWh(electr)					
C7	Levelised cost of heat at L_{GT}=0.7					8.78E-03 USD/kWh(therm)					

APPENDIX B. CASH FLOW CHART FOR AB SYSTEM

		Assumptions
1	Size of AB	124 MW(therm)
2	AB- load variation	from 0 to 100 % (equivalent thermal output: 0 to 124 MW(therm)*
3	Specific capital cost	130 USD /kW(therm).
4	Initial capital investment (DC+IC)	15 mln $
5	Nominal Interest Rate	0.1
6	Retail price index (Inflation rate)	0.01
7	Real Interest Rate	0.09
8	Cost of primary fuel	2.0 USD /GJ
9	Carbon tax	20.0 USD/tone of emitted carbon
	O&M costs	1.12 mln USD /year

	year	1	2	3	4	5	6	7	8	9	10
	FV- Disc factor	1.09	1.19	1.29	1.41	1.53	1.67	1.82	1.98	2.16	2.35
	PV-Disc factor	0.92	0.84	0.77	0.71	0.65	0.60	0.55	0.51	0.46	0.43
		Load- independent segment of annual cost									
A1	Recovery of Principal & Interest l, mln USD/y	2.44	2.44	2.44	2.44	2.44	2.44	2.44	2.44	2.44	2.44
A2	O&M cost, mln USD/y	1.33	1.45	1.58	1.72	1.87	2.04	2.22	2.42	2.63	2.86
		Load- dependent segment of annual cost at L_{GT}= 0.7 (L_{AB}=0.375*)									
A3	Fuel cost, mln USD/y	4.01E+00	4.36E+00	4.75E+00	5.17E+00	5.64E+00	6.14E+00	6.69E+00	7.28E+00	7.93E+00	8.64E+00
A4	Carbon-tax, mln USD/y	7.03E-01	7.66E-01	8.34E-01	9.08E-01	9.89E-01	1.08E+00	1.17E+00	1.28E+00	1.39E+00	1.52E+00
B1	Total cost (in FV), mln USD/y	8.48E+00	9.02E+00	9.60E+00	1.02E+01	1.09E+01	1.17E+01	1.25E+01	1.34E+01	1.44E+01	1.55E+01
B2	Total cost (in PV), mln USD/yr	7.79E+00	7.60E+00	7.43E+00	7.28E+00	7.14E+00	7.01E+00	6.89E+00	6.78E+00	6.68E+00	6.58E+00
C1	Cost of heat (in FV), USD/kWh(therm)	1.99E-02	2.12E-02	2.26E-02	2.41E-02	2.57E-02	2.75E-02	2.94E-02	3.15E-02	3.38E-02	3.63E-02
C1	Cost of heat (in PV), USD/kWh(therm)	1.83E-02	1.79E-02	1.75E-02	1.71E-02	1.68E-02	1.65E-02	1.62E-02	1.59E-02	1.57E-02	1.55E-02

C2	Cumulative PV of expenses at 70 % GT-load (10 years)	[7.12E+01] mln USD over 10 years
C3	Levelised cost of heat at 70% GT- load	1.67E-02 USD/kWh(heat)

*Relation between load of gas turbine (GT) and auxiliary boiler (AB) has been described as: $L_{AB} = (1 - L_{GTt})/(1 - L_{\min(GT)})$. L_{GT} ranges from L_{GT}=1 (at L_{AB}=0) to $L_{MIN(GT)}$=0.2 (at L_{AB}=1), thus L_{GT}=0.7 at L_{AB}= 0.375.

APPENDIX C. CASH FLOW CHART FOR MSF SYSTEM

		Assumptions
1	Size of MSF subsystem:	60000 m^3/day
2	PR	10
3	Specific heat energy consumption	61 kWh(therm) /m^3(distillate);
4	Auxiliary electricity consumption [24]	3 kWh(el)/m^3(distillate)
5	Specific capital cost	1500 USD/ m^3/day
6	Initial capital cost (DC+IC)	90 mln USD,
7	Nominal interest rate	0.1
8	Retail price index (Inflation rate)	0.01
9	Real interest rate	0.09
10	Cost of primary fuel	2.0 USD/GJ
11	Carbon tax	20.0 USD /tone of emitted carbon
12	Annual O&M costs (MSF):.	5.4 mln USD/year

	year	1	2	3	4	5	6	7	8	9	10
	FV- Disc factor	1.09	1.19	1.29	1.41	1.53	1.67	1.82	1.98	2.16	2.35
	PV-Disc factor	0.92	0.84	0.77	0.71	0.65	0.60	0.55	0.51	0.46	0.43
		Load- independent segment of annual expenses									
A1	Recovery of principal & Interest l, mln USD/y	11.99	11.99	11.99	11.99	11.99	11.99	11.99	11.99	11.99	11.99
A2	O&M cost, mln USD/y	6.43	7.00	7.62	8.30	9.04	9.85	10.72	11.68	12.72	13.85
		Load- dependent segment of annual cost at L_{GT}=1									
A3	Cost of energy, mln USD/y	1.42E+01	1.55E+01	1.68E+01	1.83E+01	2.00E+01	2.18E+01	2.37E+01	2.58E+01	2.81E+01	3.06E+01
B1	Total cost (in FV), mln USD/y	3.26E+01	3.45E+01	3.65E+01	3.86E+01	4.10E+01	4.36E+01	4.64E+01	4.95E+01	5.28E+01	5.65E+01
B2	Total cost (in PV), mln USD/yr	2.99E+01	2.90E+01	2.82E+01	2.75E+01	2.68E+01	2.61E+01	2.55E+01	2.50E+01	2.45E+01	2.40E+01
C1	Cost of water (in FV), USD/m^3(MSF)	1.82E+00	1.92E+00	2.03E+00	2.16E+00	2.29E+00	2.43E+00	2.59E+00	2.76E+00	2.95E+00	3.15E+00
C1	Cost of water (inPV), USD/m^3(MSF)	1.67E+00	1.62E+00	1.57E+00	1.53E+00	1.49E+00	1.46E+00	1.42E+00	1.39E+00	1.37E+00	1.34E+00
C2	Cumulative PV of expenses at L_{GT}=1 (10 years)					[2.67E+02] mln USD (over 10 years)					
C3	Levelised cost of water at L_{GT}=1					1.49 USD/m^3(MSF)					
		Load- dependent segment of annual cost at L_{GT}=0.7									
A3	Cost of energy mln USD/y	2.15E+01	2.35E+01	2.56E+01	2.78E+01	3.03E+01	3.30E+01	3.60E+01	3.92E+01	4.27E+01	4.65E+01

Appendix C. (Continued)

B1	Total cost (in FV), mln USD/y	4.00E+01	4.25E+01	4.52E+01	4.81E+01	5.13E+01	5.49E+01	5.87E+01	6.28E+01	6.74E+01	7.23E+01
B2	Total cost (in PV), mln USD/y	3.67E+01	3.58E+01	3.50E+01	3.42E+01	3.35E+01	3.29E+01	3.23E+01	3.17E+01	3.12E+01	3.08E+01
C1	Cost of water (in FV), USD/m^3(MSF)	2.23E+00	2.37E+00	2.52E+00	2.69E+00	2.87E+00	3.06E+00	3.27E+00	3.51E+00	3.76E+00	4.03E+00
C1	Cost of water (in PV), USD/m^3(MSF)	2.05E+00	2.00E+00	1.95E+00	1.91E+00	1.87E+00	1.83E+00	1.80E+00	1.77E+00	1.74E+00	1.72E+00
C2	Cumulative PV of expenses at L_{GT}=0.7 (10 years)					[3.34E+02] mln USD (over 10 years)					
C3	Levelised cost of water (MSF) at L_{GT}=0.7					1.86 USD/m^3(MSF)					

APPENDIX D. CASH FLOW CHART FOR RO SYSTEM

		Assumptions
1	Size of RO subsystem	60000 m^3/day;
2	Specific energy consumption	5 kWh(el)/m^3(perm)
3	Specific capital cost	1000 USD/m3/day
4	Initial capital investment (DC+IC)	60 mln USD,
5	Nominal Interest Rate	0.1
6	Retail price index (Inflation rate)	0.01
7	Real Interest Rate	0.09
8	Cost of primary fuel	2.0 USD/GJ of primary fuel
9	Carbon tax	20.0 USD/tone of emitted carbon

	year	1	2	3	4	5	6	7	8	9	10
	FV- discount factor	1.09	1.19	1.29	1.41	1.53	1.67	1.82	1.98	2.16	2.35
	PV-discount factor	0.92	0.84	0.77	0.71	0.65	0.60	0.55	0.51	0.46	0.43
		Load- independent segment of annual expenses									
A1	Recovery of principal & Interest, mln USD/y	8.14	8.14	8.14	8.14	8.14	8.14	8.14	8.14	8.14	8.14
A2	O&M cost, mln USD/y	11.98	13.05	14.21	15.48	16.86	18.36	19.99	21.78	23.72	25.83
		Load- dependent segment of annual expenses at L_{GT}=1									

A3	Cost of energy mln USD/y	3.25E+00	3.54E+00	3.85E+00	4.19E+00	4.57E+00	4.97E+00	5.42E+00	5.90E+00	6.43E+00	7.00E+00
B1	Total cost (in FV), mln USD/y	2.34E+01	2.47E+01	2.62E+01	2.78E+01	2.96E+01	3.15E+01	3.35E+01	3.58E+01	3.83E+01	4.10E+01
B2	Total cost (in PV), mln USD/yr	2.15E+01	2.08E+01	2.03E+01	1.98E+01	1.93E+01	1.89E+01	1.85E+01	1.81E+01	1.78E+01	1.74E+01
C1	Cost of water (in FV), USD/m^3(RO)	1.28E+00	1.35E+00	1.44E+00	1.52E+00	1.62E+00	1.72E+00	1.84E+00	1.96E+00	2.10E+00	2.24E+00
C2	Cost of water (in PV), USD/m^3(RO)										
C3	Cumulative PV of expenses at L_{GT}=1 (10 years)						[1.92E+02] mln USD over 10 years				
C4	Levelised cost of water at L_{GT}=1						1.05 USD/m^3(RO)				

Appendix D. (Continued)

	year	1	2	3	4	5	6	7	8	9	10
		Load- dependent segment of annual expenses at L_{GT}=0.7									
A3	Cost of energy mln USD/y	3.96E+00	4.31E+00	4.69E+00	5.11E+00	5.57E+00	6.06E+00	6.60E+00	7.19E+00	7.83E+00	8.53E+00
B1	Total cost (in FV), mln USD/y	2.41E+01	2.55E+01	2.70E+01	2.87E+01	3.06E+01	3.26E+01	3.47E+01	3.71E+01	3.97E+01	4.25E+01
B2	Total cost (in PV), mln USD/yr	2.21E+01	2.15E+01	2.09E+01	2.04E+01	1.99E+01	1.95E+01	1.91E+01	1.87E+01	1.84E+01	1.81E+01
C1	Cost of water (in FV), USD/m^3(RO)	1.32E+00	1.40E+00	1.48E+00	1.57E+00	1.67E+00	1.78E+00	1.90E+00	2.03E+00	2.17E+00	2.33E+00
C2	Cost of water (in PV), USD/m^3(RO)	1.21E+00	1.18E+00	1.15E+00	1.12E+00	1.09E+00	1.07E+00	1.05E+00	1.03E+00	1.01E+00	9.92E-01
C3	Cumulative PV of expenses at L_{GT}=0.7 (10 years)					[1.99E+02] mln USD over 10 years					
C4	Levelised cost of water at L_{GT}=0.7					1.09 USD/m^3(RO)					

In: Desalination: Methods, Cost and Technology
Editor: Irena A. Urbonienė pp. 53-116

ISBN: 978-1-61668-909-4

Chapter 2

PERFORMANCE EVALUATION AND OPTIMIZATION OF THE MSF DESALINATION PLANT

A. Sabah Abdul-Wahab**[*a]**, K V. Reddy**[b]**,
***A. Mohammed Al-Weshahi**[c]**, Salim Al-Hatmi**[d]*
***and M. Yasir Tajeldin**[a]*

[a] Department of Mechanical & Industrial Engineering, College of Engineering, Sultan Qaboos University, P.O. Box 33, Al Khoud 123, Sultanate of Oman
[b] Middle East Desalination Research Center (MEDRC), Sultanate of Oman,
[c] Shinas College of Technology, Sultanate of Oman,
[d] Barka Power and Desalination Plant, Sultanate of Oman.

ABSTRACT

In this chapter, a steady state mathematical model for multistage flash (MSF) desalination plants is developed. The model is based on basic principles of physics and chemistry that describe the physical phenomenon occurring in the process. The input plant parameters that are known to affect the performance of the MSF desalination plant are taken into account in the construction of the model. These parameters include make-up flow, brine recycle flow, seawater flow, seawater temperature, seawater concentration, top brine temperature, steam temperature and the plant load. For each stage, the developed model predicts the temperatures of the brine, distillate and cooling brine, and the flow rates of brine outlet and distillate production. For the purpose of this study, one of the MSF plant is selected for investigation. The developed model is evaluated with the MSF plant vendor simulation results and its actual operating data. Model predictions matched well with the vendor simulation results and the plant operating data. Thus, the developed model is sufficiently accurate and model predictions will be reliable over complete operating plant range. The developed model is thus used to carry out the optimization of the MSF plant performance with an objective to maximize its performance ratio or maximize its distillate production rate. It was found that the main

[*] Correspondiong author: E-mail: sabah1@squ.edu.om (Sabah A. Abdul-Wahab). Tel.: +968-2414-1360; fax: +968-2414-1316.

variables that affect the performance of the plant were the top brine temperature and the brine recycle flow rate.

This model can be used for determining the optimum controller's set points of a running MSF desalination plant at different loads to maximize the water production or minimize the energy consumption or any other required objective function. This model can be integrated with the plant control software, and the optimum set points for controllers can be evaluated off line whenever plant loads are changed or management requirements change.

NOMENCLATURE

A_H	total heat transfer area of the brine heater (m^2)
A_j	total heat transfer area based on outside diameter of cooling tube in stage j (m^2)
B_0	flashing brine entering first stage or the feed brine (ton/hr)
B_1	flashing brine leaving first stage (ton/hr)
B_{j-1}	flashing brine entering stage j (ton/hr)
B_j	flashing brine leaving stage j (ton/hr)
B_D	brine blow down (ton/hr)
B_N	flashing brine flow rate (concentrated brine) leaving stage N (ton/hr)
$C_{B,j-1}$	salt concentration of flashing brine entering stage j in (g/L) or mass fraction
$C_{B,j}$	salt concentration of flashing brine leaving stage j in (g/L) or mass fraction
$C_{B,N}$	salt concentration in the flashing brine leaving last stage N (g/L) or mass fraction
C_F	salt concentration in the feed seawater (g/L) or mass fraction
C_R	salt concentration in the cooling brine to the heat recovery section (g/L) or mass fraction
C_w	reject seawater (ton/hr)
D	distillate product (ton/hr)
$D_{i,j}$	cooling tube inside diameter in stage j (m)
D_1	distillate product leaving first stage (ton/hr)
D_{j-1}	distillate product entering stage j (ton/hr)
D_j	distillate product leaving stage j (ton/hr)
$D_{o,j}$	cooling tube outside diameter in stage j (m)
FF	fouling factor (m^2 K/ W)
F_m	makeup seawater flow rate stream (ton/hr)
F_{sea}	seawater feed to the plant (ton/hr)
$H_{l,j}$	brine level in stage j (m)

(Continued)

H_v	specific enthalpy of saturated steam (kJ/kg)
$h_{B,j-1}$	enthalpy of the flashing brine entering stage j (kJ/kg)
$h_{B,j}$	enthalpy of the flashing brine leaving stage j (kJ/kg)
$h_{D,j}$	specific enthalpy of saturated water (kJ/kg) at $T_{D,j}$
$h_{i,j}$	cooling tube inside heat transfer coefficient in stage j (kW/m^2 K)
$h_{o,j}$	cooling tube outside heat transfer coefficient in stage j or condensation coefficient on tubes in stage j (kW/m^2)
$h_{v,j}$	specific enthalpy of flashed vapor at stage j (kJ/kg)
$K_{B,j}$	thermal conductivity of brine leaving stage j (kW/m K)
$K_{D,j}$	thermal conductivity of distillate product leaving stage j (kW/m K)
$K_{F,j}$	thermal conductivity of cooling brine in stage j (kW/m K)
K_m	thermal conductivity of tube metal (kW/m K)
n	number of tubes in a column of the tube bundle
P_s	saturation pressure of the steam (atm)
P_j	pressure in stage j (atm)
Q_j	heat loss from stage j due to the loss of vapor from the stage to the vent system (kJ/s)
$S_{B,j-1}$	specific heat capacity of flashing brine entering stage j (kJ /kg K)
$S_{B,j}$	specific heat capacity of flashing brine leaving stage j (kJ /kg K)
$S_{D,j-1}$	specific heat capacity of distillate entering stage j (kJ /kg K)
$S_{D,j}$	specific heat capacity of distillate leaving stage j (kJ /kg K)
$S_{F,j}$	specific heat capacity of cooling brine leaving stage j at heat rejection section(kJ /kg K)
$S_{R,H}$	specific heat capacity of brine in the brine heater (kJ/kg K)
$S_{R,j}$	specific heat capacity of cooling brine leaving stage j at heat recovery section (kJ /kg K)
$T_{B,0}$	temperature of the flashing brine leaving the brine heater (°C). It equals to top brine temperature (TBT)
T^*	reference temperature for enthalpy (°C)
$T_{B,j-1}$	temperature of flashing brine entering stage j (°C)
$T_{B,j}$	temperature of flashing brine leaving stage j (°C)
$T_{D,j-1}$	temperature of distillate product entering stage j (°C)
$T_{D,j}$	temperature of distillate product leaving stage j (°C)
$T_{F,j}$	temperature of the cooling brine leaving stage j (°C)
$T_{F,j+1}$	temperature of the cooling brine entering stage j (°C)

(Continued)

$T_{F,1}$	temperature of the cooling brine leaving stage 1 and entering the brine heater (°C)
T_s	steam temperature (°C)
$\Delta T_{B,j}$	stage flash down temperature (°C), $(T_{B,j-1} - T_{B,j})$
$\Delta T_{BPR,j}$	boiling point elevation in stage j (°C)
$\Delta T_{DEM,j}$	temperature drop in demister in stage j (°C)
ΔT_{NONE}	non-equilibrium allowance in temperature for flashing brine in stage j (°C)
$\Delta T_{TB,j}$	temperature drop in cooling tube bundle in stage j (°C)
$\Delta T_{DEM,j}$	temperature drop in the demister and the cooling tube bundle in stage j (°C)
$U_{o,j}$	overall heat transfer coefficient in stage j (kW/m^2 K)
U_H	the overall heat transfer coefficient at the brine heater (kW/m^2 K)
$\mu_{B,j}$	viscosity of flashing brine leaving stage j (kg/ m s)
$\mu_{D,j}$	viscosity of distillate product leaving stage j (kg/ m s)
$\mu_{F,j}$	viscosity of the cooling brine in stage j (kg/ m s)
V_F	velocity of brine in cooling tubes (m/s)
W	cooling brine flow rate in the heat recovery section (ton/hr)
$W_{F,j}$	condensate flow rate in a stage j (kg/s)
W_j	brine mass flow rate per unit stage width in stage j (ton/ hr m)
W_s	steam flow rate to the brine heater (ton/hr)
$\rho_{D,j}$	density of distillate product leaving stage j (kg/m^3)
$\rho_{B,j}$	density of the flashing brine leaving stage j (kg/m^3)
$\rho_{D,j}$	distillated density (kg/m^3) in stage j
$\rho_{F,j}$	density of the cooling brine in stage j (kg/m^3)
λ_s	latent heat of vaporization of water in the brine heater (kJ/kg)
ΔP	pressure drop across the sluice gate (atm)

1. Introduction

1.1. Objective

The objective of this chapter is to develop a steady state mathematical model to study the performance of the MSF desalination plant and optimize its operating conditions at various loads.

1.2. Methodology

1. Collect the process details of the selected MSF desalination plant including its dimensional details and intake seawater conditions. The dimensional details include the number of stages, number of tubes per stage, the tube internal diameter, the effective tube length, the tube thermal conductivity, the fouling factor, the tube bundle width, the tube material, the overall heat transfer coefficients, the distillate duct height from bottom, the demister height from bottom, the stage length, the stage width, the stage height, the demister specific area and the free volume. The intake seawater parameters include the temperature and salt concentration.
2. Collect the operating data of the selected plant for various plant loads during summer and winter operations such as seawater flow rate, makeup flow rate, brine recycle flow rate, TBT, steam temperature, and steam pressure.
3. Develop the mathematical model that describing the physical phenomena occurring in the flash evaporation stages based on the basic equations of mass balance, enthalpy balance and the equilibrium relations.
4. Select appropriate methods to solve the developed model and to optimize the plant operations.
5. Check the developed model accuracy with the design/vendor simulation data of the selected MSF plant.
6. Evaluate the performance of the selected MSF plant with the operating data.
7. Identify bottlenecks and optimize the selected MSF plant operations and evaluate the economic benefits.

2. Process Description of the Selected MSF Desalination Plant

The selected Multi Stage Flash (MSF) desalination plant is conventional brine recycle MSF plant consisting of number of stages placed in series. Subsequent stages are maintained at a lower pressure by steam ejectors to facilitate the flashing of brine in each stage down the line. Each stage consists of two sections: flash chamber and the heat exchanger tube bundle. In the flash chamber, the superheated brine entering the stage (*flashing brine*) is flash vaporized. The vapor so produced condenses on the outer surface of the tube bundle placed at the top of the flash chamber. The incoming brine (*cooling brine*) flows inside the cooling tubes as a coolant, countercurrent to the flashing brine. The condensate product falls into a product tray located below the tube bundle [1,2].

The flow schematic of the selected MSF desalination plant is shown in Figure 2.1 with the symbols for parameters of each flow stream. This plant consists of three sections as shown in Figure 2.1; heat input section (brine heater), the heat recovery section (HRS) and the heat rejection section (HRJ). The heat recovery section constitutes of 16 stages (stages 1-16) and heat rejection section constitutes of 3 stages (stages 17-19). Makeup stream after deaeration mixes with the part of the brine from the last stage of the heat rejection section and flows to the heat recovery section as recycle brine (coolant). The purpose of heat rejection section is to reject the excess thermal energy from the plant, thus cooling the distillate product and the concentrated brine to the lowest possible temperatures as they exit from the plant (through stage 19). Dimensional details of the selected MSF desalination plant are given in Table 2.1.

The cooling seawater feed (F_{sea}) passes inside the tubes of the heat rejection section (i.e., from stage 19 to stage 17). Cooling seawater leaving the heat rejection section (stage 17), is split into two streams; the reject seawater (C_w) and the makeup seawater stream (F_m). The reject seawater stream goes back to the sea carrying part of the heat supplied in brine heater. The makeup stream after deaeration combines with the part of the brine stream (R) from the last stage of the heat rejection section to form the combined stream (W) called the recycle brine. This combined stream (as cooling brine feed) enters the heat recovery section.

The cooling brine (W) flows inside the tubes of the heat recovery section (entering at stage 16) and gets heated gradually by the flashing brine of each stage as it flow from stage 16 to the first stage of heat recovery section. Heated cooling brine from heat recovery section at $T_{F,1}$ is further heated in brine heater to top brine temperature, $T_{B,0}$, with low pressure steam from the low pressure steam turbine exhaust. Heated brine from brine heater enters stage 1 of the heat recovery section through an orifice, thus reducing brine pressure. Therefore this brine will become superheated in stage 1 and flashes to produce vapor. Vapor produced in the stage goes through a demister to trap the liquid droplets entrained with the vapor. Then this vapor passes over the heat exchanger tube bundle where it condenses and condensate falls into a distillate collection tray.

The flashing brine flows into the subsequent stages maintained at low pressures and the flashing process is repeated all the way down the last stage. The distillate from stage 1 flows to stage 2, and stage 2 to 3 and so on. The accumulated distillate from all stages is finally collected in a distillate box of the last stage of the heat rejection section (D_N) and is extracted by a distillate pump. The concentrated brine (B_N) leaving stage 19 is split into two streams; blow down stream (B_D) and brine for recycle (R). The brine blow down stream is discharged back to the sea and brine for recycle is mixed with the makeup seawater stream to form the combined brine stream (W). This combined stream passes through the heat recovery section inlet (stage 16) to serve as the cooling brine.

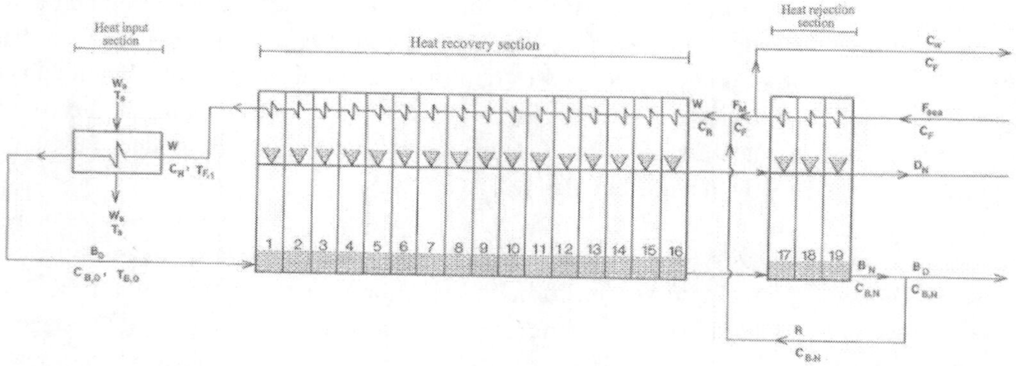

Figure 2.1. Schematic diagram of the selected MSF desalination plant

Table 2.1. Dimensional details of the selected MSF desalination plant

No.	Parameter	Recovery	Rejection	Brine heater
1	Number of stages	16	3	-
2	Number of tubes per stage	Stage 1-5: 3156 tubes Stage 6-16: 3309 tubes	Stage 17-19: 2567 tubes	3126 tubes
3	Tube internal diameter/thickness (mm)	25.9/1	26.1/1	25.9/1
4	Effective tube length	13.706	13.706	11.096
5	Fouling factor (m^2 K/W)	0.07713	0.1057	0.1720
6	Wall thickness (mm)	1.0	1.0	1.0
7	Tube material	Stage 1-3: 66/30/2/2 Cu/Ni/Fe/Mg Stage 4-5: Cu/Ni 90/10 Stage 6-16: Al Brass	Stage 17-19: 66/30/2/2 Cu/Ni/Fe/Mg	66/30/2/2 Cu/Ni/Fe/Mg

3. Model Equations

The model equations are constituted of a set of mass and energy balance equations. They are developed with the following assumptions:

The distillate produced and accumulated in any stage is salt free;
No heat losses occur at any stage of the process;
Heat of the mixing of brine solutions is negligible;
All the saline liquid droplets are retained in the demister;

The developed model includes the following:

Overall material balance equations
Material balance equations for each stage
Salt balance equation for each stage
Mixing and splitting equations
Enthalpy balance equations on flashing brine
Overall enthalpy balance for any stage
Heat transfer equation
Equilibrium relation
Brine heater model
Auxiliary equations

3.1. Overall Material Balance Equations

The overall material balance for MSF plant (Figure 3.1) is given by

$$F_{sea} = C_w + D + B_D \tag{3.1}$$

Where:

F_{sea} is the seawater feed to the plant (ton/hr)

C_w is the reject seawater (ton/hr)

D is the distillate product (ton/hr)

B_D is the brine blow down stream or rejected brine (ton/hr)

The material balance for the seawater splitter is given by

$$F_{sea} = F_m + C_w \quad (3.2)$$

Where:

F_{sea} is the seawater feed to the plant (ton/hr)

F_m is the makeup feed seawater (ton/hr)

C_w is the reject seawater (ton/hr)

Substituting equation (3.2) into equation (3.1) yields

$$F_m = D + B_D \quad (3.3)$$

As per equation (3.3), the overall material balance for MSF plant have one input stream (the makeup feed seawater, F_m), and two output streams (the distillate product, D, and the brine blow down stream, B_D). Therefore, the excess energy added to the system in the brine heater is rejected in brine blow down, distillate product and reject seawater streams.

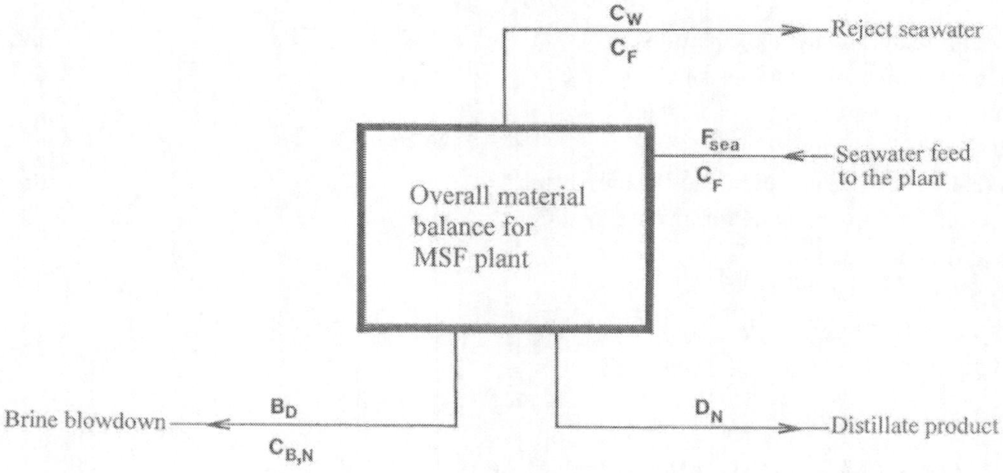

Figure 3.1. Overall material balance for MSF plant

The overall salt balance is given by

$$F_m C_F = B_D C_{B,N} \tag{3.4}$$

Where:

C_F is the salt concentration in the feed seawater (g/L)

$C_{B,N}$ is the salt concentration in the flashing brine leaving last stage N (g/L).

It should be noted that the above equation assumes that the distillate is salt free.

3.2. Material Balance Equations for Each Stage

For any stage j in the heat recovery or heat rejection sections, except for stage 1, at steady state, the sum of the stage inlet flashing brine and distillate flow rates should be equal to the sum of these steams leaving the stage (Figure 3.2):

$$B_{j-1} + D_{j-1} = B_j + D_j \tag{3.5}$$

Where:

B_{j-1} is the flashing brine entering stage j (ton/hr)

D_{j-1} is the distillate product entering stage j (ton/hr)

B_j is the flashing brine leaving stage j (ton/hr)

D_j is the distillate product leaving stage j (ton/hr)

For stage 1 (Figure 3.3), the mass balance is given by,

$$B_0 = B_1 + D_1 \tag{3.6}$$

Where:

B_0 is the flashing brine entering first stage or the feed brine (ton/hr)

B_1 is the flashing brine leaving first stage (ton/hr)

D_1 is the distillate product leaving first stage (ton/hr)

Since no distillate enters the first stage ($D_0 = 0.0$). It should be noted here that $B_0 = W$

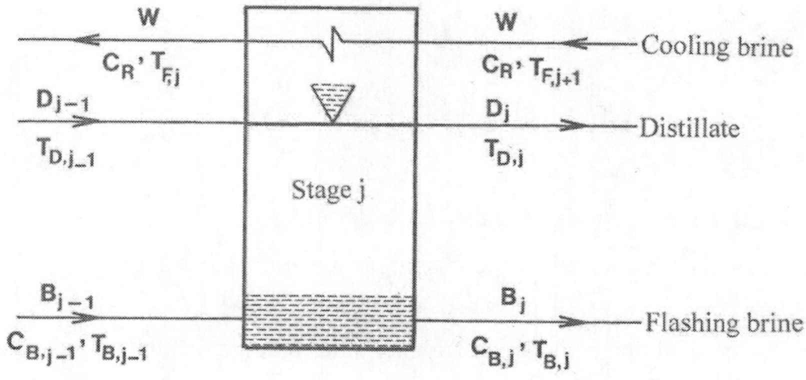

Figure 3.2. Material balance for any stage j, except stage 1

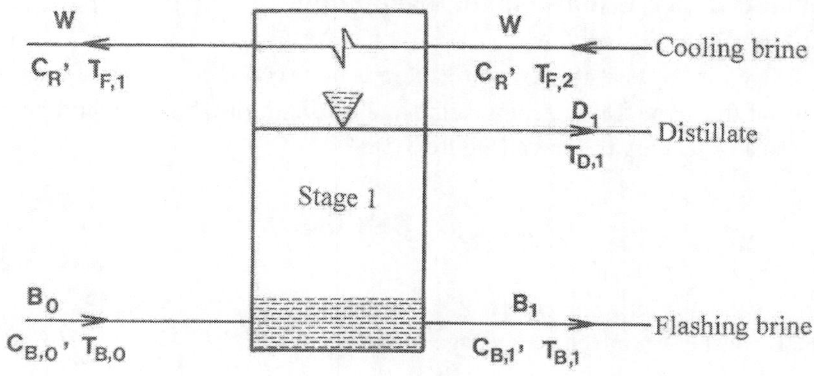

Figure 3.3.Material balance for stage 1

3.3. Salt Balance Equation for Each Stage

Similarly the salt balance equation for any stage j in the heat recovery or heat rejection sections (Figure 3.2) is given as

$$B_{j-1}\, C_{B,j-1} = B_j\, C_{B,j} \tag{3.7}$$

Where :

B_{j-1} is the flashing brine entering stage j (ton/hr)

B_j is the flashing brine leaving stage j (ton/hr)

$C_{B,j-1}$ is the salt concentration of flashing brine entering stage j (g/L)

$C_{B,j}$ is the salt concentration of flashing brine leaving stage j (g/L)

It should be noted that the flow rate as well as the concentration of the cooling brine do not change from stage to stage in either the heat recovery section or the heat rejection section.

In the heat recovery section, the flow rate and the concentration of the cooling brine are W and C_R, respectively; where W is the sum of the brine for recycle (R) and the makeup seawater (F_m) flow rates. In the heat rejection section, the flow rate and the concentration of the cooling brine are equal to that of the seawater feed to the plant $(F_{sea}$ and $C_F)$.

3.4. Mixing and Splitting Equations

The splitter divides the entering stream into more than one stream and each output stream has the same salt concentration and the specific enthalpy as the input stream. Only the mass is distributed in a particular ratio. Figure 3.4 shows the splitters and mixer in the plant. There are two splitters, blow down stream splitter and reject seawater stream splitter.

The blow down splitter (S1) can be described by,

$$B_D = B_N - R \tag{3.8}$$

Where:

B_D is the blow down flow rate (ton/hr)

B_N is the flashing brine flow rate (concentrated brine) leaving stage N (ton/hr)

R is the brine for recycle flow rate (ton/hr)

The reject seawater splitter (S2) can be described by,

$$C_W = F_{sea} - F_m \tag{3.9}$$

Where:

C_W is the rejected seawater flow rate (ton/hr)

F_{sea} is the seawater feed to the plant (ton/hr).

F_m is the makeup flow rate stream (ton/hr)

Total material balance for mixer (M) gives

$$W = R + F_m \tag{3.10}$$

And the salt balance for mixer is given by,

$$WC_R = RC_{B,N} + F_m C_F \tag{3.11}$$

Where:

W is the cooling brine flow rate to the heat recovery section (ton/hr)

R is the brine for recycle flow rate (ton/hr)

F_m is the makeup seawater flow rate (ton/hr)

C_R is the salt concentration in the cooling brine to the heat recovery section (g/L)

C_F is the salt concentration in the feed seawater (g/L)

$C_{B,N}$ is the salt concentration in the flashing brine leaving stage N (g/L)

3.5. Enthalpy Balance Equations on Flashing Brine

Referring to Figure 3.2, the enthalpy balance equation on the flashing brine at any stage j, in the heat recovery or heat rejection sections, is given by,

$$B_{j-1}\,h_{B,j-1} = B_j\,h_{B,j} + \left(B_{j-1} - B_j\right) h_{v,j} \tag{3.12}$$

Where:

B_{j-1} is the flashing brine entering stage j (ton/hr)

B_j is the flashing brine leaving stage j (ton/hr)

$h_{v,j}$ is specific enthalpy of flashing vapor at stage j (kJ/kg)

Enthalpy of flashing brine $h_{B,j-1}$ and $h_{B,j}$ in equation 3.12 are calculated by

$$\begin{aligned} h_{B,j-1} &= \left(4.185 - 5.381\text{x}10^{-3}C_{B,j-1} + 6.26\text{x}10^{-6}C_{B,j-1}{}^{2}\right)T_{B,j-1} \\ &-\left(3.055\text{x}10^{-5} + 2.774\text{x}10^{-6}C_{B,j-1} - 4.318\text{x}10^{-8}C_{B,j-1}{}^{2}\right)T_{B,j-1}{}^{2} \\ &+\left(8.844\,\text{x}10^{-7} + 6.527\text{x}10^{-8}C_{B,j-1} - 4.003\text{x}10^{-10}C_{B,j-1}{}^{2}\right)T_{B,j-1}{}^{3} \end{aligned} \tag{3.13a}$$

$$\begin{aligned} h_{B,j} &= \left(4.185 - 5.381\text{x}10^{-3}C_{B,j} + 6.26\text{x}10^{-6}C_{B,j}{}^{2}\right)T_{B,j} \\ &-\left(3.055\text{x}10^{-5} + 2.774\text{x}10^{-6}C_{B,j} - 4.318\text{x}10^{-8}C_{B,j}{}^{2}\right)T_{B,j}{}^{2} \\ &+\left(8.844\,\text{x}10^{-7} + 6.527\text{x}10^{-8}C_{B,j} - 4.003\text{x}10^{-10}C_{B,j}{}^{2}\right)T_{B,j}{}^{3} \end{aligned} \tag{3.13b}$$

Where:

$C_{B,j-1}$ is the salt concentration of flashing brine entering stage j (g/L)

$C_{B,j}$ is salt concentration of flashing brine leaving stage j (g/L)

$T_{B,j-1}$ is temperature of flashing brine entering stage j (°C)

$T_{B,j}$ is temperature of flashing brine leaving stage j (°C)

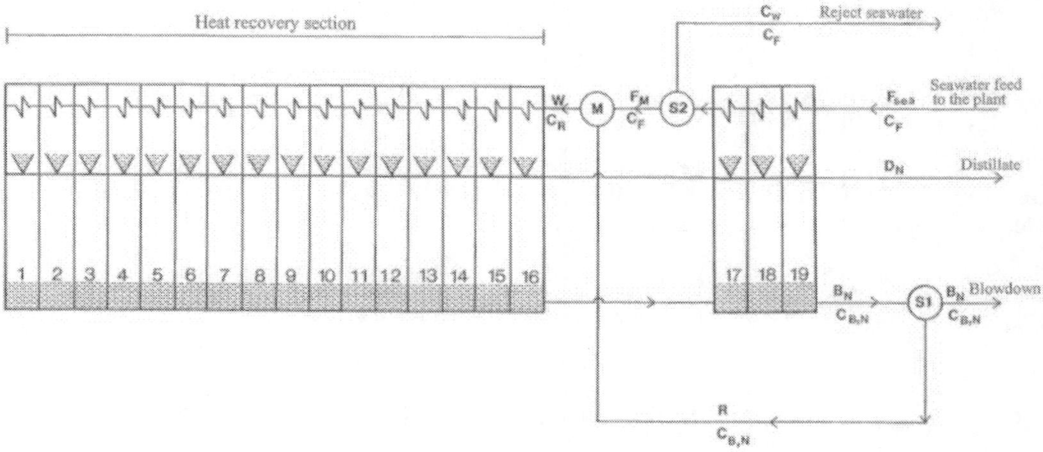

Figure 3.4. Mixing and splitting points

3.6. Overall Enthalpy Balance for Any Stage

The sum of the enthalpies of all the streams entering the stage should be equal to the sum of all those leaving the stage for a steady-state operation, since there is no accumulation or generation of energy within the stage. Considering the T^* as the reference temperature for all enthalpy evaluation, the enthalpy balance for any stage j, in the heat recovery or heat rejection sections, is given by,

$$W\ S_{R,j}\ (T_{F,j} - T_{F,j-1}) = D_{j-1}\ S_{D,j-1}\ (T_{D,j-1} - T^*) + B_{j-1}\ S_{B,j-1}\ (T_{B,j-1} - T^*) - D_j\ S_{D,j}\ (T_{D,j} - T^*) - B_j\ S_{B,j}\ (T_{B,j} - T^*) - Q_j \tag{3.14}$$

Where:

W is the flow rate of the cooling brine in the heat recovery section (ton/hr). It is substituted by seawater feed to the plant, F_{sea} , in the heat rejection section.

$T_{F,j+1}$ is the temperature of the cooling brine entering stage j (°C)

$T_{F,j}$ is the temperature of the cooling brine leaving stage j (°C)

D_{j-1} is the distillate product entering stage j (ton/hr)

$T_{D,j-1}$ is the temperature of distillate product entering stage j (°C)

$T_{B,j-1}$ is the temperature of the flashing brine entering stage j (°C)

T^* is the reference temperature for enthalpy (°C)

B_{j-1} is the flashing brine entering stage j (ton/hr)

B_j is the flashing brine leaving stage j (ton/hr)

D_{j-1} is the distillate product entering stage j (ton/hr)

$T_{D,j-1}$ is temperature of distillate product entering stage j (°C)

$T_{B,j}$ is temperature of the flashing brine leaving stage j (°C)

Q_j is the heat loss from stage j by vapor loss from the stage to the venting system (kJ/s)

Specific heat capacity of cooling brine leaving stage j (kJ/kg K) at heat recovery section inlet, $S_{R,j}$, in equation 14 is calculated by

$$S_{R,j} = 4.185 - 5.381\text{x}10^{-3}C_R + 6.26\text{x}10^{-6}C_R^{\ 2} - (3.055\text{x}10^{-5} + 2.774\text{x}10^{-6}C_R - 4.318\text{x}10^{-8}C_R^{\ 2})T_{F,j} + (8.844\text{ x}10^{-7} + 6.527\text{x}10^{-8}C_R - 4.003\text{x}10^{-10}C_R^{\ 2})T_{F,j}^{\ \ 2} \quad (3.15a)$$

Where:

C_R is the salt concentration of the cooling brine to the heat recovery section (g/L). It is constant for all the stages of the heat recovery section

$T_{F,j}$ is the temperature of the cooling brine leaving stage j (°C)

Specific heat capacity, $S_{F,j}$, at the heat rejection section in equation 3.14 is calculated by

$$S_{F,j} = 4.185 - 5.381\text{x}10^{-3}C_F + 6.26\text{x}10^{-6}C_F^{\ 2} - (3.055\text{x}10^{-5} + 2.774\text{x}10^{-6}C_F - 4.318\text{x}10^{-8}C_F^{\ 2})T_{F,j} + (8.844\text{ x}10^{-7} + 6.527\text{x}10^{-8}C_F - 4.003\text{x}10^{-10}C_F^{\ 2})T_{F,j}^{\ \ 2} \quad (3.15b)$$

Where:

C_F is the salt concentration of the feed seawater to the heat rejection section inlet (g/L). It is constant for all stages of the heat rejection section.

$T_{F,j}$ is the temperature of the cooling brine leaving stage j (°C)

Specific heat capacity of distillate entering stage j, $S_{D,j-1}$, and leaving stage j, $S_{D,j}$, (kJ/kg K) in equation 3.14 is calculated by

$$S_{D,j-1} = \begin{pmatrix} 1.001183 - 6.61666652\,x\,10^{-5}\,T_{D,j-1} \\ +1.3999989x10^{-7}\,T_{D,j-1}^{\ \ 2} + \\ 1.3333336x10^{-9}\,T_{D,j-1}^{\ \ 3} \end{pmatrix} \text{x } 4.184 \quad (3.16a)$$

$$S_{D,j} = \begin{pmatrix} 1.001183 - 6.61666652\,x\,10^{-5}\,T_{D,j} \\ +1.3999989x10^{-7}\,T_{D,j}^{\ \ 2} + \\ 1.3333336x10^{-9}\,T_{D,j}^{\ \ 3} \end{pmatrix} \text{x } 4.184 \quad (3.16b)$$

Where:

$T_{D,j-1}$ is temperature of distillate product entering stage j (°F)

$T_{D,j}$ is temperature of distillate product leaving stage j (°F)

Specific heat capacity of flashing brine entering stage j, $S_{B,j-1}$, and leaving stage j, $S_{B,j}$, (kJ/kg K) in equation 3.14 is calculated by

$$\begin{aligned} S_{B,j-1} = {} & 4.185 - 5.381\text{x}10^{-3} C_{B,j-1} + 6.26\text{x}10^{-6} C_{B,j-1}{}^2 \\ & -(3.055\text{x}10^{-5} + 2.774\text{x}10^{-6} C_{B,j-1} \ - 4.318\text{x}10^{-8} C_{B,j}{}^2) T_{B,j-1} \\ & +(8.844\ \text{x}10^{-7} + 6.527\text{x}10^{-8} C_{B,j-1} \ - 4.003\text{x}10^{-10} C_{B,j-1}{}^2) T_{B,j-1}{}^2 \end{aligned} \tag{3.17a}$$

$$\begin{aligned} S_{B,j} = {} & 4.185 - 5.381\text{x}10^{-3} C_{B,j} + 6.26\text{x}10^{-6} C_{B,j}{}^2 \\ & -(3.055\text{x}10^{-5} + 2.774\text{x}10^{-6} C_{B,j} \ - 4.318\text{x}10^{-8} C_{B,j}{}^2) T_{B,j} \\ & +(8.844\ \text{x}10^{-7} + 6.527\text{x}10^{-8} C_{B,j} C \ - 4.003\text{x}10^{-10} C_{B,j}{}^2) T_{B,j}{}^2 \end{aligned} \tag{3.17b}$$

Where:

$C_{B,j-1}$ is the salt concentration of flashing brine entering stage j (g/L)

$T_{B,j-1}$ is temperature of flashing brine entering stage j (°C)

$C_{B,j}$ is salt concentration of flashing brine leaving stage j (g/L)

$T_{B,j}$ is temperature of flashing brine leaving stage j (°C)

3.7. Heat Transfer Equation

The heat transferred due to condensation of the flashed vapor on the cooling tubes in any stage j, in the heat recovery or the heat rejection sections [3], is given by,

$$W S_{R,j}(T_{F,j} - T_{F,j+1}) = \frac{U_j A_j (T_{F,j} - T_{F,j+1})}{ln\left(\frac{T_{D,j} - T_{F,j+1}}{T_{D,j} - T_{F,j}}\right)} \tag{3.18}$$

Where:

W is the flow rate of the cooling brine to the heat recovery section (ton/hr). It is replaced by seawater feed to the plant, F_{sea} , in the heat rejection section.

$S_{R,j}$ is the specific heat capacity of the cooling brine leaving stage j in the heat recovery section (kJ/kg K). It is replaced by $S_{F,j}$ in the heat rejection section

$T_{F,j}$ is the temperature of the cooling brine leaving stage j (°C)

$T_{F,j+1}$ is the temperature of the cooling brine entering stage j (°C)

U_j is the overall heat transfer coefficient at stage j (W/m2 K)

A_j is total heat transfer area based on outside diameter of cooling tube in stage j (m2)

$T_{D,j}$ is the temperature of the distillate leaving stage j (°C)

3.8. Equilibrium Equation

The flashing brine temperature $(T_{B,j})$ and the distillate temperature $(T_{D,j})$ in any stage j are related by the following equation accounting for the boiling point elevation, the non-equilibrium allowance and the temperature loss due to pressure drops through demister and the condenser tube bundle:

$$T_{B,j} = T_{D,j} + \Delta T_{BPR,j} + \Delta T_{NONE,j} + \Delta T_{DEM,j} + \Delta T_{TB,j} \quad (3.19)$$

Where:

$T_{B,j}$ is the temperature of flashing brine leaving stage j (°C)

$T_{D,j}$ is the temperature of distillate leaving stage j (°C)

$\Delta T_{BPR,j}$ is the boiling point elevation in stage j (°C)

$\Delta T_{NONE,j}$ is the non-equilibrium allowance in temperature for flashing brine in stage j (°C)

$\Delta T_{DEM,j}$ is the temperature drop in demister in stage j (°C)

$\Delta T_{TB,j}$ is the temperature drop in cooling tube bundle in stage j (°C)

The boiling point elevation in stage j $(\Delta T_{BPR,j})$ is estimated by the correlation developed by Friedrich and Hafford [4] based on the data reported by Stoughton and Lietzke [5].

$$\Delta T_{BPR,j} = \left[\frac{565.757}{T_{B,j}} - 9.81559 + 1.54739 \ln T_{B,j} - \left(\frac{337.178}{T_{B,j}} - 6.41981 + 0.922753 \ln T_{B,j}\right) C + \left(\frac{32.681}{T_{B,j}} - 0.55368 + 0.079022 \ln T_{B,j}\right) C^2\right]\left[\frac{C}{\frac{266919.6}{T_{B,j}} - \frac{379.669}{T_{B,j}} + 0.334169}\right] \quad (3.20)$$

Where:

$$C = 19.819\, C_{B,j}/(1 - C_{B,j}) \quad (3.21)$$

$\Delta T_{BPR,j}$ is the boiling point elevation in stage j (°C)

$T_{B,j}$ is the temperature of flashing brine leaving stage j (K)

$C_{B,j}$ is the salt concentration of flashing brine leaving stage j in mass fraction

A number of empirical correlations and rigorous models are reported in the literature for the estimation of non-equilibrium allowance for the flashing brine $(\Delta T_{NONE,j})$. In the range of operation of the MSF plants, the following correlation by Burns and Roe [6] is most suitable:

$$\Delta T_{NONE,j} = (352(H_{l,j})^{1.1}(1.8\Delta T_{B,j})^{-0.25}(2.2046 \times 0.3043\, W_j)^{0.5}(1.8\, T_{D,j} + 32.0)^{-2.5})/1.8 \tag{3.22}$$

Where:

$\Delta T_{NONE,j}$ is the non-equilibrium allowance for flashing brine in stage j (°C)

$H_{l,j}$ is the brine level in stage j (m)

$\Delta T_{B,j}$ is the stage flash down temperature (°C), $(T_{B,j-1} - T_{B,j})$

W_j is the brine mass flow rate per unit stage width in stage j (ton/ hr m)

$T_{D,j}$ is the temperature of distillate leaving stage j (°C)

Both the temperature losses in the demister $(\Delta T_{DEM,j})$ and the tube bundle $(\Delta T_{TB,j})$ are considered together and an empirical equation of Omar (1981) is used to calculate the total loss,

$$\Delta T_{DEM,j} + \Delta T_{TB,j} = exp[1.885 - 0.02063(1.8\, T_{D,j} + 32)]/1.8 \tag{3.23}$$

Where:

$\Delta T_{DEM,j} + \Delta T_{TB,j}$ is the temperature drop in the demister and the cooling tube bundle in stage j (°C)

$T_{D,j}$ is the temperature of distillate leaving stage j (°C)

3.9. Brine Heater Model

Referring to Figure 3.5, the total mass balance for brine heater is given by,

$$B_0 = W \tag{3.24}$$

Where:

B_0 is the flashing brine flow rate leaving the brine heater (ton/hr)

W is the cooling brine flow rate in the heat recovery section (ton/hr). It is equal to the brine flow rate entering the brine heater.

Since $B_0 = W$, then the salt balance gives

$$C_{B,0} = C_R \tag{3.25}$$

Where:

$C_{B,0}$ is the salt concentration of the flashing brine leaving the brine heater (g/L)

C_R is the salt concentration of the cooling brine in the heat recovery section (g/L). It is the same as the salt concentration of the cooling brine that is entering the brine heater.

For making enthalpy balance of the brine heater (Figure 3.5), two assumptions are made; steam condensate does not sub-cool, and heat loss to surroundings is negligible. Therefore, the enthalpy balance can be written as

$$WS_{R,H}\left(T_{B,0} - T_{F,1}\right) = W_s\lambda_s \tag{3.26}$$

Where:

W is the cooling brine flow rate in the heat recovery section (ton/hr). It equals to B_0

$S_{R,H}$ is the specific heat capacity of brine in the brine heater (kJ/kg K)

$T_{B,0}$ is the temperature of the flashing brine leaving the brine heater (°C). It equals to top brine temperature (TBT)

$T_{F,1}$ is the temperature of the cooling brine leaving stage 1 and entering the brine heater (°C)

W_s is the steam flow rate to the brine heater (ton/hr)

λ_s is the latent heat of vaporization of water in the brine heater (kJ/kg)

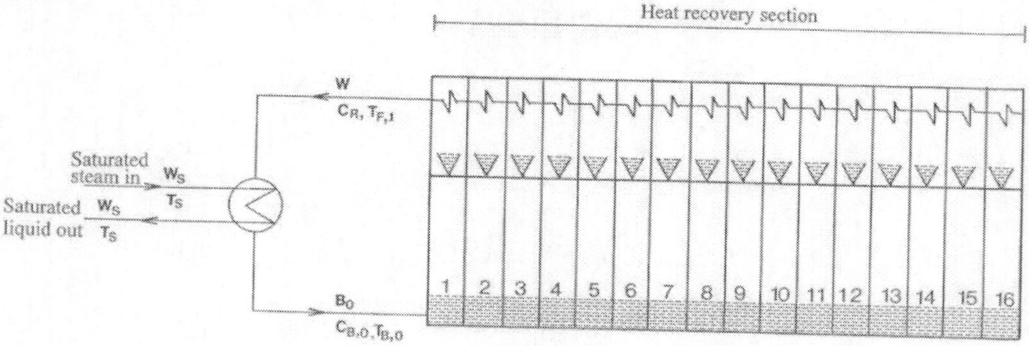

Figure 3.5.Material balance around the brine heater (input heat section)

The heat transfer equation for brine heater can be described by

$$WS_{R,H}\left(T_{B,0} - T_{F,1}\right) = \frac{U_H A_H (T_{B,0} - T_{F,1})}{ln\left(\frac{T_S - T_{F,1}}{T_S - T_{B,0}}\right)} \tag{3.27}$$

Where:

W is the cooling brine flow rate in the heat recovery section (ton/hr).

$S_{R,H}$ is the specific heat capacity of brine in the brine heater (kJ/kg K)

$T_{B,0}$ is the temperature of the brine leaving the brine heater (°C). It equals to top brine temperature (TBT)

$T_{F,1}$ is the temperature of the cooling brine leaving stage 1 and entering the brine heater (°C)

U_H is the overall heat transfer coefficient at the brine heater (W/m^2 K)

A_H is total heat transfer area of the brine heater (m^2)

T_s is the steam temperature (°C)

3.10. Auxiliary Equations

These consist of the physical, thermodynamic and transport property correlations for water, steam and brine solutions, which are applicable over the range of temperatures and concentrations encountered in the MSF plant. Moreover, correlations are needed to calculate film heat transfer coefficients inside and outside the cooling tubes. The correlations used here are taken from Perry and Green [7].

3.10.1. Properties of Pure Saturated Water (Distillate)

Specific heat capacity of pure saturated water

The specific heat capacity of pure saturated water as a function of boiling point at each stage given by Helal et al. [8] is used:

$$S_{D,j} = \begin{pmatrix} 1.001183 - 6.61666652 \times 10^{-5}\, T_{D,j} \\ +1.3999989x10^{-7}\, T_{D,j}^{\ 2} + \\ 1.3333336x10^{-9}\, T_{D,j}^{\ 3} \end{pmatrix} \times 4.184 \qquad (3.28)$$

Where:

$S_{D,j}$ is the specific heat capacity of distillate leaving stage j (kJ/kg K)

$T_{D,j}$ is the temperature of distillate product leaving stage j (°F)

Specific enthalpy of pure saturated water

To obtain the specific enthalpy of saturated water, equation 3.28 is integrated between the limits of the reference temperature of 32°F and the boiling temperature ($T_{D,j}$) in °F and resulting equation is given by

$$h_{D,j} = \left(1.001183T_{D,j} - 3.0833326x10^{-5}\, T_{D,j}^{\ 2} + 4.666663x10^{-8}\, T_{D,j}^{\ 3} + 3.333334x10^{-10}\, T_{D,j}^{\ 4} - 31.92\right)\left(\frac{4.184}{1.8}\right) \qquad (3.29)$$

Where:

$h_{D,j}$ is the specific enthalpy of saturated water (kJ/kg) at $T_{D,j}$

$T_{D,j}$ is the temperature of distillate product leaving stage j (°F)

Density of pure saturated water

The density of pure saturated water as a function of boiling point at each stage given by Mothershed [9], is used

$$\rho_{D,j} = (62.707172 - 0.43955304\text{x}10^{-2}\, T_{D,j} - 0.46076921\text{x}10^{-4} T_{D,j}{}^2)[0.4535924/0.3048^3] \tag{3.30}$$

Where:

$\rho_{D,j}$ is the density of distillate product leaving stage j (kg/m^3)

$T_{D,j}$ is temperature of distillate product leaving stage j (°F)

Viscosity of pure saturated water

The viscosity of pure saturated water as a function of boiling point at each stage given by Perry and Green [7], is used

$$\mu_{D,j} = 1.00\text{x}10^{-3}/\text{Exp}(3.244(T_{D,j} - 20)/(T_{D,j} + 109)) \tag{3.31}$$

Where:

$\mu_{D,j}$ is the viscosity of distillate product leaving stage j (kg/ m s)

$T_{D,j}$ is temperature of distillate product leaving stage j (°C)

Thermal conductivity of pure saturated water

The thermal conductivity of pure saturated water as a function of boiling point at each stage given by Perry and Green [7], is used

$$K_{D,j} = 5.756\text{x}10^{-4} + 1.526\text{x}10^{-6} T_{D,j} - 5.81\text{x}10^{-9} T_{D,j}{}^2 \tag{3.32}$$

Where:

$K_{D,j}$ is thermal conductivity of distillate product leaving stage j (kW/m C)

$T_{D,j}$ is temperature of distillate product leaving stage j (°C)

3.10.2. Properties of Brine

Specific heat capacity of brine

The specific heat capacity of flashing brine at each stage given by Perry and Green [7], is used

$$S_{B,j} = 4.185 - 5.381\text{x}10^{-3}C_{B,j} + 6.26\text{x}10^{-6}C_{B,j}{}^{2}$$
$$-(3.055\text{x}10^{-5} + 2.774\text{x}10^{-6}C_{B,j} \quad - 4.318\text{x}10^{-8}C_{B,j}{}^{2})T_{B,j}$$
$$+(8.844\ \text{x}10^{-7} + 6.527\text{x}10^{-8}C_{B,j} \quad - 4.003\text{x}10^{-10}C_{B,j}{}^{2})T_{B,j}{}^{2} \quad (3.33)$$

Where:

$C_{B,j}$ is salt concentration of flashing brine leaving stage j (g/L)

$T_{B,j}$ is temperature of flashing brine leaving stage j (°C)

Specific enthalpy of brine

The specific enthalpy of brine at each stage given by Perry and Green [7], is used

$$h_{B,j} = (4.185 - 5.381\text{x}10^{-3}C_{B,j} + 6.26\text{x}10^{-6}C_{B,j}{}^{2})\ T_{B,j}$$
$$-(3.055\text{x}10^{-5} + 2.774\text{x}10^{-6}C_{B,j} \quad - 4.318\text{x}10^{-8}C_{B,j}{}^{2})T_{B,j}{}^{2}$$
$$+(8.844\ \text{x}10^{-7} + 6.527\text{x}10^{-8}C_{B,j} \quad - 4.003\text{x}10^{-10}C_{B,j}{}^{2})T_{B,j}{}^{3} \quad (3.34)$$

Where:

$C_{B,j}$ is salt concentration of flashing brine leaving stage j (g/L)

$T_{B,j}$ is temperature of flashing brine leaving stage j (°C)

Density of brine

The brine density correlation reported by Mothershed [9], is used. It is valid for 0.0 to 26.0% salt concentration and 40 to 300 °F.

$$\rho_{B,j} =$$
$$(62.707172 + 49.364088C_{B,j} - 0.43955304\text{x}10^{-2}T_{B,j} - 0.032554667C_{B,j}T_{B,j} -$$
$$0.46076921\text{x}10^{-4}T_{B,j}{}^{2} + 0.63240299\text{x}10^{-4}C_{B,j}T_{B,j}{}^{2})[0.45359924/(0.3048)^{3}] \quad (3.35)$$

Where:

$\rho_{B,j}$ is density cf the flashing brine leaving stage j (kg/m^3)

$C_{B,j}$ is salt concentration of flashing brine leaving stage j in mass fraction

$T_{B,j}$ is temperature of flashing brine leaving stage j (°F)

It should be noted that the above equation can also be used to calculate the pure water density (distillate) by equating $C_{B,j}$ to zero

Viscosity of brine

The viscosity of brine at each stage given by Perry and Green [7], is used

$$\mu_{B,j} = 1.002x10^{-3} + 1.55x10^{-6}C_{B,j} + 8.5x10^{-9}C_{B,j}{}^{2}/Exp(3.244(T_{B,j} - 20)/(T_{B,j} + 109)) \quad (3.36)$$

Where:

$\mu_{B,j}$ is viscosity of flashing brine leaving stage j (kg/ m s)

$C_{B,j}$ is salt concentration of flashing brine leaving stage j (g/L)

$T_{B,j}$ is temperature of flashing brine leaving stage j (°C)

Thermal conductivity of brine

The thermal conductivity of brine at each stage given by Perry and Green [7], is used

$$K_{B,j} = 5.756x10^{-4} - 3.464x10^{-5}C_{B,j} + 7.286X10^{-6}C_{B,j}{}^{2} + (1.526 + 0.4662C_{B,j} - 0.2268xC_{B,j}{}^{2} + 0.02867xC_{B,j}{}^{3})x10^{-6}T_{B,j} - (5.81 + 20.55C_{B,j} - 9.916C_{B,j}{}^{2} + 1.464C_{B,j}{}^{3})x10^{-9}T_{B,j}{}^{2} \quad (3.37)$$

Where:

$K_{B,j}$ is thermal conductivity of brine leaving stage j (kW/m C)

$T_{B,j}$ is temperature of flashing brine leaving stage j (°C)

$C_{B,j}$ is salt concentration of flashing brine leaving stage j in mass fraction

It should be noted that the above equations described for the flashing brine can also be used for the cooling brine by replacing $C_{B,j}$ by C_R and $T_{B,j}$ by $T_{F,j}$.

3.10.3. Properties of steam used in the brine heater

Saturation pressure of steam

The vapor pressure of saturated steam is estimated using the empirical correlation of Griffin and Keller [10]

$$log_{10}\left(\frac{P}{218.167}\right) = \frac{-(647.27-T)}{T[3.2437814+5.86826x10^{-3}(647.27-T)+1.1702379x10^{-8}(647.27-T)^{3}]}[1 + 2.1878462x10^{-3}(647.27 - T)^{3}] \quad (3.38)$$

Where:

P is the saturation pressure of the steam (atm)

T is saturation temperature of the steam (K)

Latent heat of vaporization of water

The latent heat of vaporization given by Perry and Green [7], is used

$$\lambda_s = 2495 - 2.132T - 2.632\text{x}10^{-3}T^2 \quad (3.39)$$

Where:

λ_s is the latent heat of vaporization of water in the brine heater (KJ/kg)

T is the saturation temperature of water (°C)

Specific enthalpy of saturated steam

The specific enthalpy of saturated steam given by Perry and Green [7], is used

$$H_v = 2499.15 + 1.955T - 1.927\text{x}10^{-3}T^2 \quad (3.40)$$

Where:

H_v is the specific enthalpy of saturated steam (kJ/kg)

T is temperature of saturated steam (°C)

3.10.4. Heat Transfer Coefficients

Tube side heat transfer coefficient

The tube inside film heat transfer coefficient in the heat recovery or heat rejection section is estimated using the Sieder-Tate correlation for turbulent flow in circular tubes [3].

$$h_{i,j} = 0.027\frac{K_{F,j}}{D_{i,j}}\left(\frac{D_{i,j}V_F\rho_{F,j}}{u_{F,j}}\right)^{0.8}\left(\frac{S_{F,j}u_{F,j}}{K_{F,j}}\right)^{1/3} \quad (3.41)$$

Where:

$h_{i,j}$ is the cooling tube inside heat transfer coefficient in stage j (kW/m^2 K)

$D_{i,j}$ is the cooling tube inside diameter in stage j (m)

V_F is the velocity of brine in cooling tubes (m/s)

$\rho_{F,j}$ is is density of the cooling brine in stage j (kg/m^3)

$\mu_{F,j}$ is the viscosity of the cooling brine in stage j (kg/ m s)

$S_{F,j}$ is the specific heat capacity of the cooling brine in stage j (kJ/kg K)

$K_{F,j}$ is the thermal conductivity of cooling brine in stage j (kW/m K)

It should be noted that the above equations described for the flashing brine can also be used for the calculation of the cooling brine properties ($\rho_{F,j}, u_{F,j}, S_{F,j}, K_{F,j}$) by replacing $C_{B,j}$ by C_R and $T_{B,j}$ by $T_{F,j}$.

Condensation coefficient on tubes

The condensation coefficient outside the tube is estimated [3] using

$$h_{o,j} = 1.64327\left[\frac{k_{D,j}^{3}\rho_{D,j}^{2}}{\frac{u_{D,j}W_{F,j}D_{o,j}}{A_j}}\right]^{1/3}(n)^{-1/6} \tag{3.42}$$

Where:

$h_{o,j}$ is the cooling tube outside heat transfer coefficient in stage j or condensation coefficient on tubes in stage j (kW/m^2 K)

$k_{D,j}$ is the distillate thermal conductivity in stage j (kW/m K) calculated by Eq. (3.32)

$\mu_{D,j}$ is the distilled viscosity in stage j (kg/ m s)calculated by Eq. (3.31)

$\rho_{D,j}$ is the distilled density (kg/m^3) in stage j calculated by Eq. (3.30)

$D_{o,j}$ is the cooling tube outside diameter in stage j (m)

A_j is the total heat transfer area based on outside diameter of cooling tube in stage j (m^2)

n is the number of tubes in a column of the tube bundle

$W_{F,j}$ is the condensate flow rate in a stage j (kg/s) and calculated by

$$W_{F,j} = D_j - D_{j-1} \tag{3.43}$$

Overall Heat Transfer coefficient

The overall heat transfer coefficient, based on outside area, is then calculated using the individual coefficients, fouling factor and the tube wall resistance [3] as,

$$U_{o,j} = \frac{1}{\frac{1}{h_{i,j}}\frac{D_{o,j}}{D_{i,j}} + FF + \frac{D_{o,j}}{2K_m}\ln\frac{D_{o,j}}{D_{i,j}} + \frac{1}{h_{o,j}}} \tag{3.44}$$

Where:

$U_{o,j}$ is overall heat transfer coefficient in stage j (kW/m^2 K)

$D_{o,j}$ is the cooling tube outside diameter (m)

$D_{i,j}$ is the cooling tube inside diameter (m)

FF is the fouling factor (m^2 K/W)

K_m is the thermal conductivity of tube metal (kW/m K)

$h_{i,j}$ is the cooling tube inside heat transfer coefficient in stage j (kW/m^2 K) calculated by Eq. (3.41).

$h_{o,j}$ is the cooling tube outside heat transfer coefficient in stage j or condensation coefficient on tubes in stage j (kW/m^2 K) calculated by Eq. (3.42)

3.10.5. Orifice Flow Equation

For evaluation of liquid level in each stage, the following orifice equation [3] is used:

$$B_j = C_D\, A_{o,j} \sqrt{2\,\rho_B \Delta P} \tag{3.45}$$

Where:

ΔP = Pressure drop across the sluice gate and it is calculated by

$$\Delta P = \frac{P_j - P_{j+1}}{\rho_B g} + H_{l,j} - C_c H_{l,j+1} \tag{3.45a}$$

Where:

P_j is pressure in stage j

C_c = Contraction coefficient and it is calculated by

$$C_c = 0.61 + 0.18R_{op} - 0.58R_{op}^2 + 0.7R_{op}^3 \tag{3.45b}$$

C_D = Discharge coefficient and it is calculated by

$$C_D = \frac{C_c}{\sqrt{1 - C_c R_{op}}} \tag{3.45c}$$

Where

$$R_{op} = \frac{H_{o,j}}{\frac{P_j - P_{j+1}}{\rho_B g} + H_{1,j}} \tag{3.45d}$$

4. TDM FORMULATION

The model equations described in the previous section for the MSF plant constitute a set of sparse, nonlinear, algebraic equations. Linearization and tri-diagonal matrix arrangement, suggested by Helal et al. [8], is selected since this approach has the following advantages:

The Thomas algorithm can be used to solve the tri-diagonal matrix saving considerable computer storage [7].
The number of iterations required is small.
The algorithm has high convergence stability and results in shorter run times compared to calculation approaches which follow stage wise calculations.

A detailed linearization and tri-diagonal matrix arrangement is described in the following sections.

4.1. Stage j

For any stage j in the recovery or the rejection section, except a few stages dealt with later, the following treatment applies:

Rewrite the overall enthalpy balance Eq. (3.14) as,

$$T_{F,j} - T_{F,j-1} = a_{1,j}T_{D,j-1} + a_{2,j}T_{B,j-1} - a_{3,j}T_{D,j} - a_{4,j}T_{B,j} + a_{5,j} \quad (4.1)$$

Where:

$$a = WS_{R,j}$$

$$a_{1,j} = \frac{D_{j-1}S_{D,j-1}}{a}$$

$$a_{2,j} = \frac{B_{j-1}S_{B,j-1}}{a}$$

$$a_{3,j} = \frac{D_jS_{D,j}}{a}$$

$$a_{4,j} = \frac{B_jS_{B,j}}{a}$$

$$a_{5,j} = T^*(a_{3,j} + a_{4,j} - a_{1,j} - a_{2,j}) - Q_j/a$$

Now, rewrite the heat transfer Eq. (3.18) for the adjacent stages j and $(j-1)$ as follows:

$$T_{D,j} = \frac{1}{1-\alpha_j}(T_{F,j+1} - \alpha_j T_{F,j}) \quad (4.2)$$

$$T_{D,j-1} = \frac{1}{1-\alpha_{j-1}}(T_{F,j} - \alpha_{j-1}T_{F,j-1}) \quad (4.3)$$

Where:

$$\alpha_j = 3.6\ Exp\ (\frac{U_jA_j}{WS_{R,j}})$$

$$\alpha_{j-1} = 3.6\ Exp\ (\frac{U_{j-1}A_{j-1}}{WS_{R,j-1}})$$

Next, consider the equilibrium relation given by Eq. (3.19) for stages j and $(j-1)$ as,

$$T_{B,j} = T_{D,j} + z_j \tag{4.4}$$

$$T_{B,j-1} = T_{D,j-1} + z_{j-1} \tag{4.5}$$

Where:

$$z_j = \Delta T_{BPR,j} + \Delta T_{NONE,j} + \Delta T_{DEM,j} + \Delta T_{TB,j}$$

$$z_{j-1} = \Delta T_{BPR,j-1} + \Delta T_{NONE,j-1} + \Delta T_{DEM,j-1} + \Delta T_{TB,j-1}$$

Substituting Eq. (4.2) to Eq. (4.5) into Eq. (4.1) and rearranging,

$$Ay_j\ T_{F,j-1} + Be_j\ T_{F,j} + Ce_j\ T_{F,j+1} = De_j \tag{4.6}$$

The above equation relates the cooling brine temperatures in the three consecutive stages $(j-1), j$ and $(j+1)$, in which coefficients of the equation are as follows:

$$Ay_j = C_{1,j}$$

$$Be_j = C_{2,j} - 1$$

$$Ce_j = C_{3,j} + 1$$

$$De_j = -b_{3,j}$$

$$C_{1,j} = -\frac{b_{1,j}\alpha_{j-1}}{(1-\alpha_{j-1})}$$

$$C_{2,j} = \frac{b_{1,j}}{(1-\alpha_{j-1})} - \frac{b_{2,j}\alpha_j}{(1-\alpha_j)}$$

$$C_{3,j} = \frac{b_{2,j}}{(1-\alpha_j)}$$

$$b_{1,j} = a_{1,j} + a_{2,j}$$

$$b_{2,j} = -(a_{3,j} + a_{4,j})$$

$$b_{3,j} = (a_{5,j} + a_{2,j}z_{j-1} - a_{4,j}z_j)$$

If all the coefficients, a's, b's and c's as well as the z_j and z_{j-1}, in Eq. (4.6) are known, it becomes a linear equation. In other words, all the physical properties involved in the said parameter must be evaluated in advance, which requires temperatures and concentrations of the concerned streams to be assumed or known.

4.2. Stage j=1

This is the first stage in heat recovery section. If the flashing brine inlet temperature, $T_{B,0}$ is fixed, Eq. (3.14) can be written using the procedure followed for stage j as

$$Be_1\, T_{F,1} + Ce_1\, T_{F,2} = De_1 \tag{4.7}$$

Where:

$$Be_1 = C_{2,1} - 1$$

$$Ce_1 = C_{3,1} + 1$$

$$De_1 = -b_{3,1}$$

$$C_{2,1} = -\frac{b_{2,1}\alpha_1}{(1-\alpha_1)}$$

$$C_{3,1} = \frac{b_{2,1}}{(1-\alpha_1)}$$

$$b_{2,1} = -(a_{3,1} + a_{4,1})$$

$$b_{3,1} = a_{5,1} - a_{4,1}z_1$$

$$a = W\, S_{R,1}$$

$$a_{2,1} = \frac{B_0\, S_{B,0}}{a}$$

$$a_{3,1} = \frac{D_1\, S_{D,1}}{a}$$

$$a_{4,1} = \frac{B_1 S_{B,1}}{a}$$

$$a_{5,1} = a_{2,1} T_{B,0} + T^*(a_{3,1} + a_{4,1} - a_{2,1}) - Q_1/a$$

4.3. Stage j=NR

This is the last stage in the heat recovery section. The overall enthalpy balance equation for this stage can be written as,

$$T_{F,NR} - T_{B,N} = a_{1,NR} T_{D,NR-1} + a_{2,NR} T_{B,NR-1} - a_{3,NR} T_{D,NR} - a_{4,NR} T_{B,NR} + a_{5,NR} \tag{4.8}$$

where

$$a = W\, S_{R,NR}$$

$$a_{1,NR} = \frac{D_{NR-1}\, S_{D,NR-1}}{a}$$

$$a_{2,NR} = \frac{B_{NR-1}\, S_{B,NR-1}}{a}$$

$$a_{3,NR} = \frac{D_{NR}\, S_{D,NR}}{a}$$

$$a_{4,NR} = \frac{B_{NR}\, S_{B,NR}}{a}$$

$$a_{5,NR} = T^*(a_{3,NR} + a_{4,NR} - a_{1,NR} - a_{2,NR}) - Q_{NR}/a$$

Eliminating $T_{B,NR-1}$, $T_{B,NR}$, and $T_{D,NR-1}$, $T_{D,NR}$ in Eq. (4.8) using the equilibrium relation and heat transfer equation, respectively, we get

$$\frac{b_{1,NR} T_{F,NR}}{1-\alpha_{NR-1}} - \frac{b_{1,NR}\, \alpha_{NR-1} T_{F,NR-1}}{1-\alpha_{NR-1}} + \frac{b_{2,NR} T_{B,N}}{1-\alpha_{NR}} - \frac{b_{2,NR} \alpha_{NR}\, T_{F,NR}}{1-\alpha_{NR}} - T_{F,NR} + T_{B,N} + b_3 = 0 \tag{4.9}$$

Where:

$$b_{1,NR} = a_{1,NR} + a_{2,NR}$$

$$b_{2,NR} = -(a_{3,NR} + a_{4,NR})$$

$$b_{3,NR} = (a_{5,NR} + a_{2,NR} z_{NR-1} - a_{4,NR} z_{NR})$$

This stage receives the recycle brine on the tube side, hence $T_{F,NR+1} = T_{B,N}$, for which a relation can be derived by making an overall enthalpy balance of heat rejection section:

$$F_{sea} S_{R,NR+1} (T_{F,NR+1} - T^*) + B_N S_{B,N} (T_{B,N} - T^*) + D_N S_{D,N} (T_{D,N} - T^*) + \sum_{j=NR+1}^{N} Q_j - D_{NR} S_{D,NR} (T_{D,NR} - T^*) + B_{NR} S_{B,NR} (T_{B,NR} - T^*) - F_m S_F(T_F - T^*) - F_{sea} S_{sea}(T_{sea} - T^*) = 0 \quad (4.10)$$

Eq. (4.10) can be rewritten as

$$q_1 T_{F,NR+1} + q_2 T_{B,N} + q_3 T_{D,N} - q_4 T_{D,NR} - q_5 T_{B,NR} + q_7 = 0 \quad (4.11)$$

Where:

$$q_1 = F_{sea} S_{R,NR+1}$$

$$q_2 = B_N S_{B,N}$$

$$q_3 = D_N S_{D,N}$$

$$q_4 = D_{NR} S_{D,NR}$$

$$q_5 = B_{NR} S_{B,NR}$$

$$q_6 = F_m S_F (T_F - T^*) + F_{sea} S_{sea}(T_{sea} - T^*)$$

$$q_7 = T^*(-q_1 - q_2 - q_3 + q_4 + q_5) - q_6$$

From equilibrium relation for stages N and NR,

$$T_{D,N} = T_{B,N} - z_N \quad (4.12)$$

$$T_{B,NR} = T_{D,NR} + z_{NR} \quad (4.13)$$

Where:

$$z_N = \Delta T_{BPR,N} + \Delta T_{NONE,N} + \Delta T_{DEM,N} + \Delta T_{TB,N}$$

$$z_{NR} = \Delta T_{BPR,NR} + \Delta T_{NONE,NR} + \Delta T_{DEM,NR} + \Delta T_{TB,NR}$$

Substituting $T_{D,N}$ and $T_{B,NR}$ from Eq. (4.12) and Eq. (4.13) into Eq. (4.11) we get

$$q_2 T_{B,N} + q_3 T_{B,N} + q_{45} T_{D,NR} + q_1 T_{F,NR+1} + q_8 = 0 \tag{4.14}$$

Where:

$$q_{45} = -(q_4 + q_5)$$

$$q_8 = -q_3 z_N - q_5 z_{NR} + q_7$$

From the heat transfer Eq. (3.18) for stage NR we get

$$T_{D,NR} = \frac{1}{1-\alpha_{NR}} (T_{B,N} - \alpha_{NR} T_{F,NR}) \tag{4.15}$$

Where:

$$\alpha_{NR} = Exp\left(\frac{U_{NR}\, A_{NR}}{R\, S_{R,NR}}\right)$$

Eq. (4.15) can be written as,

$$T_{D,NR} = x\, T_{B,N} - x\, \alpha_{NR}\, T_{F,NR} \tag{4.16}$$

Where:

$$x = \frac{1}{1-\alpha_{NR}}$$

Substituting $T_{D,NR}$ value from Eq. (4.16) into Eq. (4.14) and on rearranging, the required relation between $T_{B,N}$ and $T_{F,NR}$, $T_{F,NR+1}$ is obtained as

$$T_{B,N} = x^{11}\, T_{F,NR} - x_q\, T_{F,NR+1} - x^1 q_8 \tag{4.17}$$

Where:

$$x^1 = \frac{1}{q_2 + q_3 + x q_{45}}$$

$$x^{11} = x^1 x\, q_{45} \alpha_{NR}$$

$$x_q = x^1 q_1$$

Substituting the value of $T_{B,N}$ from Eq. (4.17) into Eq. (4.9), and rearranging we get

$$Ay_{NR}\ T_{F,NR-1} + Be_{NR}\ T_{F,NR} + Ce_{NR} T_{F,NR+1} = De_{NR} \tag{4.18}$$

Where:

$$Ay_{NR} = -\frac{b_{1,NR}\ \alpha_{NR-1}}{1-\ \alpha_{NR-1}}$$

$$Be_{NR} = \frac{b_{1,NR}}{1-\alpha_{NR-1}} - x\ b_{2,NR}\ \alpha_{NR} + x\ x^{11} b_{2,NR} + x^{11} - 1$$

$$Ce_{NR} = -x_q\left(1 + x\ b_{2,NR}\right)$$

$$De_{NR} = x^1 q_8\left(x b_{2,NR} + 1\right) - b_{3,NR}$$

4.4. Stage j=NR+1

This is the first stage in heat rejection section. For this stage, the enthalpy balance Eq. (3.14) is written as,

$$T_{F,NR+1} - T_{F,NR+2} = \\ a_{1,NR+1} T_{D,NR} + a_{2,NR+1} T_{B,NR} - a_{3,NR+1} T_{D,NR+1} - a_{4,NR+1}\ T_{B,NR+1} + a_{5,NR+1} \tag{4.19}$$

Where:

$$a = F_{sea}\ S_{R,NR+1}$$

$$a_{1,NR+1} = \frac{D_{NR}\ S_{D,NR}}{a}$$

$$a_{2,NR+1} = \frac{B_{NR}\ S_{B,NR}}{a}$$

$$a_{3,NR+1} = \frac{D_{NR+1}\ S_{D,NR+1}}{a}$$

$$a_{4,NR+1} = \frac{B_{NR+1}\ S_{B,NR+1}}{a}$$

$$a_{5,NR+1} = T^*\left(a_{3,NR+1} + a_{4,NR+1} - a_{1,NR+1} - a_{2,NR+1}\right) - \left(\frac{Q_{NR+1}}{a}\right)$$

Eliminating $T_{B,NR}$, $T_{B,NR+1}$, $T_{D,NR}$ and $T_{D,NR+1}$ using the heat transfer Eq. (3.18) and equilibrium Eq. (3.19) respectively we get,

$$T_{F,NR+1} - T_{F,NR+2} = \frac{b_{1,NR+1}}{1-\alpha_{NR}} T_{B,N} - \frac{b_{1,NR+1}\alpha_{NR}}{1-\alpha_{NR}} T_{F,NR} + \frac{b_{2,NR+1}}{1-\alpha_{NR+1}} (T_{F,NR+2} - \alpha_{NR+1} T_{F,NR+1}) + b_{3,NR+1} \quad (4.20)$$

Where:

$$b_{1,NR+1} = a_{1,NR+1} + a_{2,NR+1}$$

$$b_{2,NR+1} = -(a_{3,NR+1} + a_{4,NR+1})$$

$$b_{3,NR+1} = (a_{5,NR+1} + a_{2,NR+1} z_{NR} - a_{4,NR+1} z_{NR+1})$$

Substituting the value of $T_{B,N}$ from Eq. (4.17) into Eq. (4.20) and rearranging we get

$$Ay_{NR+1}\, T_{F,NR} + Be_{NR+1}\, T_{F,NR+1} + Ce_{NR+1} T_{F,NR+2} = De_{NR+1} \quad (4.21)$$

Where:

$$Ay_{NR+1} = -x\, b_{1,NR+1} \alpha_{NR} + x\, x^{11} b_{1,NR+1}$$

$$Be_{NR+1} = -\frac{b_{2,NR+1}\alpha_{NR+1}}{1-\alpha_{NR+1}} - x\, x_q\, b_{1,NR} - 1$$

$$Ce_{NR+1} = \frac{b_{2,NR+1}}{1-\alpha_{NR+1}} + 1$$

$$De_{NR+1} = x\, x^1 q_8 b_{1,NR} - b_{3,NR}$$

4.5. Stage j=N

This is the last stage in the heat rejection. The enthalpy balance Eq. (3.14) for this stage is written as

$$T_{F,N} - T_{Sea} = a_{1,N} T_{D,N-1} + a_{2,N} T_{B,N-1} - a_{3,N} T_{D,N} - a_{4,N} T_{B,N} + a_{5,N} \quad (4.22)$$

Where:

$$a = F_{sea}\, S_{sea}$$

$$a_{1,N} = \frac{D_{N-1} S_{D,N-1}}{a}$$

$$a_{2,N} = \frac{B_{N-1} S_{B,N-1}}{a}$$

$$a_{3,N} = \frac{D_N\, S_{D,N}}{a}$$

$$a_{4,N} = \frac{B_N\, S_{B,N}}{a}$$

$$a_{6,N} = \frac{F_m\, S_m}{a}$$

$$a_{5,N} = T^*\left(a_{3,N} + a_{4,N} - a_{1,N} - a_{2,N} - a_{6,N}\right) - a_{6,N} T_m - Q_N/a$$

A similar procedure followed earlier is used to eliminate $T_{B,N}$,$T_{D,N}$, $T_{B,N-1}$, $T_{D,N-1}$ in Eq. (4.22) to get

$$Ay_N\; T_{F,N-1} + Be_N\; T_{F,N} = De_N \tag{4.23}$$

Where:

$$Ay_N = -\frac{b_{1,N}\alpha_{N-1}}{1-\alpha_{N-1}}$$

$$Be_N = \frac{b_{1,N}}{1-\alpha_{N-1}} - \frac{b_{2,N}\,\alpha_N}{1-\alpha_N} - 1$$

$$De_{NR+1} = -\frac{b_{2,N} T_{sea}}{1-\alpha_N} - T_{sea} - b_{3,N}$$

$$b_{1,N} = a_{1,N} + a_{2,N}$$

$$b_{2,N} = -(a_{3,N} + a_{4,N})$$

$$b_{3,NR+1} = (a_{5,N} + a_{2,} z_{N-1} - a_{4,N} z_N)$$

4.6. TDM

Finally, the linearized enthalpy balance equations for all the stages, in terms of T_F, is arranged in the following tri-diagonal matrix form:

$$\begin{bmatrix} Be_1 Ce_1 & & & & & & & \\ Ay_2 Be_2 Ce_2 & & & & & & & \\ & Ay_3 Be_3 Ce_3 & & & & & & \\ & & \cdots \cdots & & & & & \\ & & \cdots & \cdots \cdots & & & & \\ & & & Ay_i Be_i Ce_i & & & & \\ & & & & \cdots \cdots \cdots & & & \\ & & & & & \cdots \cdots \cdots & & \\ & & & & & & Ay_N Be_N & \end{bmatrix} \begin{bmatrix} T_{F,1} \\ T_{F,2} \\ T_{F,3} \\ \cdots \\ \cdots \\ T_{F,i} \\ \cdots \\ \cdots \\ T_{F,N} \end{bmatrix} = \begin{bmatrix} De_1 \\ De_2 \\ De_3 \\ \cdots \\ \cdots \\ De_i \\ \cdots \\ \cdots \\ De_N \end{bmatrix}$$

5. COMPUTATION PROCEDURE

Evaluation of production rate, temperature and flow rate profiles for the specified operating conditions in a given plant is the basic objective to be fulfilled by a computation module for any optimization study. The stepwise procedure for such a module is described below:

Step 1: Specify the following data:

Dimensional parameters for the flash chambers, heat exchanger bundle and brine heater (number of stages, number of tubes per stage, tubes thermal conductivity, fouling factors, tubes materials) (see Table 2.1).

Seawater: Temperature (Tsea), salt concentration (CF) and flow rate (F_{sea})

Makeup flow rate (F_m)

Recycle brine flow rate (W)

Steam to brine heater: temperature (Ts) and pressure (Ps)

Flashing brine first stage inlet temperature (TBT or $T_{B,0}$)

These data can be obtained from vendor simulation manual as well as from plant operation data such steam temperature (T_s) and pressure (P_s) and top brine temperature (TBT).

Step 2: Assume suitable brine levels in all stages.

Step 3: Estimate initial temperature profiles for the flashing brine ($T_{B,j}$), distillate ($T_{D,j}$), and cooling brine ($T_{F,j}$), in all stages. An initial guess for flashing brine temperature

($T_{B,j}$) is obtained by linear interpolation between the top brine temperature (TBT) and the seawater temperature (T_{sea}) as follows:

$$T_{B,j} = TBT - j\Delta TB \quad (5.1)$$

Where:

$$\Delta TB = \frac{TBT - T_{sea}}{N}$$

N is the total number of stages

A profile for cooling brine ($T_{F,j}$) is obtained using the following interpolation relations [12]

$$T_{F,j} = T_{B,N} + (N - i)\Delta TB - (j - 1)\Delta TB \quad (5.2)$$

This is for cooling brine in heat recovery section, where i is the number of stages in the heat rejection section. The cooling brine temperature at heat rejection section is calculated as follows [12]

$$T_{F,j} = T_{sea} + (N + 1 - j)\Delta T_{diff} \quad (5.3)$$

Where:

$$\Delta T_{diff} = \frac{T_{B,N} - T_{sea}}{i}$$

The flashing brine flow rate profile is calculated using enthalpy balance Eq. (3.12) and then distillate flow rate profile is obtained from material balance Eq. (3.5)

Step 4: Assume the makeup feed temperature (T_F). The makeup feed enters the flashing chamber of the last stage. Part of the brine from the last stage of heat rejection section is discharged to sea as blow down and the other part is mixed with the makeup feed and this mixed stream is pumped to the last stage of heat recovery section as cooling brine. It is worth noting that the temperature of the exiting makeup feed from first stage of heat rejection should be same as stream entering the last stage of heat recovery section.

Step 5: Calculate the flashing brine flow rate in all stages (B_j) using flashing brine enthalpy balance, starting from the first stage. The flashing brine flow rate is calculated using the enthalpy balance (Eq. (3.12)). The flashing brine flow rate for the stage 0 or brine heater is equal to the recycle flow rate. Whereas for the last stage, flashing brine flow rate (B_N) is equal to the remaining brine (R) plus blow down (B_D) flow rates.

Step 6: Calculate distillate flow rate in all stages (D_j) using material balance for flashing brine. It should be noted that there is no distillate at brine heater as there is no production. The amount of distillate in last stage (D_N) is the accumulated sum of the distillate produced in each separately.

Step 7: Update the salt concentration in flashing brine by salt balance in all stages.

Step 8: Evaluate the coefficients of TDM matrix and solve it to update the cooling brine temperature profile ($T_{F,j}$).

Step 9: Use the new ($T_{F,j}$) profile obtained in the above step to update the distillate temperature profile ($T_{D,j}$).

Step 10: Update the flashing brine temperature profile ($T_{B,j}$) using Newton-Raphson method by minimizing [11],

$$FUN = T_{B,j} - T_{D,j} - \Delta T_{BPR,j} - \Delta T_{NONE,j} - \Delta T_{DEM,j} - \Delta T_{TB,j}$$

Step 11: If the convergence criteria for $T_{B,j}$ profile is satisfied, go to step 12, otherwise repeat the procedure from step 5.

Step 12: If the makeup temperature (T_F) is equal to exit cooling brine temperature from first rejection stage then calculate brine levels, otherwise repeat from step 5.

Figure 5.1 shows the logic diagram for calculation procedure presented above.

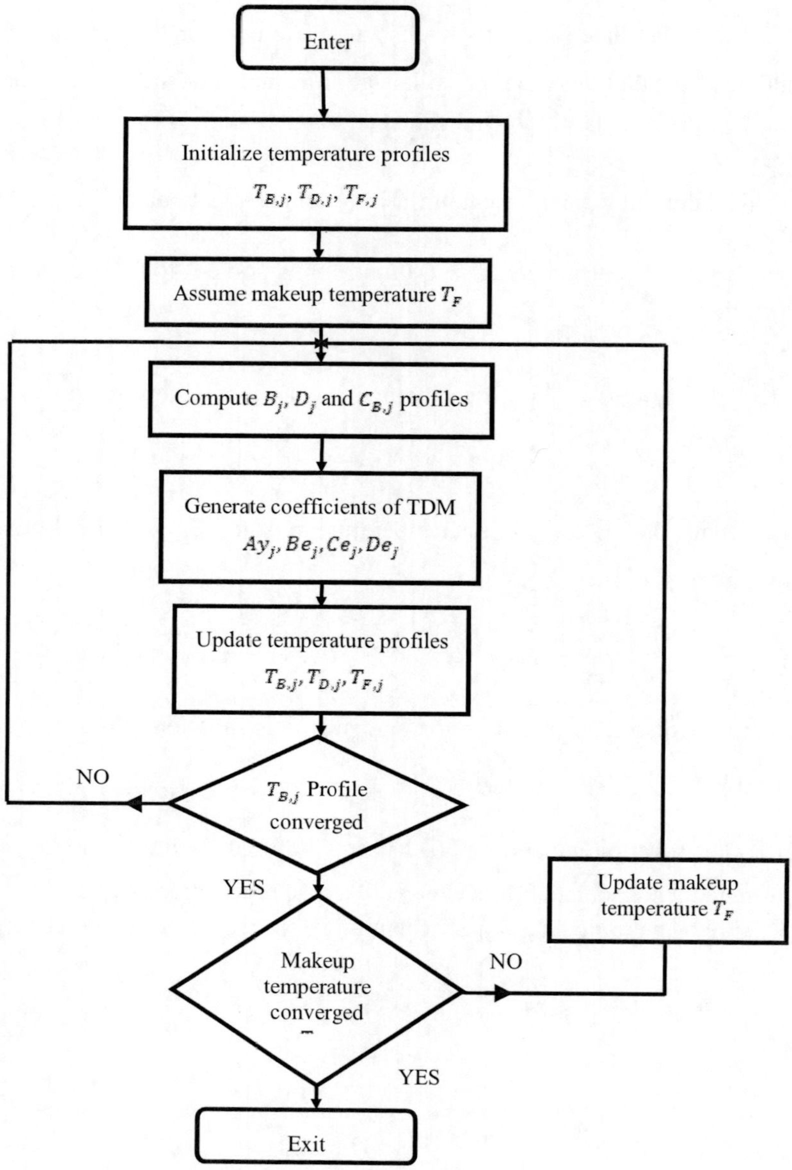

Figure 5.1. The logic diagram for the solution of the developed model (TDM-logic)

6. Results of MSF Process Simulation

To check the accuracy of the developed model, the selected MSF desalination plant data is used. The dimensional data of the selected plant were given in Table 2.1. Four sets of vendor simulated data of the plant is given in Tables 6.1-6.4 are used to check the accuracy of the developed model predictions. The model predicted results for the same vendor simulated conditions are presented in Tables 6.5-6.8. A comparison of the vendor simulated results with the developed model predictions indicates that the agreement between the two is excellent (Table 6.9 and Figures 6.1-6.20).

Further, one set of actual MSF desalination plant operating data at 99% load is collected and it is given in Table 6.10. Model predictions for this operating condition are given in Tables 6.11 and 6.12. Again comparison of the model prediction with operating data clearly indicates that the accuracy of the developed model predictions is excellent (Table 6.12 and Figure 6.21).

Table 6.1. Vendor simulation results for summer 100% load

Makeup	= 3320 ton/hr
Recycle (W)	= 12873 ton/hr
Seawater feed to the plant	= 10555 ton/hr
Seawater temperature	= 35 °C
Seawater salt concentration	= 40 g/L
TBT	= 105.0 °C
Steam temperature	= 121.0 °C
Distillate production	= 1270.20 ton/hr

j	$T_{B,j}$	$T_{D,j}$	$T_{F,j}$	B_j	D_j
1	101.33	99.95	98.21	12792	79.10
2	97.8	96.43	94.67	12713	156.80
3	94.3	92.94	91.16	12636	233.10
4	90.74	89.38	87.68	12559	309.70
5	87.21	85.87	84.14	12484	385.30
6	83.56	82.22	80.61	12406	462.80
7	79.92	78.56	76.95	12330	539.10
8	76.3	74.98	73.3	12255	614.30
9	72.69	71.38	69.67	12181	688.20
10	69.11	67.8	66.05	12108	761.00
11	65.52	64.23	62.46	12036	832.10
12	62.02	60.72	58.91	11966	901.10
13	58.59	57.2	55.38	11899	966.40
14	55.24	53.76	51.93	11833	1031.70
15	51.98	50.38	48.56	11771	1094.60
16	48.87	47.07	45.27	11711	1154.00
17	46.88	44.67	42.51	11673	1190.70
18	44.76	42.53	40.14	11633	1230.70
19	42.11	40.11	37.6	11593	1270.20

Table 6.2. Vendor simulation results for winter 100% load

Makeup	= 3196 ton/hr
Recycle (W)	= 12245 ton/hr
Seawater feed to the plant	= 10555 ton/hr
Seawater temperature	= 24 °C
Seawater salt concentration	= 40 g/L
TBT	= 99 °C
Steam temperature	= 116 °C
Distillate production	= 1274.40 ton/hr

j	$T_{B,j}$	$T_{D,j}$	$T_{F,j}$	B_j	D_j
1	94.99	93.64	91.76	12161	81.70
2	91.12	89.78	87.88	12080	161.90
3	87.29	85.94	84.04	12001	240.50
4	83.41	82.07	80.23	11922	318.90
5	79.58	78.25	76.37	11845	396.20
6	75.61	74.29	72.53	11766	475.20
7	71.68	70.36	68.56	11688	552.70
8	67.77	66.46	64.61	11612	628.80
9	63.89	62.58	60.69	11538	703.30
10	60.05	58.73	56.81	11465	776.40
11	56.26	54.9	52.95	11393	847.10
12	52.61	51.13	49.17	11325	915.00
13	49.07	47.46	45.5	11260	980.00
14	45.62	43.9	41.93	11197	1043.00
15	42.29	40.43	38.46	11137	1099.50
16	39.13	37.04	35.1	11081	1156.00
17	36.93	34.33	31.83	11042	1194.10
18	34.65	31.98	29.32	11001	1234.20
19	31.88	29.54	26.71	10961	1274.40

Table 6.3. Vendor simulation results for summer 60% load

Makeup	= 2008 ton/hr
Recycle (W)	= 9535 ton/hr
Seawater feed to the plant	= 10555 ton/hr
Seawater temperature	= 35 °C
Seawater salt concentration	= 40 g/L
TBT	= 91°C
Steam temperature	= 104 °C
Distillate production	= 771.20 ton/hr

j	$T_{B,j}$	$T_{D,j}$	$T_{F,j}$	B_j	D_j
1	87.95	86.65	85.63	9486	47.90
2	85.02	83.72	82.69	9438	95.00
3	82.1	80.81	79.77	9392	141.30
4	79.16	77.87	76.88	9350	187.70
5	76.24	74.96	73.95	9300	233.70
6	73.23	71.96	71.02	9252	280.60
7	70.23	68.97	68.01	9206	326.90
8	67.24	65.99	65.01	9160	372.70
9	64.27	63.20	62.01	9115	417.90
10	61.3	60.06	59.03	9070	462.60

Table 6.3. (Continued)

j	$T_{B,j}$	$T_{D,j}$	$T_{F,j}$	B_j	D_j
11	58.38	57.90	56.05	9026	505.80
12	55.56	54.18	53.15	8985	547.60
13	52.78	51.35	50.132	89.44	588.30
14	50.08	48.56	47.52	8905	627.70
15	47.44	45.85	44.8	8866	665.80
16	44.9	43.19	42.5	8830	762.30
17	42.96	40.97	39.43	8802	729.30
18	41.23	39.12	37.73	8778	753.80
19	39.58	37.25	36.19	8761	771.20

Table 6.4. Vendor simulation results for winter 60% load

Makeup	= 1992 ton/hr
Recycle (W)	= 8925 ton/hr
Seawater feed to the plant	= 10555 ton/hr
Seawater temperature	= 24 °C
Seawater salt concentration	= 40 g/L
TBT	= 85 °C
Steam temperature	= 98.96 °C
Distillate production	= 773.80 ton/hr

j	$T_{B,j}$	$T_{D,j}$	$T_{F,j}$	B_j	D_j
1	81.61	80.33	79.23	8875	49.50
2	78.34	77.07	75.95	8826	98.20
3	75.1	73.83	72.7	8777	145.90
4	71.83	70.57	69.48	8730	193.60
5	68.61	67.35	65.23	8683	240.60
6	65.29	64.04	63.	8635	288.30
7	62	60.78	59.68	8588	335.40
8	58.72	57.48	56.37	8541	381.80
9	55.46	54.21	53.09	8496	427.40
10	52.29	50.96	49.02	8452	471.40
11	49.19	47.77	46.64	8410	513.70
12	46.18	44.7	43.54	8368	554.80
13	43.25	41.67	40.51	8328	594.40
14	40.4	38.73	37.67	8290	632.60
15	37.64	35.87	34.7	8254	669.30
16	34.99	33.08	31.92	8218	704.30
17	32.8	30.59	28.82	8190	732.30
18	30.95	28.54	27.02	8166	756.40
19	29.21	26.66	25.47	8149	773.80

Table 6.5. Model predictions for summer 100% load

Makeup	= 3320 ton/hr
Recycle (W)	= 12873 ton/hr
Seawater feed to the plant	= 10555 ton/hr
Seawater temperature	= 35 °C
Seawater salt concentration	= 40 g/L
TBT	= 105.0 °C
Steam temperature	= 121.0 °C
Distillate production	= 1269.50 ton/hr

j	$T_{B,j}$	$T_{D,j}$	$T_{F,j}$	B_j	D_j
1	101.01	99.57	98.62	12783.55	89.45
2	97.46	96.00	95.03	12705.12	167.88
3	93.83	92.35	91.38	12626.26	246.74
4	90.18	88.67	87.70	12548.06	324.94
5	86.52	84.97	84.00	12470.66	402.34
6	82.83	81.25	80.29	12394.00	479.00
7	79.15	77.53	76.56	12318.37	554.63
8	75.46	73.79	72.83	12243.84	629.16
9	71.78	70.07	69.10	12170.53	702.47
10	68.12	66.35	65.39	12098.53	774.47
11	64.48	62.65	61.70	12027.96	845.04
12	60.86	58.97	58.03	11958.91	914.09
13	57.29	55.33	54.40	11891.48	981.52
14	53.76	51.72	50.81	11825.80	1047.20
15	50.29	48.17	47.27	11761.95	1111.05
16	46.42	44.20	43.79	11691.79	1181.21
17	45.03	42.76	42.22	11666.47	1206.53
18	43.15	40.84	40.26	11632.96	1240.04
19	41.44	39.09	38.20	11603.50	1269.50

Table 6.6. Model predictions for winter 100% load

Makeup	= 3196 ton/hr
Recycle (W)	= 12245 ton/hr
Seawater feed to the plant	= 10555 ton/hr
Seawater temperature	= 24 °C
Seawater salt concentration	= 40 g/L
TBT	= 99 °C
Steam temperature	= 116 °C
Distillate production	= 1275.80 ton/hr

j	$T_{B,j}$	$T_{D,j}$	$T_{F,j}$	B_j	D_j
1	94.60	93.13	92.09	12152.21	92.79
2	90.70	89.20	88.15	12071.33	173.67
3	86.73	85.19	84.14	11990.31	254.69
4	82.74	81.17	80.12	11910.32	334.68
5	78.76	77.14	76.09	11831.53	413.47
6	74.77	73.10	72.06	11753.86	491.14
7	70.79	69.07	68.04	11677.63	567.37
8	66.83	65.05	64.02	11602.93	642.07
9	62.90	61.05	60.04	11529.89	715.11
10	59.01	57.09	56.09	11458.62	786.38
11	55.17	53.17	52.18	11389.24	855.76
12	51.39	49.30	48.33	11321.86	923.14
13	47.68	45.50	44.55	11256.63	988.37
14	44.05	41.77	40.84	11193.65	1051.35
15	40.51	38.13	37.22	11133.08	1111.92
16	36.74	34.22	33.71	11069.25	1175.75
17	34.97	32.39	31.73	11039.36	1205.64
18	32.72	30.06	29.41	11002.24	1242.76
19	30.70	27.98	27.10	10969.20	1275.80

Table 6.7. Model predictions for summer 60% load

Makeup	= 2008 ton/hr
Recycle (W)	= 9535 ton/hr
Seawater feed to the plant	= 10555 ton/hr
Seawater temperature	= 35 °C
Seawater salt concentration	= 40 g/L
TBT	= 91 °C
Steam temperature	= 104 °C
Distillate production	= 771.13 ton/hr

j	$T_{B,j}$	$T_{D,j}$	$T_{F,j}$	B_j	D_j
1	87.60	86.09	85.38	9480.41	54.59
2	84.62	83.08	82.37	9433.21	101.79
3	81.58	80.01	79.30	9385.66	149.34
4	78.54	76.93	76.23	9338.51	196.49
5	75.50	73.86	73.15	9291.92	243.08
6	72.46	70.78	70.08	9245.82	289.18
7	69.43	67.70	67.01	9200.40	334.60
8	66.42	64.64	63.96	9155.68	379.32
9	63.42	61.60	60.92	9111.71	423.29
10	60.46	58.58	57.90	9068.54	466.46

Table 6.7. (Continued)

j	$T_{B,j}$	$T_{D,j}$	$T_{F,j}$	B_j	D_j
11	57.52	55.58	54.91	9026.20	508.80
12	54.62	52.62	51.96	8984.74	550.26
13	51.76	49.69	49.04	8944.20	590.80
14	48.94	46.81	46.17	8904.61	630.39
15	46.19	43.98	43.35	8866.01	668.99
16	43.30	41.01	40.59	8826.76	708.24
17	41.42	39.07	38.74	8795.79	739.21
18	40.29	37.91	37.55	8779.15	755.85
19	39.01	36.60	36.25	8763.87	771.13

Table 6.8. Model predictions for winter 60% load

Makeup = 1992 ton/hr
Recycle (W) = 8925 ton/hr
Seawater feed to the plant = 10555 ton/hr
Seawater temperature = 24 °C
Seawater salt concentration = 40 g/L
TBT = 85 °C
Steam temperature = 98.96 °C
Distillate production = 773.03 ton/hr

j	$T_{B,j}$	$T_{D,j}$	$T_{F,j}$	B_j	D_j
1	81.17	79.61	78.83	8867.29	57.71
2	77.85	76.25	75.47	8817.87	107.13
3	74.47	72.83	72.05	8768.25	156.75
4	71.11	69.42	68.65	8719.45	205.55
5	67.76	66.02	65.25	8671.51	253.49
6	64.42	62.63	61.88	8624.37	300.63
7	61.12	59.26	58.52	8578.24	346.76
8	57.85	55.93	55.19	8533.16	391.84
9	54.62	52.63	51.91	8489.21	435.79
10	51.45	49.38	48.67	8446.44	478.56
11	48.33	46.19	45.49	8404.91	520.09
12	45.28	43.05	42.37	8364.70	560.30
13	42.29	39.98	39.32	8325.88	599.12
14	39.40	36.99	36.35	8288.51	636.49
15	36.59	34.09	33.47	8252.67	672.33
16	33.81	31.21	30.68	8217.38	707.62
17	31.44	28.75	28.31	8188.78	736.22
18	29.94	27.18	26.75	8171.53	753.47
19	28.34	25.53	25.19	8151.97	773.03

Table. 6.9. Comparison of various parameters of vendor simulation results with model predictions

No.	parameter	Summer 100%		Summer 60%		Winter 100%		Winter 60%	
		Vendor	Simulation	Vendor	Simulation	Vendor	Simulation	Vendor	Simulation
1	Steam flow rate to brine heater (ton/hr)	153.9	146.23	87.7	93.24	154.3	149.55	87.6	95.17
2	Brine heater duty* (kJ/hr)	3.44×10^8	3.22×10^8	2.01×10^8	2.09×10^8	3.49×10^8	3.31×10^8	2.02×10^8	2.15×10^8
3	Performance ratio	8.30	8.68	8.80	8.20	8.30	8.53	8.80	8.12
4	Blow down (ton/hr)	2049.8	2050.5	1236.1	1236.9	1921.6	1920.2	1218.2	1219.0
5	Blow down salt concentration (g/L)	66.60	65.10	65.40	64.30	67.00	65.40	65.70	65.57

*Brine heater duty = $W\ 1000\left(TBT - T_{F,1}\right) S_{R,H}$

Table 6.10. MSF desalination plant actual operating data for 99% load

Makeup	= 3150 ton/hr
Recycle (w)	= 12800 ton/hr
Seawater feed to the plant	= 9700 ton/hr
Seawater temperature	= 26.8 °C
Seawater salt concentration	= 40 g/L
TBT	= 97.6°C
Steam temperature	= 112.6 °C
Distillate production	= 1271 ton/hr

j	$T_{B,j}$	$T_{D,j}$	$T_{F,j}$	B_j	D_j
1	93.4	-	-	-	-
2	89.2	-	-	-	-
3	86.7	-	-	-	-
4	82.8	-	-	-	-
5	78.9	-	-	-	-
6	75.2	-	-	-	-
7	71.6	-	-	-	-
8	68.5	-	-	-	-
9	65.4	-	-	-	-
10	61.0	-	-	-	-
11	57.5	-	-	-	-
12	53.8	-	-	-	-
13	50.6	-	-	-	-
14	47.3	-	-	-	-
15	43.4	-	-	-	-
16	40.8	-	-	-	-
17	38.3	-	-	-	-
18	35.9	-	-	-	-
19	33.8	-	-	-	-

Table 6.11. Model predictions for actual plant operating conditions of plant at 99% load

Makeup	= 3150 ton/hr
Recycle (W)	= 12800 ton/hr
Seawater feed to the plant	= 9700 ton/hr
Seawater temperature	= 26.8 °C
Seawater salt concentration	= 40 g/L
TBT	= 97.6°C
Steam temperature	= 112.6 °C
Distillate production	= 1270.68 ton/hr

j	$T_{B,j}$	$T_{D,j}$	$T_{F,j}$	B_j	D_j
1	94.39	92.92	91.89	12705.11	94.89
2	90.61	89.10	88.07	12623.08	176.92
3	86.77	85.23	84.20	12541.23	258.77
4	82.94	81.36	80.34	12460.81	339.19
5	79.13	77.51	76.49	12381.92	418.08
6	75.33	73.66	72.66	12304.46	495.54
7	71.56	69.84	68.85	12228.70	571.30
8	67.82	66.04	65.06	12154.70	645.30
9	64.13	62.28	61.32	12082.53	717.47
10	60.48	58.57	57.61	12012.27	787.73
11	56.89	54.90	53.96	11943.98	856.02
12	53.35	51.29	50.37	11877.75	922.25
13	49.89	47.74	46.84	11813.66	986.34
14	46.51	44.27	43.39	11751.77	1048.23
15	43.22	40.88	40.03	11692.19	1107.81
16	39.61	37.16	36.75	11627.86	1172.14
17	38.31	35.81	35.19	11604.18	1195.82
18	36.17	33.60	32.95	11567.72	1232.28
19	33.98	31.34	30.60	11529.32	1270.68

Table 6.12. Actual plant operating data and model predictions for various parameters

No.	parameter	99% load	
		Actual plant data	Model simulation data
1	Steam flow rate to brine heater (ton/hr)	151.50	153.40
2	Brine heater duty* (kJ/hr)	3.54×10^8	3.36×10^8
3	Performance ratio	8.19	8.28
4	Blowdown (ton/hr)	1879.0	1879.3
5	Blowdown salt concentration (g/L)	-	65.19

*Brine heater duty = $W\ 1000\left(TBT - T_{F,1}\right) S_{R,H}$

Actual $T_{F,1}$= 90.7 °C

Simulated $T_{F,1}$= 91.89 °C

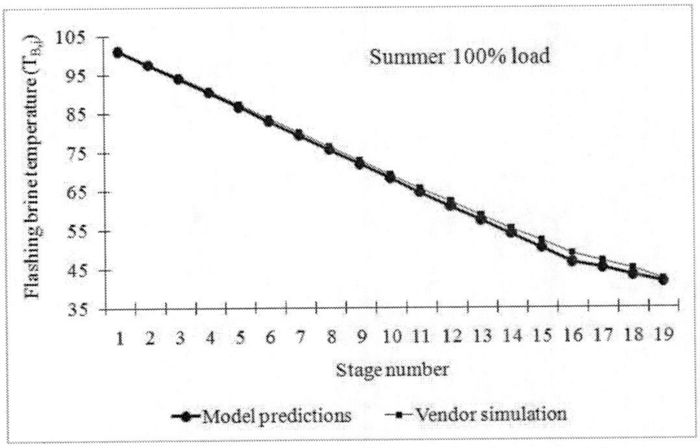

Figure 6.1. Vendor simulated and model predicted flashing brine temperature profile (°C) for summer 100% load

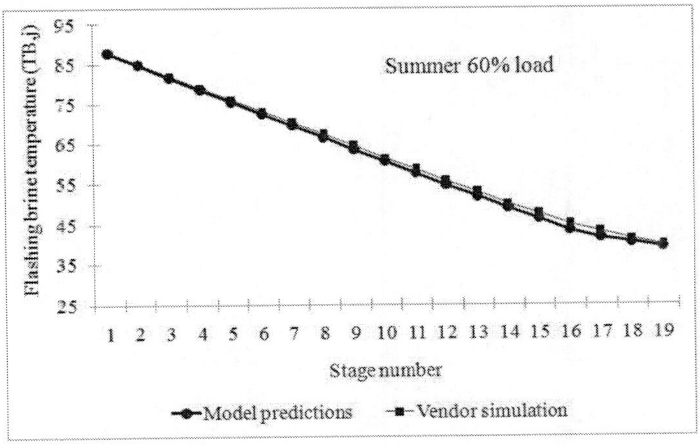

Figure 6.2. Vendor simulated and model predicted flashing brine temperature profile (°C) for summer 60% load

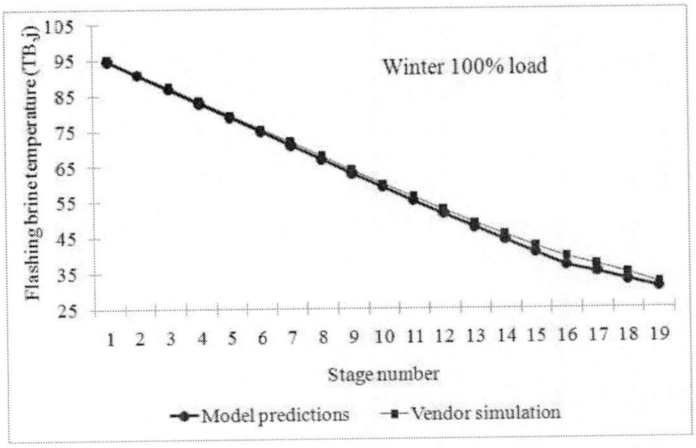

Figure 6.3. Vendor simulated and model predicted flashing brine temperature profile (°C) for winter 100% load

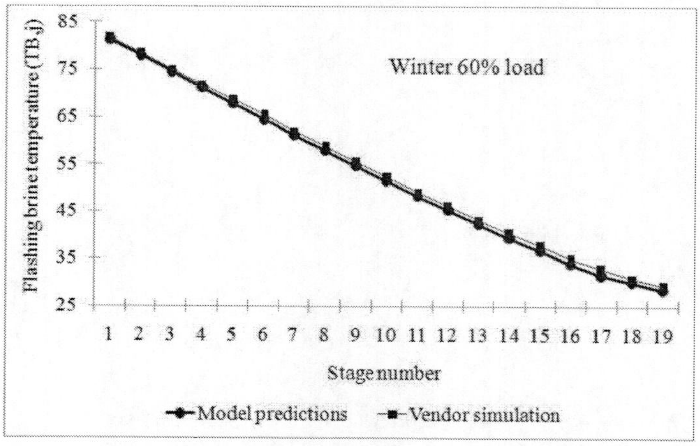

Figure 6.4. Vendor simulated and model predicted flashing brine temperature profile (°C) for winter 60% load

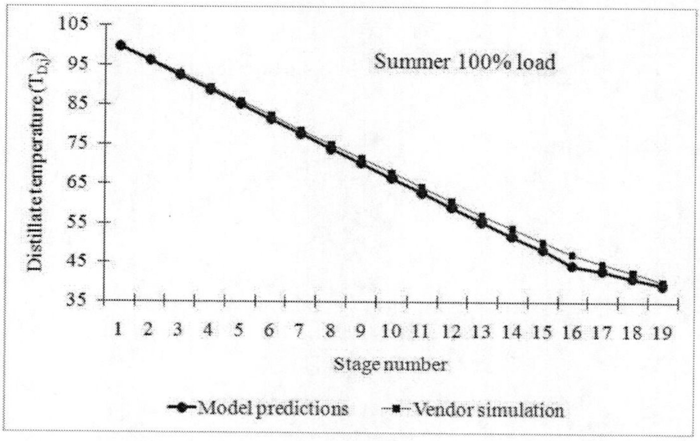

Figure 6.5. Vendor simulated and model predicted distillate temperature profile (°C) for summer 100% load

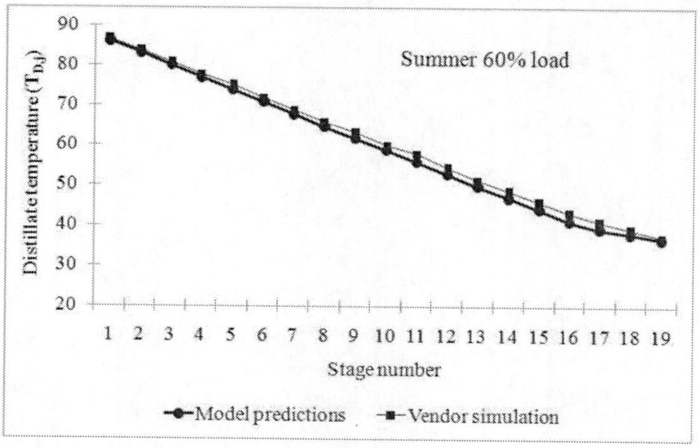

Figure 6.6. Vendor simulated and model predicted distillate temperature profile (°C) for summer 60% load

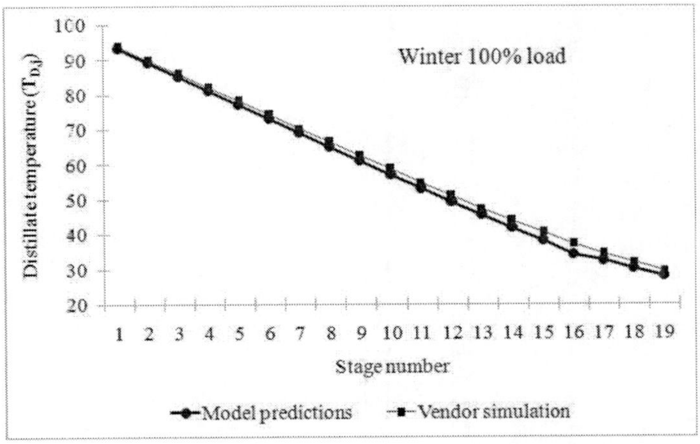

Figure 6.7. Vendor simulated and model predicted distillate temperature profile (°C) for winter 100% load

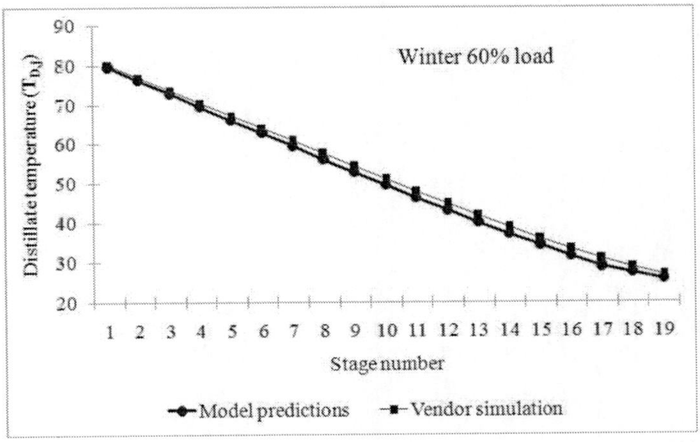

Figure 6.8. Vendor simulated and model predicted distillate temperature profile (°C) for winter 60% load

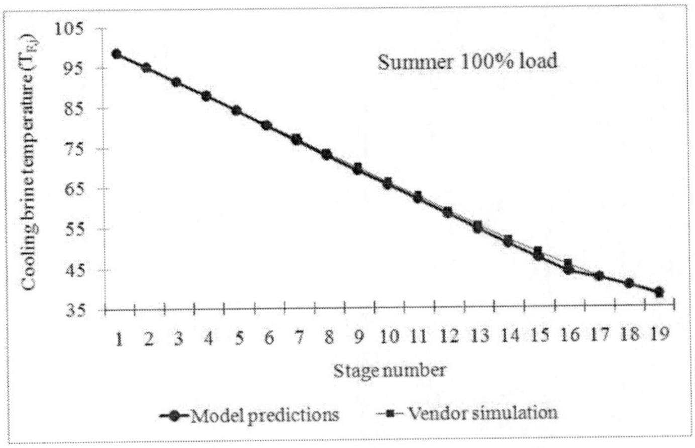

Figure 6.9. Vendor simulated and model predicted cooling brine temperature profile (°C) for summer 100% load

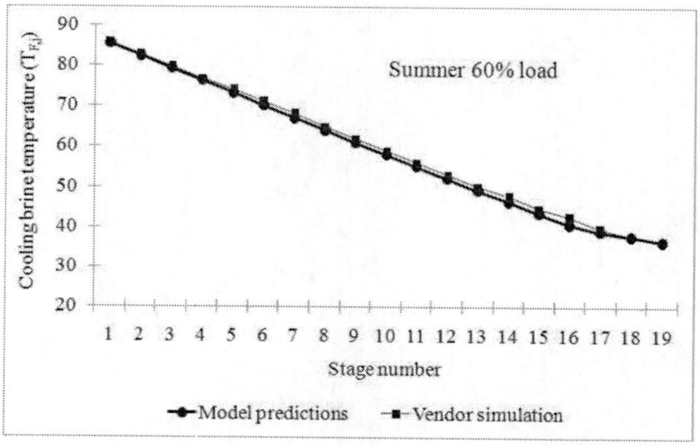

Figure 6.10: Vendor simulated and model predicted cooling brine temperature profile (°C) for summer 60% load.

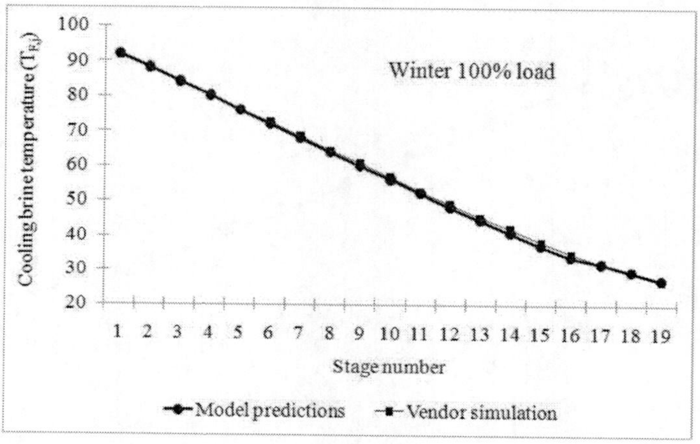

Figure 6.11. Vendor simulated and model predicted cooling brine temperature profile (°C) for winter 100% load

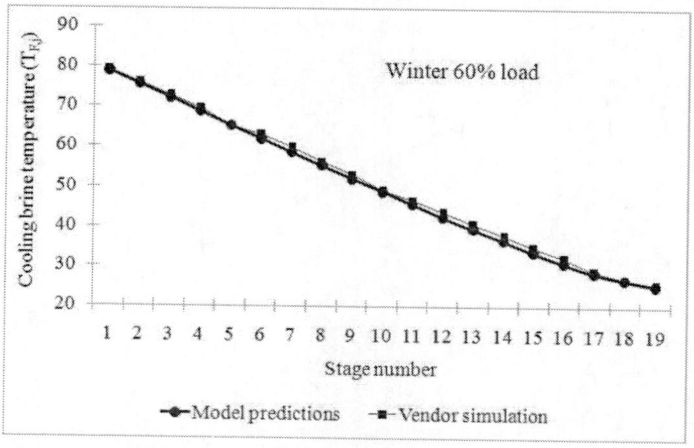

Figure 6.12. Vendor simulated and model predicted cooling brine temperature profile (°C) for winter 60% load

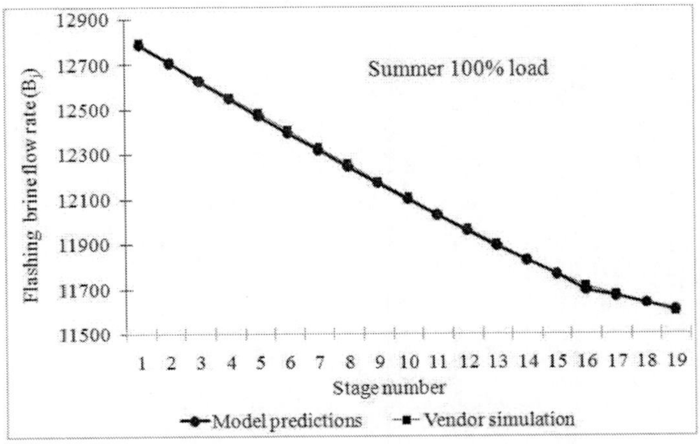

Figure 6.13. Vendor simulated and model predicted flashing brine flow rate profile (ton/hr) for summer 100% load

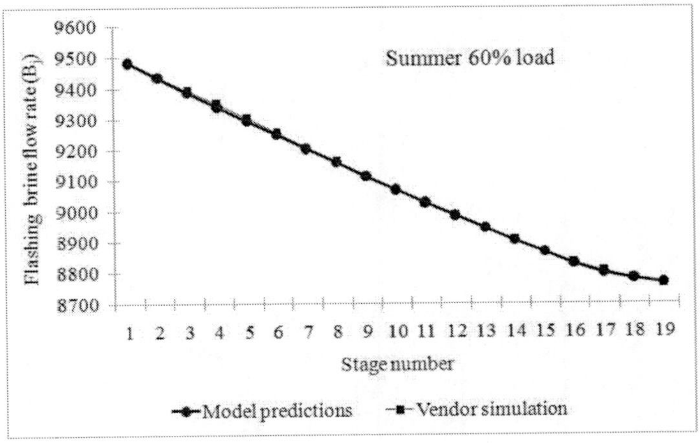

Figure 6.14. Vendor simulated and model predicted flashing brine flow rate profile (ton/hr) for summer 60% load

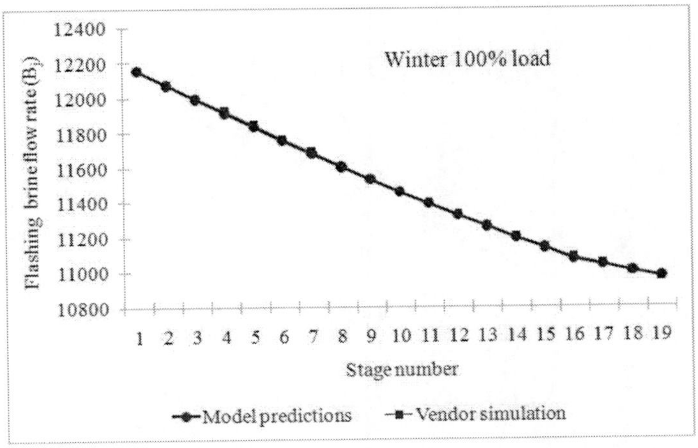

Figure 6.15. Vendor simulated and model predicted flashing brine flow rate profile (ton/hr) for winter 100% load

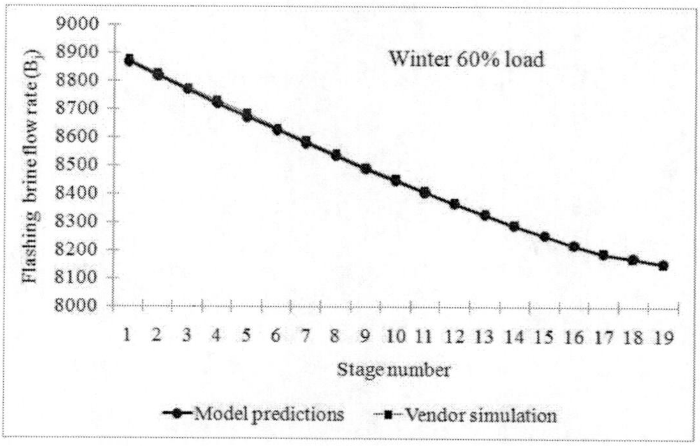

Figure 6.16. Vendor simulated and model predicted flashing brine flow rate profile (ton/hr) for winter 60% load

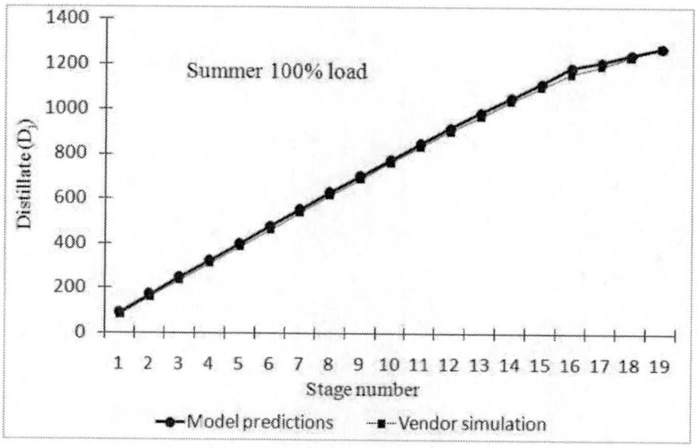

Figure 6.17. Vendor simulated and model predicted distillate flow rate profile (ton/hr) for summer 100% load

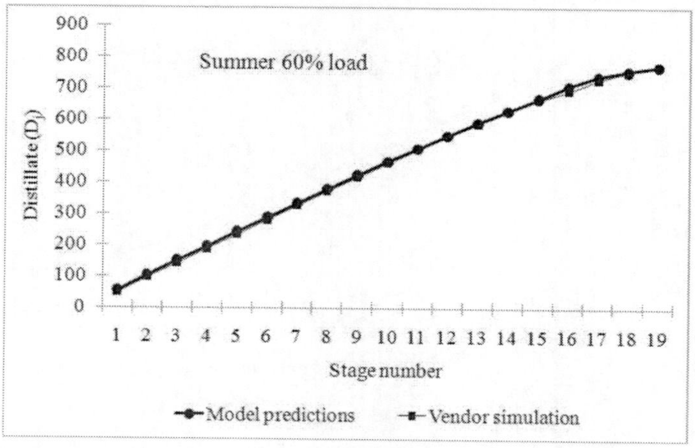

Figure 6.18. Vendor simulated and model predicted distillate flow rate profile (ton/hr) for summer 60% load

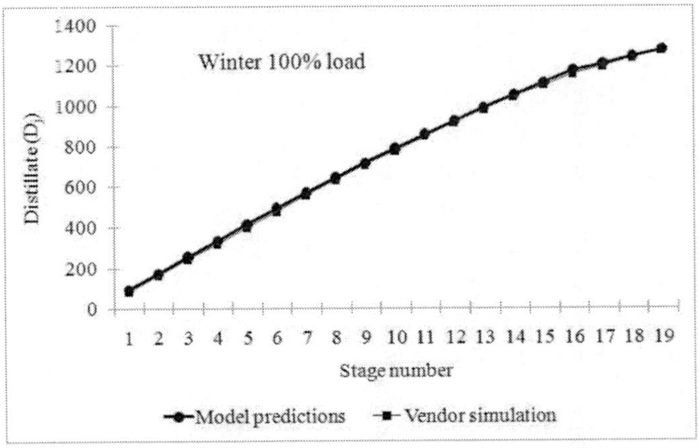

Figure 6.19. Vendor simulated and model predicted distillate flow rate profile (ton/hr) for winter 100% load

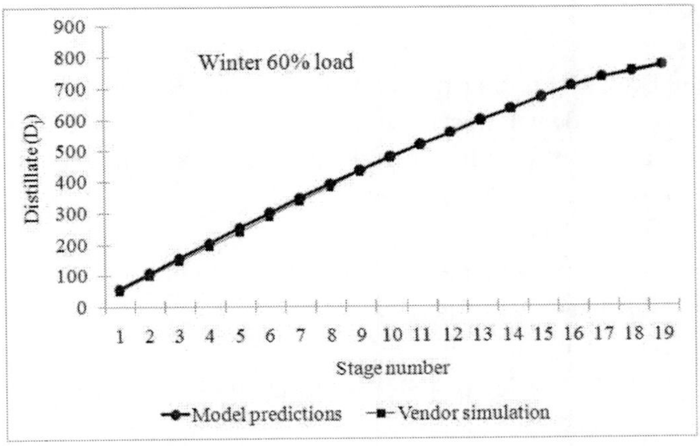

Figure 6.20. Vendor simulated and model predicted distillate flow rate profile (ton/hr) for winter 60% load

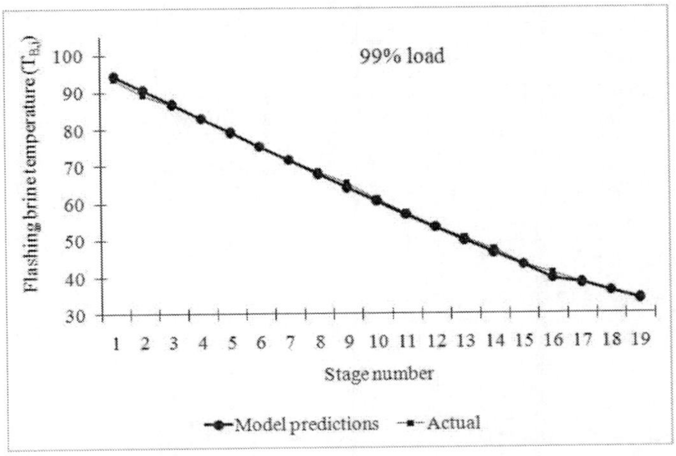

Figure 6.21. Actual measured and model predicted flashing brine temperature profile (°C) for the 99% load

7. MSF Process Optimization

The selected MSF desalination plant was commissioned in 2003 to produce 3800 ton/hr of distillate water by consuming low pressure steam, which is extracted from combined cycle power plant. In this section, the optimization of the MSF desalination plant is carried out by investigating two different modes of operation: maximization of the performance ratio for a target production and maximization of the distillate production rate. Optimum conditions are obtained by series of simulation runs by varying different operating variables at a time within the maximum and minimum constrains of certain variable of the plant operation and equipment like pump limitations etc.

First mode: Maximization of the performance ratio for a target production

In some situations, the distillate product flow rate is fixed due to limitation on storage capacity. The objective, therefore, is to maximize the performance ratio at the desired distillate production.

Second mode: Maximization of the distillate production rate

In the summer season, demand for the water increases. The situation may arise to produce maximum possible amount of distillate at any performance ratio whatsoever. The objective, therefore, is to maximize the distillate production rate.

In both modes, the maximization was carried out to satisfy all the constraints of the design and the operational conditions given in Table 7.1.

Table 7.1. Plant and operational constraints of the desalination plant

Maximum capacity of recycle pump (ton/hr)	2 pumps, each 6567
Maximum capacity of intake pump (ton/hr)	12873
Maximum capacity of distillate pump (ton/hr)	1500
Maximum capacity of blow down pump (ton/hr)	1957
Maximum capacity of condensate pump (ton/hr)	180
Brine velocity limits inside condenser tubes in different plant sections	2.2
Maximum value for top brine temperature (°C)	105
Maximum blow down concentration (ppm)	67500
Limits of reject coolant temperature (°C)	45
Maximum steam temperature (°C)	121
*Cooling brine velocity in the tubes (m/s)	1.8-2.2
Vapor velocity (m/s)	3.5-7.0
Water vapor loss in ejectors	2.5%
**Last stage level (mm)	200-680

* This limits were selected to minimize the tube erosion at high velocities and minimize the scaling/fouling at lower velocities.

** Last stage level is determined in order to prevent the brine flooding in the stages.

The main variables that contribute to the performance of MSF plant are the top brine temperature (TBT) and the recycle stream flow rate (W). Therefore, these two variables will be manipulated to achieve the maximum in both modes without violating any plant constraints listed in Table 7.1. TBT plays a critical role on the performance of the MSF plant. By increasing TBT, both the production rate and the performance ratio will increase. The production rate will increases because of an increase in the flashing range. But the performance ratio increases due to decrease in the latent heat of vaporization of water at higher temperature. At the same time, higher TBT leads to increased scale formation. Increase in the recycle stream flow rate increases the distillate production but adversely affects the performance ratio. An increase in the former is due to the drop in the flashing brine temperature from stage to stage. The maximum rate for the recycle flow is limited by the maximum allowable velocity in the cooling tubes, as higher velocities cause erosion of the tube material.

7.1. Maximization of the Performance Ratio

Two specific loads were considered for maximizing the performance ratio:

Summer 70 % load (Distillate product flow rate = 893 ton/hr)
Winter 78% load (Distillate product flow rate = 985 ton/hr)

Summer 70 % load (893 ton/hr)

The operational parameters associated for this load, as simulated by the vendor, are given in Table 7.2. For the same operating conditions, simulation run was carried out using the developed model. The results are shown in Table 7.3. The optimization is then done by using the developed model. The objective of the optimization is to maximize the performance ratio for targeted distillate production of 893 ton/hr. This will be done mainly by varying the values of the TBT and the brine recycle rate (W). First TBT is incremented by 2 $^{\circ}$C from 95 $^{\circ}$C to maximum allowed value of 105°C. By increasing TBT, more steam will be produced and thus distillate product will go up. However, the distillate product is required to maintain at 893 ton/hr. To achieve this, brine recycle flow was reduced gradually to achieve desired production rate of 893 ton/hr. This resulted in reducing the steam consumption rate, which contributed in increasing the performance ratio. The new operating conditions will be selected to ensure the guarantee of the maximization without the violation of any one of the constraints given in Table 7.1. The effect of variation of the TBT and the recycle flow rate on the performance ratio is given in Table 7.4. By examining the results in Table 7.4, we observe that the maximum performance ratio is 8.95) at TBT=105 °C and W=8749 ton/hr. Therefore, simulation run is made at this operating condition using the developed model and the result of simulation are given in Table 7.5.

Table 7.2. Vendor data for summer 70% load (893 ton/hr)

Makeup	= 2460 ton/hr
Recycle (W)	= *10373 ton/hr*
Seawater feed to the plant	= 10555 ton/hr
Seawater temperature	= 35 °C
Seawater salt concentration	= 40 g/L
TBT	= *95 °C*
Steam temperature	= 116 °C
Distillate production	= *893 ton/hr*

Table 7.3. Model simulation results for summer 70% load (893 ton/hr) (Before optimization)

Makeup	= 2460 ton/hr
Recycle (W)	= *10373 ton/hr*
Seawater feed to the plant	= 10555 ton/hr
Seawater temperature	= 35 °C
Seawater salt concentration	= 40 g/L
TBT	= *95 °C*
Steam temperature	= 116 °C
Distillate production	= *893.20 ton/hr*

j	$T_{B,j}$	$T_{D,j}$	$T_{F,j}$	B_j	D_j
1	91.40	89.91	89.13	10309.10	63.90
2	88.23	86.72	85.93	10253.71	119.29
3	85.00	83.46	82.67	10197.93	175.07
4	81.76	80.19	79.40	10142.74	230.26
5	78.52	76.91	76.13	10088.21	284.79
6	75.27	73.62	72.85	10034.26	338.74
7	72.04	70.34	69.57	9981.13	391.87
8	68.81	67.07	66.30	9928.85	444.15
9	65.60	63.81	63.05	9877.50	495.50
10	62.42	60.57	59.82	9827.14	545.86
11	59.27	57.36	56.61	9777.84	595.16
12	56.15	54.18	53.44	9729.66	643.34
13	53.07	51.03	50.30	9682.66	690.34
14	50.05	47.93	47.22	9636.92	736.08
15	47.08	44.88	44.18	9592.49	780.51
16	43.90	41.62	41.21	9545.57	827.43
17	42.19	39.86	39.45	9520.35	852.65
18	40.77	38.40	37.99	9500.13	872.87
19	39.34	36.94	36.52	9479.80	893.20

Steam flow rate to brine heater (ton/hr) = 107.56

Performance ratio = 8.30

Table 7.4. Effect of the TBT and the recycle flow rate on the performance ratio (Summer 70 %)

No.	TBT (°C)	Recycle flow rate or W (ton/hr)	Steam flow rate to brine heater (ton/hr)	Performance ratio
1	95	10555	107.56	8.30
2	97	10002	105.85	8.44
3	99	9656	104.22	8.57
4	101	9333	102.66	8.70
5	103	9031	101.18	8.83
6	105	8749	99.76	8.95

Table 7.5. Maximization of the performance ratio for the summer 70% load

Makeup	= 2460 ton/hr
Recycle (W)	= *8749 ton/hr*
Seawater feed to the plant	= 10555 ton/hr
Seawater temperature	= 35 °C
Seawater salt concentration	= 40 g/L
TBT	= *105 °C*
Steam temperature	= 116 °C
Distillate production	= *893.20 ton/hr*

j	$T_{B,j}$	$T_{D,j}$	$T_{F,j}$	B_j	D_j
1	100.83	99.39	98.56	8685.78	63.72
2	97.11	95.65	94.80	8629.81	119.69
3	93.31	91.83	90.97	8573.62	175.88
4	89.49	87.97	87.13	8517.90	231.60
5	85.66	84.10	83.25	8462.82	286.68
6	81.80	80.21	79.37	8408.30	341.20
7	77.94	76.30	75.46	8354.58	394.92
8	74.08	72.40	71.56	8301.72	447.78
9	70.24	68.50	67.67	8249.81	499.69
10	66.42	64.62	63.80	8198.92	550.58
11	62.63	60.77	59.95	8149.14	600.36
12	58.87	56.94	56.13	8100.55	648.95
13	55.16	53.15	52.36	8053.23	696.27
14	51.51	49.42	48.64	8007.26	742.24
15	47.93	45.75	44.98	7962.72	786.78
16	44.15	41.86	41.40	7916.33	833.17
17	42.00	39.65	39.23	7890.23	859.27
18	40.49	38.10	37.71	7872.36	877.14
19	39.11	36.69	36.32	7856.30	893.20

Steam flow rate to brine heater (ton/hr) = 99.76

Performance ratio = 8.95

Winter 78% load (985 ton/hr)

Similar procedure for optimization of performance ratio as used for 893 ton/hr is performed for load of 795 ton/hr. The operational parameters associated with this load, as carried out by the vendor, is shown in Table 7.6. For the same operating conditions, simulation run is made using the developed model. The results are shown in Table 7.7.

Table 7.6. Vendor data for winter at 78% load (985 ton/hr)

Makeup	= 2742 ton/hr
Recycle (W)	= *10346 ton/hr*
Seawater feed to the plant	= 10555 ton/hr
Seawater temperature	= 24 °C
Seawater salt concentration	= 40 g/L
TBT	= *92 °C*
Steam temperature	= 116 °C
Distillate production	= *985 ton/hr*

Table 7.7. Model simulation results for winter 78% load (989.14 ton/hr) (Before optimization)

Makeup	= 2742 ton/hr
Recycle (W)	= 10346 ton/hr
Seawater feed to the plant	= 10555 ton/hr
Seawater temperature	= 24 °C
Seawater salt concentration	= 40 g/L
TBT	= 92 °C
Steam temperature	= 116 °C
Distillate production	= 989.14 ton/hr

j	$T_{B,j}$	$T_{D,j}$	$T_{F,j}$	B_j	D_j
1	87.84	86.33	85.42	10273.09	72.91
2	84.20	82.65	81.73	10210.10	135.90
3	80.49	78.91	77.99	10146.94	199.06
4	76.78	75.15	74.24	10084.70	261.30
5	73.08	71.40	70.50	10023.49	322.51
6	69.38	67.65	66.76	9963.22	382.78
7	65.71	63.92	63.04	9904.17	441.83
8	62.07	60.22	59.34	9846.40	499.60
9	58.46	56.54	55.67	9790.00	556.00
10	54.90	52.91	52.05	9735.06	610.94
11	51.40	49.33	48.49	9681.66	664.34
12	47.96	45.80	44.98	9629.90	716.10
13	44.60	42.34	41.54	9579.87	766.13
14	41.32	38.97	38.19	9531.66	814.34
15	38.13	35.68	34.92	9485.37	860.63
16	34.87	32.30	31.77	9438.45	907.55
17	32.73	30.08	29.54	9407.74	938.26
18	30.89	28.17	27.64	9382.64	963.36
19	29.05	26.27	25.77	9356.86	989.14

Steam flow rate to brine heater (ton/hr) = 120.51

Performance ratio = 8.23

The effect of variation of the TBT and the recycle flow rate on the performance ratio is shown in Table 7.8. The result of the model at the maximum performance ratio of 9.10 at TBT=105 °C and W=8508 ton/hr is shown in Table 7.9.

Table 7.8. Effect of the TBT and the recycle flow rate on the performance ratio (Winter 78 %)

No	TBT (°C)	Recycle flow rate or W (ton/hr)	Steam flow rate to brine heater (ton/hr)	Performance ratio
1	92	10346	120.51	8.23
2	95	9859	117.59	8.41
3	97	9558	115.95	8.53
4	99	9277	114.11	8.65
5	101	9011	112.91	8.76
6	103	8761	111.48	8.87
7	105	8508	108.70	9.10

Table. 7.9. Maximization of the performance ratio for the winter 78% load

Makeup	= 2742 ton/hr
Recycle (W)	= *8508 ton/hr*
Seawater feed to the plant	= 10555 ton/hr
Seawater temperature	= 24 °C
Seawater salt concentration	= 40 g/L
TBT	= *105 °C*
Steam temperature	= 116 °C
Distillate production	= *989.14 ton/hr*

j	$T_{B,j}$	$T_{D,j}$	$T_{F,j}$	B_j	D_j
1	100.19	98.74	97.78	8436.61	70.89
2	95.87	94.40	93.42	8374.37	133.13
3	91.48	89.97	88.99	8313.01	194.49
4	87.05	85.51	84.54	8252.45	255.05
5	82.61	81.03	80.05	8192.86	314.64
6	78.16	76.52	75.56	8134.15	373.35
7	73.71	72.02	71.06	8076.56	430.94
8	69.27	67.52	66.57	8020.18	487.32
9	64.86	63.04	62.09	7965.10	542.40
10	60.49	58.59	57.65	7911.40	596.10
11	56.17	54.18	53.25	7859.18	648.32
12	51.90	49.82	48.91	7808.55	698.95
13	47.71	45.53	44.63	7759.62	747.88
14	43.61	41.31	40.44	7712.53	794.97
15	39.61	37.19	36.34	7667.44	840.06
16	35.62	33.06	32.35	7626.20	881.30
17	32.24	29.55	29.01	7578.00	929.50
18	30.33	27.57	27.10	7544.17	963.33
19	28.64	25.83	25.43	7518.36	989.14

Steam flow rate to brine heater (ton/hr) = 108.70
Performance ratio = 9.10

Values of the various parameters before and after maximization of the performance ratio for both summer 70% and winter 78% loads are listed in Table 7.10.

7.2. Maximization of Distillate Production Rate

Maximization of the distillate production rate will be obtained when TBT and brine recycle flow rate are maximum. For maximization of production rate, summer 100% load (Table 6.5) and winter 100% load (Table 6.6) are selected.

The result of the maximization of the distillate production rate for summer is shown in Table 7.11 and for winter in Table 7.12. It can be seen from Table 7.11 that the distillate production rate in summer is increased from 1269.5 ton/hr (100%) to 1305.8 ton/hr (102.9%). In winter as shown in Table 7.12, it can be seen that the distillate production rate is increased from 1275.8 ton/hr (100%) to 1482.33 ton/hr (116.2%), which is 16.2% increase. Table 7.13 illustrates various parameters before and after maximization of the distillate production rate for summer 100% and winter 100% loads.

Table 7.10. Various parameters before and after maximization of performance ratio

No.	parameter	Summer 70% load		Winter 78% load	
		Before optimization	After optimization	Before optimization	After optimization
1	Steam flow rate to brine heater (ton/hr)	107.56	99.76	120.51	108.70
2	Brine heater duty* (kJ/hr)	$2.38x10^8$	$2.21\ x10^8$	$2.66\ x10^8$	$2.41x10^8$
3	Performance ratio	8.30	8.95	8.23	9.10
4	Blowdown (ton/hr)	1566.8	1566.8	1752.9	1752.9
5	Blowdown salt concentration (g/L)	64.49	65.20	64.90	65.78

*Brine heater duty = $W\ 1000\left(TBT - T_{F,1}\right) S_{R,H}$

Table 7.11. Maximization of the distillate production rate for summer (Seawater temperature = 35 °C)

Makeup	= 3320 ton/hr
Recycle (*W*)	= *13134 ton/hr*
Seawater feed to the plant	= 10555 ton/hr
Seawater temperature	= 35 °C
Seawater salt concentration	= 40 g/L
TBT	= *105 °C*
Steam temperature	= 121 °C
Distillate production	= *1305.76 ton/hr*

j	$T_{B,j}$	$T_{D,j}$	$T_{F,j}$	B_j	D_j
1	100.98	99.54	98.57	13041.95	92.05
2	97.39	95.93	94.95	12961.24	172.76
3	93.73	92.25	91.26	12880.11	253.89
4	90.05	88.54	87.55	12799.67	334.33
5	86.36	84.81	83.82	12720.08	413.92
6	82.64	81.06	80.08	12641.26	492.74
7	78.92	77.30	76.32	12563.51	570.49
8	75.21	73.54	72.56	12486.92	647.08
9	71.50	69.78	68.80	12411.60	722.40
10	67.80	66.03	65.06	12337.65	796.35
11	64.14	62.30	61.34	12265.18	868.82
12	60.50	58.60	57.64	12194.30	939.70
13	56.90	54.93	53.98	12125.11	1008.89
14	53.35	51.30	50.37	12057.73	1076.27
15	49.85	47.72	46.81	11992.27	1141.73
16	45.94	43.70	43.31	11919.95	1214.05
17	44.63	42.36	41.81	11895.79	1238.21
18	42.73	40.40	39.82	11861.08	1272.92
19	40.86	38.49	37.73	11828.24	1305.76

Steam flow rate to brine heater (ton/hr) = 150.40
Performance ratio = 8.68
Original load = 1269.5 ton/hr (100%), See Table 6.5
New load = (1305.76/1269.5)*100% =102.9%

Table 7.12. Maximization of the distillate production rate for the winter (Seawater temperature = 24 °C)

Makeup	= 3196 ton/hr
Recycle (W)	= 13134 ton/hr
Seawater feed to the plant	= 10555 ton/hr
Seawater temperature	= 24 °C
Seawater salt concentration	= 40 g/L
TBT	= 105 °C
Steam temperature	= 121 °C
Distillate production	= 1482.33 ton/hr

j	$T_{B,j}$	$T_{D,j}$	$T_{F,j}$	B_j	D_j
1	100.32	98.88	97.73	13027.08	106.92
2	96.13	94.66	93.51	12933.19	200.81
3	91.87	90.37	89.21	12839.22	294.78
4	87.58	86.05	84.89	12746.36	387.64
5	83.28	81.71	80.55	12654.82	479.18
6	78.97	77.35	76.19	12564.51	569.49

Table 7.12. (Continued)

j	$T_{B,j}$	$T_{D,j}$	$T_{F,j}$	B_j	D_j
7	74.66	72.98	71.84	12475.82	658.18
8	70.36	68.63	67.49	12388.86	745.14
9	66.09	64.29	63.16	12303.78	830.22
10	61.85	59.98	58.86	12220.71	913.29
11	57.66	55.71	54.60	12139.82	994.18
12	53.53	51.48	50.39	12061.25	1072.75
13	49.46	47.32	46.25	11985.15	1148.85
14	45.48	43.23	42.19	11911.68	1222.32
15	41.60	39.24	38.22	11841.02	1292.98
16	37.37	34.87	34.37	11765.05	1368.95
17	35.79	33.24	32.49	11736.80	1397.20
18	33.27	30.62	29.88	11692.47	1441.53
19	30.91	28.19	27.26	11651.67	1482.33

Steam flow rate to brine heater (ton/hr) = 169.9
Performance ratio = 8.72
Original load = 1275.8 ton/hr (100%), See Table 6.6
New load = (1482.33/1275.8)*100% =116.2%

Table 7.13. Various parameters before and after maximization of the distillate rate

No.	parameter	Summer 100% load		Winter 100% load	
		Before optimization	After optimization	Before optimization	After optimization
1	Steam flow rate to brine heater (ton/hr)	146.23	150.4	149.55	169.9
2	Brine heater duty* (kJ/hr)	$3.22x10^8$	$3.31x10^8$	$3.31x10^8$	$3.74x10^8$
3	Performance ratio	8.68	8.68	8.53	8.72
4	Blowdown (ton/hr)	2050.5	2014.2	1920.2	1713.7
5	Blowdown salt concentration (g/L)	65.10	65.99	65.40	65.12

*Brine heater duty = $W\ 1000\left(TBT - T_{F,1}\right) S_{R,H}$

8. Conclusion

MSF desalination plant performance evaluation and optimization through modeling and simulation technique was the subject of this chapter. The selected plant was commissioned in 2003 to produce 3800 m^3/hr of distillate water from three identical desalination units.

For this purpose, a steady state mathematical model was developed to study the performance of the selected MSF desalination plant and to optimize its operating conditions at various loads. The developed model is based on basic principles of physics and chemistry to describe the physical phenomenon occurring in the process. It was constructed after considering all limitations of the desalination unit from construction, operation and capability angles.

The developed model predictions were compared with vendor simulation results and the actual plant operating data, and matching was good. Thus, the developed model was considered to be sufficiently accurate. It is safe to assume that model predictions would be reliable over complete operating plant range since the models are based on basic principles of physics and chemistry; not empirical model. The developed model was thus used to carry out the optimization of the MSF plant performance.

The plant optimization studies were then carried out for two different modes of operation: maximization of the performance ratio for a targeted production and maximization of the distillate production rate. In both modes, the optimum conditions were obtained by series of simulation runs by varying at a time different operating variables within the maximum and the minimum constrains of the design and operational limits of the plant.

It was found that both the performance ratio and the production rate were affected considerably by only the top brine temperature and the brine recycle flow rate. Hence, these two operating variables were manipulated for maximization of plant performance in both the above modes without violating any plant constraints.

The developed software thus can be used to evaluate the optimum controller's set points offline for any desired plant loads or operating conditions. Furthermore, the developed software can be integrated with the control software for online calculation of controller set points for optimal conditions and automatically adjust the controller settings.

ACKNOWLEDGMENTS

The funding provided by the Middle East Desalination Research Center (MEDRC Project 08-AS-007) is gratefully acknowledged.

REFERENCES

[1] Abdul-Wahab, S. A., Elkamel, A., Al-Weshahi, M. A. & Al Yahmadi, A. S. (2007). Troubleshooting the brine heater of the MSF plant fuzzy logic-based expert system. *Desalination*, *217*, 100-117.

[2] Abdul-Wahab, S. A. & Jupp, B. P. (2009). Levels of heavy metals in subtidal sediments in the vicinity of thermal power/desalination plants: A case study. *Desalination*, *244*, 261-282.

[3] Gengel, Y. A. (2007). *Heat and Mass Transfer: A practical approach*, 3rd Ed., McGraw-Hill, New York.

[4] Friedrich, R. O. & Hafford, J. A. (1971). Report ORNL-TM-3489.

[5] Stoughton, R. W. & Lietzke, M. H. (1965). Calculation of some thermodynamic properties of sea salt solutions at elevated temperatures from data on NaCl solutions. *Journal of Chemical and Engineering Data*, *10*, 254-260.

[6] Burns & Roe, Inc. (1983). Final Report, MSF Module Studies, O.S.W. San Diego Test Facility, O.S.W. Contract No. 14-30-3014, Page 31.

[7] Perry, R. H. & Green, D. W. (1999). *Perry's Chemical Engineers' Handbook*. 7th Ed., McGraw-Hill, New York.

[8] Helal, A. M., Medani, M. S. & Soliman, M. A. (1986). A tridiagonal matrix model for multistage flash desalination plants. *Computers & Chemical Engineering*, *10*, 327-342.

[9] Mothershed, C. T. (1966). Report ORNL-TM-1560.

[10] Griffin, W. L. & keller, R. M. (1965). Report ORNL-TM-1299.

[11] Chapra, S. C. & Canale, R. P. (1998). *Numerical Methods for Engineers*, 3th Ed., McGraw-Hill, Singapore.

[12] El-Dessouky, H. T. & Ettouney, H. M. (2002). *Fundamentals of Salt Water Desalination*, 1st Ed., Elsevier Science B.V., Amsterdam, Netherlands.

In: Desalination: Methods, Cost and Technology
Editor: Irena A. Urbonienė pp.117-142

ISBN: 978-1-61668-909-4

Chapter 3

TiO_2-Assisted Photocatalytic Degradation of Herbicides in Aqueous Solution: A Review

Biljana Abramović [*] ***and Daniela Šojić***

Faculty of Sciences, Department of Chemistry, Biochemistry and Environmental Protection, Trg D. Obradovića 3, 21000 Novi Sad, Serbia.

Abstract

In today's life, environmental pollution has far reaching negative consequences on the lives of humans. As is known, because of their stability, many pesticides after their application may be leached and thus contaminate surface and waste waters. The majority of herbicides are resistant to chemical and/or photochemical degradation under typical environmental conditions. Recent investigations in the area of chemical treatment of both waste waters and drinking water are directed toward improving efficiency of degradation of organic compounds dissolved or dispersed in water by applying catalytic and photochemical methods. Since their early applications in the 1970's, advanced oxidation processes (AOPs) have shown great potential in degrading organic pollutants. Heterogeneous TiO_2 photocatalysis, ozonation and photo-Fenton's reaction are the most intensively investigated technologies. The aim of this review is to address the fundamentals of one kind of AOP, namely, photocatalytic processes and their application for herbicide degradation in aqueous suspensions using TiO_2 as catalyst. Since AOPs rely on the generation and subsequent reaction of highly reactive oxygen radicals with herbicides, there are many factors that can affect the efficiency of this process. Hence, this review will attempt to summarize and highlight the effect of a variety of conditions on TiO_2-photocatalyzed removal of herbicides, such as the amount of catalyst, reaction pH, light intensity, concentration of herbicide, and the presence of some additives. The review will also summarize the pathways of herbicides degradation, including some of the intermediates that are generated during the degradation process.

Keywords: Photocatalytic degradation, Herbicides, TiO_2, Kinetics, Photo-intermediates and products

[*] Corresponding author: E-mail: biljana.abramovic@dh.uns.ac.rs

1. Introduction

The availability of a wider range of chemicals and international marketing of pesticides has led to a global growth in their use. Total global sales of pesticides had declined slightly in 2002 to approximately $25 million per annum, but had increased again in 2003 to $26.71 million. This is composed of almost 50% herbicides, 25% insecticides and 21.6% fungicides, and the remainder to other products (Matthews, 2006). Because of their stability, many pesticides, i.e. herbicides that are used in agriculture may be leached and thus contaminate surface and waste waters (Donald et al., 2007; Gray, 1996). Besides, their uptake by the plant root may cause pesticide accumulation in plants and, through the food chain, become a threat to the living beings.

In the recent several years, the environmental protection criteria have become more stringent, with the intention to reduce emission of pollutants (Chiron et al., 2000). However, a special problem represents the already contaminated waters, air, and soils, whose natural remediation can be hindered, if not impossible. Because of that it is desirable to have at the disposal methods that would enable complete elimination of pesticides from contaminated waters.

Photocatalytic degradation is a promising and widely used method for treatment of waste waters and air contaminated either by organic or inorganic pollutants. In the recent decades, many researchers (Burrows et al., 2002; Kabra et al., 2004; Calza et al., 2005) have studied the processes of removal of persistent contaminants from water, including pesticides, which exhibit chemical stability and resistance to biodegradation. Numerous studies have been based on the application of water suspensions of semiconductors in the presence of UV radiation to achieve degradation of the pollutants. Such processes have a number of advantages over the conventional techniques of waste water treatment such as adsorption on activated carbon, chemical oxidation, various types of biological treatments, etc. Namely, treatment with active carbon means only transition of the pollutants from one form to another without their degradation, which causes new problems when they are to be removed. The process of chemical oxidation leads to the mineralization of all organic substances and it is only economically suitable for removal of pollutants present at high concentrations (Devipriya and Yesodharan, 2005). Also, detoxication yields the degradation of particular molecules causing at the same time formation of toxic end products (Konstantinou and Albanis, 2003). When biological treatments are concerned, the main shortcomings are slow reaction rate of contaminants removal and the need for a strict control of the pH and temperature (Devipriya and Yesodharan, 2005). In this context, photocatalytic processes possess many advantages in the removal of pollutants from water when these are present at low concentrations (Konstantinou and Albanis, 2003), which assumes:

complete oxidation of organic pollutants during several hours;
absence of the formation of polycyclic products;
availability of active and cheap catalysts applied in specially designed reactor systems;
oxidation of pollutants whose concentration is of the order of ppb, etc.

As is known, recent investigations in the area of chemical water treatment of both waste waters and drinking water, are directed towards the improvement of the efficiency of degradation of organic compounds dissolved or dispersed in water, by applying catalytic and photochemical methods (Chiron et al., 2000). Photocatalytic degradation appeared as a very efficient method for mineralization of a large number of pesticides, as well as herbicides, even those that contain nitrogen and chlorine. Methods based on the formation of highly reactive radicals that may mineralize organic substances to non-toxic compounds, such as water and carbon dioxide, are commonly called Advanced Oxidation Processes (AOPs) (Prousek, 1996). These processes can be classified into two groups:

homogeneous reactions involving the presence of hydrogen peroxide, ozone or Fenton reagent in the system with or without UV radiation, and
heterogeneous reactions involving the application of photoactive semiconductors as photocatalysts.

AOPs represent a very attractive method for removing from water organic pollutants such as herbicides, as these processes are capable of destroying them, and not just of transferring them to another form, as is the case with the treatment of air by active carbon (Konstantinou and Albanis, 2003). Processes of photocatalytic oxidation represent the way of degradation of toxic pollutants at the concentration level of the order of μg dm^{-3} or mg dm^{-3}, as an alternative to some reactions of oxidation such as ozonolysis or oxidation with the aid of H_2O_2 using UV radiation, etc. (Ollis et al., 1991; Vidal et al., 1999). It is known that TiO_2–UV process has many advantages (Herrmann et al., 1999; Wang et al., 1999). Heterogeneous photocatalytic oxidation of organic compounds in aqueous solutions in the presence of different semiconductors as photocatalysts, especially TiO_2, as well as near UV radiation, offers the possibility for effective treatment of waste waters, drinking water, surface waters and ground waters, as well the preparation of ultra-pure water for the pharmaceutical industry and microlectronics. Many organic substances and their intermediates can be thus mineralized completely (Burrows et al., 2002; Kabra et al., 2004; Calza et al., 2005), producing, in dependence of the structure of the parent compound, CO_2, water, nitrate, halides, phosphates, and other inorganic ions. A very important task in this represents the choice of an appropriate semiconducting material. For a semiconductor photocatalyst to be efficient, the different interfacial electron processes involving e^- and h^+ must compete effectively with the major deactivation processes involving e^-–h^+ recombination, which may occur either in the bulk or at the surface (Figure 1) (Mills and Le Hunte, 1997). Ideally, a semiconductor photocatalyst should be chemically and biologically inert, photocatalytically stable, easy to produce and use, efficiently activated by sunlight, able to efficiently catalyze reactions, cheap, and without risks for the environment or humans. TiO_2 (with sizes ranging from clusters, via colloids to powders and large single crystals) is close to being an ideal photocatalyst, displaying almost all the above properties. The only exception is that it does not absorb visible light.

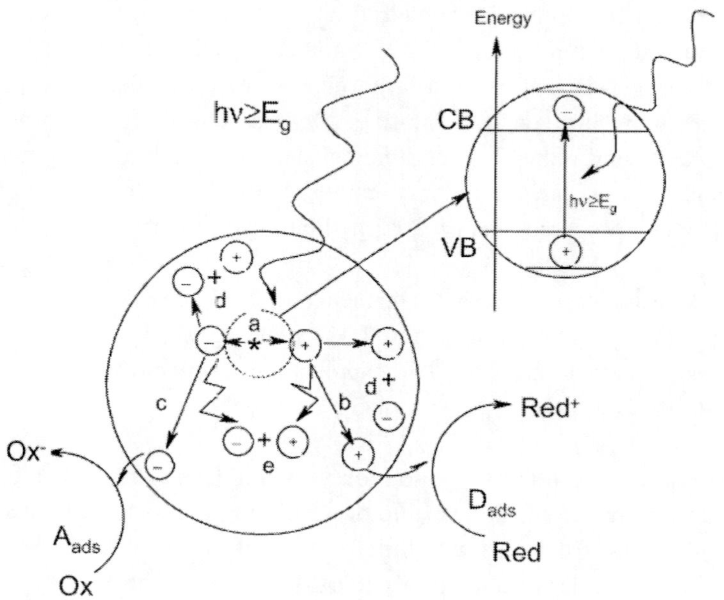

Figure 1. Main processes occurring on a semiconductor particle: (a) e^{-}–h^{+} generation; (b) oxidation of donor (D); (c) reduction of acceptor (A); (d) and (e) e^{-}–h^{+} recombination at surface and in bulk, respectively (from Mills and Le Hunte, 1997, with permission from Elsevier)

Table 1. Effect of catalysts loading on the photocatalytic degradation of herbicides

Herbicide	Spectral region	Photocatalyst	Tested range of catalyst (g dm^{-3})	Optimum dosage (g dm^{-3})	Reference
Bromacil	UV	TiO_2 P-25	0.5–5	5	Singh et al. (2003)
Isoproturon	UV	TiO_2 P-25	0.5–3	3	Haque and Muneer (2003)
Paraquat	UV	TiO_2 P-25	0.5–3	2	Moctezuma et al. (2003)
MCPP	UV	TiO_2 nanosized	1–4	2	Topalov et al. (2004)
Imazethapyr	UV	TiO_2 P-25	0.5–2.5	1.5	Ishiki et al. (2005)
Propyzamide	UV	TiO_2 P-25	0–1.67	1.33	Yano et al. (2005)
Imazapyr	UV	TiO_2 P-25 PC-500	0.1–0.7 0.2–1.2	0.2 1	Pizarro et al. (2005)
Chlorsulfuron	UV	TiO_2 P-25	0.5–2	1.75–2	Fresno et al. (2005)
MCPP	Vis	TiO_2 P-25	0.5–16	8	Abramović et al. (2007b)
Prometryn	UV	TiO_2 P-25	0.050–1	0.1	Evgenidou et al. (2007)
Imazamox	UV	TiO_2 P-25	0.1–2	1	Harir et al. (2008)
Clopyralid	UV	TiO_2 P-25	0.5–8	4	Šojić et al. (2009)

In view of the wide utilization of herbicides, the aim of the present chapter is to review the effects of operating parameters, such as the amount of catalyst, reaction pH, light intensity, concentration of herbicide, and the presence of some additives, on the photocatalytic degradation of herbicides using TiO_2 photocatalysts. The review will also summarize the pathways of herbicides degradation, including some of the intermediates that are generated during the degradation.

2. Factors Influencing Photocatalytic Processes

Numerous investigations have demonstrated that the amount of catalyst, light intensity, concentration of herbicide, presence of some additives, reaction pH and the temperature, are the main factors that influence the degradation rate of organic substrates, i.e. herbicides. The effect of each of these parameters will be discussed in the following sections.

2.1. Amount of Catalyst

Catalyst concentration is also an important factor in the photocatalytic processes taking place in TiO_2 suspensions (Dionysiou et al., 2000; Rachel et al., 2002; Herrmann et al., 2002). It has been found that the initial reaction rate of a photocatalytic reaction is directly proportional to the titania concentration (Herrmann, 1999). Such behavior is a consequence of an increase in the number of illuminated particle surfaces, which produce a larger amount of reactive e^-–h^+ pairs. However, with increase in the catalyst mass concentration the degradation rate increases up to an optimum value; a further increase causes no effect since all the available light has already been used. In some case, too high concentration of the catalyst may lead to the aggregation of TiO_2 particles, a consequence being a decrease in the number of active sites on the catalyst surface (Dionysiou et al., 2000; Topalov et al., 2004). The increase in the solution turbidity and light dispersion on the particles may be another reason for the decrease in the degradation rate at higher catalyst concentrations (Vione et al., 2005). Also, at the mass concentrations exceeding the optimal one, the intensity of UV radiation decreases because of the less efficient light penetration caused by the presence of a larger amount of the catalyst, yielding a decrease in the efficiency of the overall process (Wong and Chu, 2003). Optimal mass concentration of TiO_2 in the processes of herbicide degradation varies in a wide span, from 0.1 to 8 g dm^{-3} (Table 1), depending on the photocatalized system, type and shape of the reactor, and type and light source geometry (Chen and Ray, 1999; Dillert et al., 1999; Herrmann et al., 1999).

Pizarro et al. (2005) studied the photocatalytic oxidation of imazapyr as a function of catalyst concentration for two different titania materials: TiO_2 Degussa P-25 (TiO_2 P-25) and Millennium PC-500. They found that TiO_2 P-25 was much more efficient in capturing the photons emitted from the same lamp. Thus, while 0.2 g dm^{-3} could be considered as the optimum concentration of TiO_2 P-25, the content of 1 g dm^{-3} was optimal for Millennium PC-500. The effect of the amount of TiO_2 P-25 on the photodegradation rate was also examined in the case of degradation of the herbicide imazamox. Different amounts of the catalyst, ranging from 0.1 to 2 g dm^{-3}, at two concentration of the substrate (0.032 and 0.2 mmol dm^{-3}) were employed (Harir et al., 2008). For both substrate concentrations, the maximum degradation rate constant was observed at a catalyst content of 1 g dm^{-3}, which was selected for the further studies. In the work of Ishiki et al. (2005), a dispersion containing imazethapyr at natural pH (ca. 4.4) was used to study the catalyst effect while varying the TiO_2 P-25 content from 0.5 to 2.5 g dm^{-3}. The optimum TiO_2 P-25 concentration for the photocatalytic oxidation at the natural pH appeared to be 1.5 g dm^{-3}. These results revealed that the photons were completely absorbed by the available active sites on the catalyst surface when the concentration was in the range from 0.5 to 1.5 g dm^{-3}. However, at higher TiO_2 P-

25 concentrations, the particles may reflect the incident radiation and decrease the rate constant since not all of the photocatalyst particles can receive enough energy to promote the valence band electron to the conduction band and start the oxidation of the imazethapyr molecule. Such behavior may also be explained by the potential aggregation and precipitation of the photocatalyst particles, as has already been mentioned.

In our previous works, great attention has been paid to the study of the effect of catalyst content on the kinetics of photodegradation of the herbicides *RS*-2 (4-chloro-*o*-tolyloxy)propionic acid (mecoprop, MCPP) and 3,6-dichloropyridine-2-carboxylic acid (clopyralid). The photodegradation of 0.9 mmol dm^{-3} MCPP in the oxygenated aqueous colloids was examined in the TiO_2 concentration range of 1–4 g dm^{-3} with the aim to optimize the catalyst dosage during the irradiation process (Topalov et al., 2004). This effect was examined by determining the chloride released during the photodegradation by potentiometric titrations at different illumination times. It was found that the concentration of TiO_2 nanoparticles had a pronounced effect on the degradation rate. The optimal catalyst loading found in this study was around 2 g dm^{-3}. Up to about 2 g dm^{-3} of TiO_2 nanoparticles, the degradation rate increased simultaneously with increase in the concentration of TiO_2. Further increase of TiO_2 concentration resulted in a decrease of degradation rate. Abramović et al. (2007b), also studied the influence of catalyst loading on the photodecomposition efficiency of MCPP, but in the presence of visible light. The photodegradation of 0.9 mmol dm^{-3} MCPP in the oxygenated aqueous suspension was examined in the TiO_2 P-25 concentration range of 0.5–16 g dm^{-3}. Since the reaction conditions at higher catalyst concentrations were not the same as at the lower ones, the reaction rates of mecoprop degradation were not comparable. This is why the amount of 8 g dm^{-3} could conditionally be taken as optimal. Upon comparing these results with those in the presence of UV light, it could be noted that the effect of catalyst loading in the case of visible-light-induced photocatalytic degradation was significantly different. This was most likely due to the different photodegradation mechanism in the presence of visible and UV radiation. Obviously, in the visible light-induced photocatalytic degradation, compound-surface interaction is a critical factor in determining the visible photocatalytic degradation activity.

Based on the above, it can be concluded that the mass of catalyst used in the photoreactor has to be properly selected since it plays an important role in the global efficiency of the reaction.

2.2. Light Intensity

The rate of reaction is proportional to the radiant flux (Herrmann, 1999). This confirms the photo-induced nature of the activation of the catalytic process, with the participation of photo-induced electrical charges (electrons and holes) in the reaction mechanism. However, above a certain value, estimated to be ca. 25 mW cm^{-2} in laboratory experiments, the reaction rate becomes proportional to the square root of the intensity. The optimal light power utilization corresponds to the domain where the rate of reaction is proportional to the radiant flux. Garcia and Takashima (2003) submitted TiO_2 suspension of herbicide imazaquin to the artificial and solar radiation sources in order to determine the extent of light absorption by

TiO_2. When the mercury lamp power was changed from 125 to 250 W, without bulbs in both cases, the observed increase was 16%. This was attributed to the UV radiation intensity used to determine the amount of photons absorbed by the catalyst. Previous investigations have shown that the relationship between the degradation rate and intensity would be of half order, that is, directly proportional to the square root of the intensity. This dependence of the reaction rate may arise from enhanced band gap recombination at higher light intensities, where the surface-bound $^{\bullet}OH$ radical would be the principal hole trap and the primary initiator of the oxidation of electron-donating substrates when the intensity was increased by about 25 mW cm^{-2} (Herrmann, 1999).

Parra et al. (2004) investigated the dependence of atrazine degradation rate on the intensity of the incident light received from the Suntest solar simulator. The dose of 50 mW cm^{-2} corresponded to about 50% of the full light intensity of the solar radiation. A faster degradation was obtained with 90 mW cm^{-2} as compared to a light dose of 50 mW cm^{-2}. Accelerated degradation was obtained when a medium pressure mercury arc lamp with the main radiation centered at 366 nm (6×10^{19} photons s^{-1}) was used as the radiation source.

In the case of iodosulfuron, experiments were performed using a photonic flux varying from 0.4 to 4.6×10^{16} photon $s^{-1}cm^{-2}$ in TiO_2 aqueous suspension (Sleiman et al., 2007). The results obtained showed that for photonic fluxes below 10^{16} photon s^{-1} cm^{-2}, apparent rate constant was directly proportional to the photonic flux, indicating that most of the incident photons were efficiently converted into active species that acted in the degradation mechanism. For higher photonic fluxes, the rate constant varied as the square root of the flux. This means that the efficiency of the photocatalytic process was limited above 10^{16} photon s^{-1} cm^{-2} due to the e^-–h^- recombination, which were predominant.

Vulliet et al. (2003) carried out experiments with sulfonylureas herbicides using different radiant fluxes and corresponding to efficient photon fluxes varying from 0.4×10^{16} to 12.6×10^{16} photons s^{-1} (from 1.3 to 39.8 mW cm^{-2}). For low fluxes ($\leq 4.2 \times 10^{16}$ photons s^{-1}), the reaction rate was proportional to the photonic flux. This means that the process worked in a good photocatalytic regime: the incident photons were efficiently converted into active species that acted in the degradation mechanism. For higher fluxes, the rate varied as the square root of the radiant flux, at least to 40 mW cm^{-2}. This indicates that the process was still photocatalytic, but the e^-–h^+ recombinations predominated, which has been demonstrated by Herrmann (1999). This means that part of the light absorption and e^-–h^+ pairs are lost because of the recombination.

The above results show that, depending on the nature of the herbicides, the reaction rate is linearly proportional to the photon flux to a certain limit, and above it, the rate varies as the square root of the radiant flux.

Table 2. Influence of initial concentration herbicides on their photocatalytic degradation rate using UV radiation

Herbicide	Photocatalyst	Tested concentration range (mmol dm^{-3})	Initial concentration corresponding to maximum rate (mmol dm^{-3})	Reference
Imazaquin	TiO_2 P-25	0.005–0.1	0.005	Garcia and Takashima (2003)
Bromacil	TiO_2 P-25	0.12–0.96	0.96	Singh et al. (2003)
Isoproturon	TiO_2 P-25	0.25–0.75	0.75	Haque and Muneer (2003)
Paraquat	TiO_2 P-25	0.39–1.55	0.39	Moctezuma et al. (2003)
MCPP	TiO_2 nanosized	0.9–2.7	0.9	Topalov et al. (2004)
Imazethapyr	TiO_2 P-25	0.005–0.1	0.005	Ishiki et al. (2005)
Picloram Dicamba Floumeturon	TiO_2 P-25 TiO_2 P-25 TiO_2 P-25	0.25–1.0 0.2–0.77 0.25–0.4	0.5 0.56 0.34	Atiqur Rahman and Muneer (2005)
Triclopyr Daminozid	TiO_2 P-25 TiO_2 P-25	0.25–1.0 0.50–1.5	0.75 0.75	Qamar et al. (2006)
Chlorotoluron	TiO_2 P-25	0.15–0.45	0.45	Haque et al. (2006)
Prometryn	TiO_2 P-25	0.02-0.16	-	Evgenidou et al. (2007)
Imazamox	TiO_2 P-25	0.016–0.2	0.016	Harir et al. (2008)
Clopyralid	TiO_2 P-25	0.5–3.0	1.5	Šojić et al. (2009)

2.3. Concentration of Herbicide

The initial concentration of the reactant has a significant influence on the efficiency of the photocatalytic process. According to many researchers, the influence of the initial concentration of the solute on the photocatalytic degradation rate of the most organic compounds can be described by pseudo-first order kinetics which is rationalized in terms of the Langmuir–Hinshelwood (L–H) model, modified to accommodate reactions occurring at a solid–liquid interface (Turchi and Ollis, 1990; Poulios et al., 2000; Evgenidou et al., 2005). The L–H model describes the dependence of the reaction rate on the initial concentration of the reactant. It may be used in the cases when photocatalytic reaction takes place on the particle surface, so that the rate of the reaction is greatly influenced by adsorption (Chen and Ray, 1999). Studies showed a good agreement of the results for the majority of the investigated systems, including also herbicides. The degradation rate increases up to about 1 mmol dm^{-3}, to become constant at higher herbicide concentrations. However, in some cases, the increase in the initial concentration of the reactant yields a decrease in the rate of its degradation (Atiqur Rahman and Muneer, 2005; Qamar et al., 2006). Table 2 shows the effect of initial concentration of herbicides on their photocatalytic degradation.

Evgenidou et al. (2007) used various initial concentrations of prometryn (5–40 mg dm^{-3}) and the degradation rates obtained independently from the degradation curves based on the linear fit of the experimental data obtained only for the first 10 min of illumination in order to minimize variations arising as a result of competitive effects of intermediates, pH changes, etc. Atiqur Rahman and Muneer (2005) found that the rate of herbicides degradation increases with increase in the concentration from 0.25 to 0.5 mmol dm^{-3} (picloram), from 0.2 to 0.56 mmol dm^{-3} (dicamba), and from 0.25 to 0.34 mmol dm^{-3} (floumeturon). A further increase in the initial concentration of the reactant yielded a decrease in the rate of degradation. According to these authors, this phenomenon may be a consequence of the increase in the initial concentration of the reactant by enhanced adsorption of the organic compound molecules on the TiO_2 surface. Namely, since the light intensity, illumination time, and mass concentration remained constant, the concentrations of $O_2^{\bullet -}$ and $^{\bullet}OH$ on the TiO_2 surface did not increase, a consequence being a decrease in the photodegradation efficiency with the further increase in the herbicide concentration. Similar results were obtained by Qamar et al. (2006) in their study of the effect of the initial concentrations of triclopyr and daminozid on the rate of their degradation.

A simplified form of the L–H equation is usually used to interpret results of a great number of laboratory studies. Some authors think that in determining reaction order, some other factors such as light intensity, catalyst concentration, oxygen concentration, etc. (Turchi and Ollis, 1990; Mills and Moris, 1993) should also be taken into account. However, some authors indicated that the kinetics of photocatalytic degradation of herbicides corresponds to a half-order reaction. Namely, in the case when the initial concentrations of the reactant are different, the rate constant can not be described by the equation holding for a first-order reaction. Such a case was observed by Vulliet et al. (2003) in the photocatalytic degradation of sulfonylurea herbicides in aqueous suspension of TiO_2, which can be described as a half-order reaction. Some other authors (Vulliet et al., 2002; Fresno et al., 2005) arrived at the same conclusion. The same model has also been used in the case of photocatalytic degradation of fluroxypyr in aqueous suspension of TiO_2 (Aramendía et al., 2005), indicating the possibility of the occurrence of reactant adsorption on the catalyst surface via two active sites. Then, when two active sites participate in the adsorption, the Θ (degree of catalyst surface coverage with the substrate) can be expressed as:

$$\Theta = (Kc)^{1/2}/[1+ (Kc)^{1/2}] \tag{2.1}$$

where K is the constant of adsorption at equilibrium and c is instantaneous concentration of herbicide. On the other hand, $r = k\,\Theta$ (Langmuir–Hinselwood expression of the rate).

At a high dilution, $Kc^{1/2} << 1$, hence:

$$r = k\Theta = k[(Kc)^{1/2}/(1+ (Kc)^{1/2})] \approx k(Kc)^{1/2} = k_{app}\, c^{1/2} \tag{2.2}$$

where k is reaction rate constant and k_{app} is the apparent reaction rate constant. By integrating this equation from $t = 0$ (c_0) to t (c), one obtains:

$$r = -\, dc/dt = k_{app}\, c^{1/2} \Rightarrow (c_0^{1/2} - c^{1/2}) = k_{app}\, t/2 \tag{2.3}$$

where c_0 is initial concentration of herbicide. According to equation (2.3), there is a linear dependence between $(c_0^{1/2} - c^{1/2})$ and t.

It is evident from the above, the study of the influence of the initial herbicide concentration is a complex issue when dealing with the process of photocatalytic degradation, as one has to take into account a number of factors that influence the order of the reaction.

2.4. Electron Acceptors and Their Role in Recombination

One of the shortcomings of the application of TiO_2 as photocatalyst is certainly the possibility of recombination of the e^-–h^+ pairs within the particles, which can essentially decrease the quantum yield of the reaction. This phenomenon is especially pronounced in the absence of the appropriate electron acceptors or electron donors (Qamar et al., 2006).

In the case of TiO_2, the conduction and valence bands originate from the O 2p and Ti 3d states. At very low temperatures, the recombination of e^-–h^+ pairs is slow. Thus, for example, at 4.2 K, the photogenerated electrons and holes become trapped inside and on the surface of hydrated polycrystalline of anatase TiO_2 (Thompson and Yates, 2006). Namely, electrons get trapped on the localized Ti^{4+} sites, creating thus Ti^{3+} in the interior of the material, whereas the holes get trapped in the form of O^- ions beneath the surface, according to equation (2.4) (Kitano et al., 2007). At 90 K, the photogenerated electrons and holes have longer lifetime, lasting hours, and there is a direct linear dependence between the concentration of photogenerated electrons (Ti^{3+} ions) and holes (O^- ions) in the excited charge-transfer complex $(Ti^{3+} – O^-)^*$ (Karunakaran and Dhanalakshmi, 2005):

$$(Ti^{4+} - O^{2-}) \rightleftharpoons (Ti^{3+} - O^-)^* \tag{2.4}$$

However, at room temperature, the process of recombination of e^-–h^+ pairs is very pronounced and the reactants are adsorbed on the semiconductor's surface, making the photocatalysis process more effective (Thompson and Yates, 2006). The holes react with the adsorbed organic molecules, causing formation of reactive intermediates (Tamaki et al., 2006), which, through a series of reactions, yield formation of CO_2 and H_2O (Sene et al., 2003; Iliev et al., 2006; Iliev et al., 2007; Teoh et al., 2007). The conduction band containing adsorbed oxygen molecules yields the formation of highly reactive $O_2^{\bullet-}$ radicals (Thompson and Yates, 2006). In aqueous medium, besides $O_2^{\bullet-}$, some other highly reactive species are formed such as $^\bullet OH$ radicals (reduction potential 2.80 V), $HO_2^\bullet$ (hydrogen-peroxide radical) and H_2O_2, which can also oxidize organic molecules.

Recombination is usually prevented by separating free charge carriers and capturing electrons with the aid of the appropriate electron acceptors. Fast reactions with photogenerated electrons suppress recombination of e^-–h^+ pairs, increasing thus efficiency of the oxidative photodegradation process. Reports mentioned in this section show that the nature of the oxidizing agent has a great influence on the photocatalytic degradation of herbicides. The most frequently used electron acceptor, molecular oxygen, causes formation the $O_2^{\bullet-}$ radical anion:

$$O_2 + e^- \rightarrow O_2^{\bullet-} \tag{2.5}$$

In acidic medium, the superoxide anion gives:

$$O_2^{\bullet-} + H^+ \rightarrow HO_2^\bullet \tag{2.6}$$

The newly formed radical, by a recombination reaction, gives most often H_2O_2, which, in the absence of oxygen, may be a new source of $^{\bullet}OH$ radicals:

$$2\,HO_2^{\bullet} \rightarrow H_2O_2 + O_2 \tag{2.7}$$

$$e^- + H_2O_2 \rightarrow OH^- + {}^{\bullet}OH \tag{2.8}$$

In this way, it is actually possible to influence the photodegradation efficiency. It was found that the photocatalytic activity of TiO_2 P-25 in the degradation of clopyralid in the absence of oxygen is suppressed to a great extent (Šojić et al., 2009).

Besides oxygen, for the same purpose one may use also H_2O_2, $KBrO_3$ and $(NH_4)_2S_2O_8$ (Muneer et al., 1999; Muneer and Bahnemann, 2002; Atiqur Rahman and Muneer, 2005; Qamar et al., 2006). Apart from the efficient suppression of the recombination of e^-–h^+ pairs, by effectively taking over the electrons from the conduction band and thus contributing to the charge separation, H_2O_2 represents also a source of additional $^{\bullet}OH$ radicals that are formed in the reaction with electrons from the conduction band of the semiconducting particle (equations (2.8) and (2.9)). The reaction of the other two electron acceptors with electrons yields the corresponding potent oxidizing radicals $SO_4^{\bullet-}$ and $BrO_2^{\bullet}$ (equations (2.10) – (2.12)), which also increase the quantum yield of $^{\bullet}OH$ radicals in the reactions in which TiO_2 is used as photocatalyst (Muneer et al., 1999):

$$H_2O_2 + O_2^{\bullet-} \rightarrow {}^{\bullet}OH + OH^- + O_2 \tag{2.9}$$

$$S_2O_8^{2-} + e^- \rightarrow SO_4^{2-} + SO_4^{\bullet-} \tag{2.10}$$

$$SO_4^{\bullet-} + H_2O \rightarrow SO_4^{2-} + {}^{\bullet}OH + H^+ \tag{2.11}$$

$$BrO_3^- + 2H^+ + e^- \rightarrow BrO_2^{\bullet} + H_2O \tag{2.12}$$

$$BrO_3^- + 6H^+ + 6e^- \rightarrow [\,BrO_2^-, HOBr] \rightarrow Br^- + 3H_2O \tag{2.13}$$

Also, it appeared that the excess of H_2O_2 may in some cases lead to the formation of a peroxo-complex, causing thus an opposite effect, i.e. lowering of the efficiency of the photodegradation process (Poulios et al., 1998). For the same reason it is desirable to establish an optimal concentration of the electron acceptor in order to achieve a maximum effect. Thus, these authors reported that the increase in the concentration of H_2O_2 was accompanied by an increase in the efficiency of degradation of triclopyr to the optimum level of 1.5–3.0 mmol dm^{-3}, which decreased with the further increase in the H_2O_2 concentration. Osajima et al. (2008) in their study of the effect of H_2O_2 concentration (0.1–1.0 mmol dm^{-3}) on the photocatalytic degradation of imazapyr found that, after reaching a maximum degradation rate at the H_2O_2 concentration of 0.5 mmol dm^{-3}, a further increase yielded a decrease in the reaction rate. Also, the same authors found that optimum concentration of

$S_2O_8^{2-}$ in the photocatalytic degradation of imazapyr was 5.0 mmol dm^{-3}. Based on the above results one can conclude that the effect of investigated electron acceptors depends on both the nature of the herbicide and concentration of the electron acceptor.

Haque and Muneer (2003) studied the photocatalytic degradation of isoproturon in the presence of various electron acceptors (H_2O_2, $(NH_4)_2S_2O_8$ and $KBrO_3$). All the additives showed a beneficial effect on the degradation of the mentioned herbicide. Šojić et al. (2009) demonstrated that H_2O_2, $(NH_4)_2S_2O_8$ and $KBrO_3$ exhibited individual influence on the degradation of clopyralid. Namely, in the beginning of the reaction, all the electron acceptors accelerated the degradation process, and the influence of these electron acceptors diminished in the order: $(NH_4)_2S_2O_8 > H_2O_2 > KBrO_3$. However, in the course of the further degradation process in the presence of electron acceptors, a decrease in the degradation rate was observed. Such an effect of the electron acceptors authors explained by the change in the reaction mechanism of the photocatalytic degradation of clopyralid. This phenomenon was most pronounced in the case of $KBrO_3$. Inhibition of the degradation by $KBrO_3$ may be due to the scavenging of $OH^{\bullet}$ and holes by BrO_3^- and UV screening of TiO_2 particles due to the presence of BrO_3^- (Burns et al., 1999). Also, when H_2O_2 or $(NH_4)_2S_2O_8$ were used as an electron acceptor, a decrease of the reaction rate was observed only after a longer reaction time (t > 120 min and 170 min, respectively). In some cases, H_2O_2 may act as a hole or $OH^{\bullet}$ scavenger, or react with TiO_2 to form peroxo compounds which are detrimental to the photocatalytic action (Poulios et al., 1998).

Putting all together, there is a serious need to consider the effect of the nature of the oxidizing agents in the photocatalytic degradation of herbicides.

2.5. pH of the Solution

The pH has a great effect on the photodegradation efficiency of herbicides and it is one of the most important factors in heterogeneous photocatalytic reactions. Since it influences surface charges of the photocatalyst and size of its aggregates, it has been subject of numerous studies (Muneer and Bahnemann, 2002; Atiqur Rahman and Muneer, 2005; Qamar et al., 2006; Singh et al., 2007). Table 3 summarizes the pH influence on the photodegradation of various herbicides. The table reveals that different herbicides exhibit different activities in photocatalytic reactions. Some are photocatalytically degraded at lower pH, while the others do so at higher pH. All these may be attributed to the nature of the pollutant to be degraded. Therefore, it is important to study the nature of the pollutants to be photocatalytically degraded and determine the right pH of the process.

Table 3. pH influence on the photocatalytic degradation of various herbicides

Herbicide	Spectral region/Light source	Photocatalyst	Tested pH range	Optimum pH	Reference
Thiobencarb	Xenon	TiO_2 P-25	2–9	9	Nishida and Ohgaki (1994)
Atrazine	Solar	TiO_2 P-25	3.26–11.47	11.47	Sullivan et al. (1994)
Triclopyr	UV	TiO_2 P-25	2–13	3.5–4	Poulios et al. (1998)
Bromacil	UV	TiO_2 (P-25, UV100, PC 500)	3–9	9	Singh et al. (2003)
Imazaquin	UV	TiO_2 P-25	3–11	3	Garcia and Takashima (2003)
Alachlor	UV	TiO_2 P-25	3–11	11	Wong and Chu (2003)
Dicamba	UV	TiO_2 P-25	3–11	8	Chu and Wong (2004)
Picloram	UV	Hombikat UV100	2–11	2	Atiqur Rahman and Muneer (2005)
Dicamba	UV	TiO_2 P-25	2–11	2	
Floumeturon	UV	TiO_2 P-25	2–11	2	
Imazethapyr	UV	TiO_2 P-25	2–11	4.4	Ishiki et al. (2005)
Fluroxypyr	UV	TiO_2 P-25	2–10	8	Aramendía et al. (2005)
Imazapyr	UV	TiO_2 P-25	2.1–8.4	3.8	Carrier et al. (2006)
Chlorotoluron	UV	TiO_2 P-25	3–9.1 1.2–11.2	3 1.2	Haque et al. (2006)‡
Imazamox	UV	TiO_2 P-25	2.04–11.03	5.03	Harir et al. (2008)
Clopyralid	UV	TiO_2 P-25	1–10.5	3.2	Šojić et al. (2009)

‡Results obtained by different catalyst concentrations.

The effect of the pH on the photocatalytic reaction is a very complex issue because of the involvement of a number of factors such as electrostatic interactions between the semiconductor's surface, solvent molecules, reactants and the radicals formed during the reaction process. As is known, the isoelectric point of TiO_2 P-25 is at pH 6.25 (Qamar et al., 2006), which means that at higher pH values the particle surface has a negative charge (TiO_2^-), while at lower pH a positive one (TiO_2^+). In view of the fact that the surface functional groups on TiO_2 particles are amphoteric, one can write the following acid-base equilibrium (Wang et al., 2000; Muneer and Bahnemann, 2002):

$$-TiOH + H^+ \longrightarrow TiOH_2^+ \quad pH < 6.25$$

$$-TiOH + OH^- \longrightarrow TiO^- + H_2O \quad pH > 6.25$$

This phenomenon influences the adsorption of the reactant and the rate of its degradation. Thus, for example, fluroxypyr ($pK_a = 2.2$) contains in its molecule a hydroxyl group and two atoms of nitrogen, one in the pyridine ring and the other in the amino group. According to Aramendía et al. (2005), in the case of this herbicide the bond with the catalyst surface can be realized via the carboxylic group, as well as via both of the two nitrogen atoms (Figure 2).

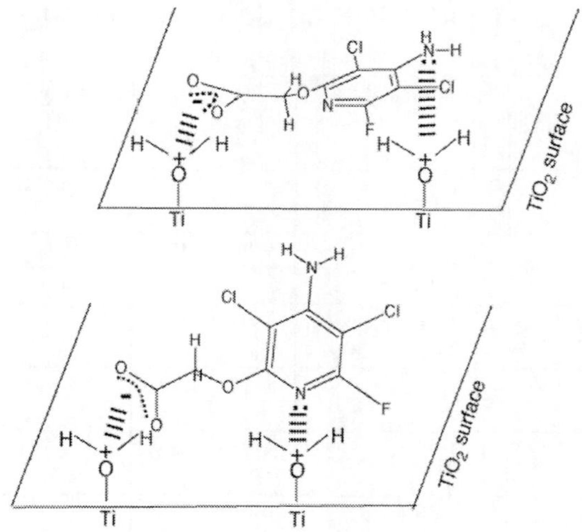

Figure 2. Suggested interactions between active sites on TiO_2 surface and fluroxypyr molecule (from Aramendía et al., 2005, with permission from Elsevier)

On the other hand, Poulios et al. (1998) pointed out the occurrence of only one type of interaction, namely via the carboxylic group, in the case of photocatalytic degradation of triclopyr (pK_a = 2.93), from the same class of pyridine acid herbicides. In their study of the degradation of imazaquin, a herbicide of the imidazolinone class, Garcia and Takashima (2003) examined the effect of the initial pH (3–11) on the degradation rate. They found that the most favorable value for the process was pH 3, because at this pH (<pK_a) the interaction between $TiOH_2^+$ and imazaquin occurred through one or more atoms with higher negative charge densities (Figure 3). Similar results obtained also Haque et al. (2006) in their study of the effect of the initial pH on the degradation rate of chlorotoluron at the mass concentrations of TiO_2 P-25 of 1 and 2 g dm^{-3}. In both cases, slightly higher degradation rates were observed under acidic pH; the rates slowly decreased with increase in pH (in the range of 3–9.1 and 1.2–11.2, respectively). Atiqur Rahman and Muneer (2005) examined the effect of pH on the photocatalytic degradation of three herbicides: picloram, dicamba, and fluometuron. In all three cases the degradation proceeded faster in acidic medium than in the basic one because of the dominant anionic form of the reactant. Since our previous study showed that clopyralid solution is stable in the presence and absence of daylight in the pH interval 1.0–9.0 during two months (Abramović et al., 2007a), we studied the effect of the initial pH on the degradation rate of clopyralid in the pH range from 1.0 to 10.5 in the presence of UV irradiation (Šojić et al., 2009). The fastest reaction was observed at pH 3.2 because of the influence of various factors (electrostatic interactions between the semiconductor surface, solvent molecules, substrate and charged radicals formed during the reaction process) at the different pHs of the system.

As can be seen from the above, the efficiency of degradation of herbicides at the different pH of the solution depends on numerous factors, so that its elucidation represents a very complex task.

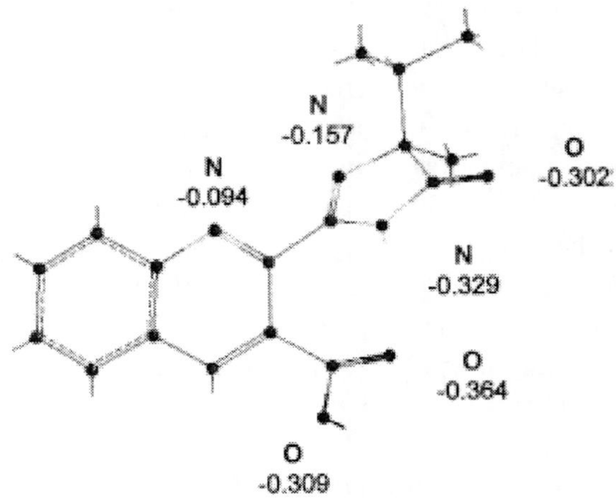

Figure 3. Charge densities on some atoms in neutral imazaquin (from Garcia and Takashima, 2003, with permission from Elsevier)

2.6. Temperature

In the majority of cases, the dependence of the herbicide photodegradation reaction rate on temperature can be described by the Arrhenius equation. The small effect of temperature on the rate of photocatalytic degradation is usually explained in terms of low activation energy compared to common heat reactions.

Haque et al. (2006) investigated the effect of temperature on the photocatalytic degradation of herbicide chlorotoluron in the presence of TiO_2 P-25 in the range of 20–50 ^{0}C as a function of time. The degradation rate was found to be more or less similar in the temperature range studied within the experimental error limits.

In the majority of cases the increase in temperature is accompanied by the increase in the herbicide photodegradation rate although this effect was not very pronounced (Garcia and Takashima, 2003; Topalov et al., 2004; Parra et al., 2004; Osajima et al., 2008; Šojić et al., 2009). Thus, the value of the overall apparent activation energy of photocatalytic oxidation of herbicides is in the range from 7.74 to 53.79 kJ mol^{-1}. Garcia and Takashima (2003) studied the effect of temperature on the photocatalytic degradation of imazaquin in the range of 20–40 ^{0}C with temperatures kept constant (±0.3 ^{0}C), and found that the rate constants increased with increase in temperature. By plotting the natural logarithm of the rate constant as a function of reciprocal absolute temperature they obtained a linear dependence (r = 0.9881) with a deviation at 40 ^{0}C. This may suggest a decrease in the adsorption of imazaquin molecules on the TiO_2 surface at this temperature. The activation energy was relatively low and was estimated at 24.8 kJ mol^{-1}. It was also demonstrated by Osajima et al. (2008) that the degradation rate constant of imazapyr increased by 38% upon the increase in the temperature from 20.0 to 50.0 ^{0}C, and that it displayed a non-Arrhenius behavior. The rate constant for imazapyr degradation increased with the temperature increase analogously to imazaquin (Garcia and Takashima, 2003), but unlike that of the imazethapyr, where the rate constant decreased with the temperature elevation from 20 to 40 ^{0}C (Ishiki et al., 2005). When the degradation of atrazine was concerned, an increase in temperature had a beneficial effect on the rate of degradation (Parra et al., 2004). The activation energy was determined by following the degradation of different atrazine concentration at 20, 40, 55 and 70 ^{0}C, under Suntest illumination, and it was found to be 10.9 kJ mol^{-1}. This value suggests that ion-molecular or radical-molecular reactions, requiring activation energy, occurred during the degradation, and not just only radical reactions, which proceed without any activation energy. Šojić et al. (2009) found that the rate constant of clopyralid degradation increased with increase in temperature from 25 to 50 ^{0}C, which was ascribed to an increased frequency of molecular collisions in the solution, although this effect was not very pronounced. On the basis of the corresponding Arrhenius plot the authors determined the overall apparent activation energy to be only 7.74 kJ mol^{-1}.

Topalov et al. (2004) studied the effect of temperature on mecoprop degradation rate in the presence of synthesized nanosized anatase TiO_2 and found that the overall apparent activation energy was 53.79 kJ mol^{-1}. This higher value compared to that obtained for photocatalytic degradation of other herbicides using TiO_2 P-25 the authors explained in terms of the very small efficiency of photocatalytic degradation. For TiO_2 photocatalyst, irradiation is the primary source of e^-–h^+ pair generation at ambient temperature, as the band gap energy is too high to be overcome by thermal activation. Therefore, the increase in the rate constant was most likely due to the increasing collision frequency of molecules in the solution, which increased with increasing temperature.

The low values of activation energy are usually characteristic for medium temperatures, from 20 to 80 ^{0}C. However, at very low temperatures (from –40 to 0 ^{0}C), the activation energies are higher (Herrmann, 2005). In contrast to that, at the "high" temperatures ≥ 70–80 ^{0}C, the activities for different types of photocatalytic reactions show a decrease, and the value of apparent activation energy becomes negative. This behavior Herrmann explained within the frame of the L–H mechanism. Namely, lowering of temperature favors adsorption of the of the final reaction product, which is a spontaneous exothermal process, and also enhances adsorption of the end product, whose desorption inhibit the reaction. In contrast to this, with increase in temperature above 80 ^{0}C, the exothermic adsorption of reactants becomes lower, which practically hinders the reaction. Therefore, the temperature optimum is in the range from 20 to 80 ^{0}C, and no heating is needed when photocatalytic processes take place in aqueous systems, which means that no further energy would be needed when water treatment is concerned.

3. Mechanism of Photodegradation of Various Herbicides

In order the photocatalytic process with the application of TiO_2 could be proposed as a universal method for efficient removal of organic pollutants it is necessary to consider in detail the chemistry of various classes of compounds under these conditions (Sturini et al., 1996). Since the chemistry of such processes is complex, a careful choice of different analytical techniques is of essential significance in monitoring and controlling all the phases of transformation of the starting compound in order the potentially formed harmful intermediates could be identified and the reaction mechanism interpreted. It is known that the disappearance of the pollutant in the first phase of photodegradation is not sufficient, as the process of heterogeneous catalysis may yield the formation of various organic intermediates. The newly formed compounds may be more toxic and in some cases more stable than the starting compound (Parra et al., 2000), so that it is necessary to monitor the process of photodegradation to the complete removal of all the intermediates formed, i.e. to the complete mineralization.

Because of the complexity of the radical formation reactions that take place during the photocatalytic processes it is difficult to propose a reaction scheme that would encompass all the intermediates formed (Konstantinou and Albanis, 2003). Also, little information are available when concerning the reaction mechanism of photocatalytic degradation of herbicides by applying simulated or natural sunlight. Because of that we give here a survey of particular classes of herbicides concerning the identification of products formed and type of the mechanism of herbicide degradation (Table 4).

In order to assess the degree of mineralization reached during the photocatalytic treatment, the formation of CO_2 and inorganic ions is generally estimated. In general, at low reactant levels, or for compounds which do not form important intermediates, complete mineralization and reactant disappearance proceed with similar half lives, but at higher reactant levels, where important intermediates occur, the mineralization process is slower than the degradation of the parent compound (Konstantinou and Albanis, 2003). Until now, total mineralization has been observed for the photocatalytic degradation of most herbicides although for some of them longer irradiation times are needed. Only in the case of s-triazine herbicides, the mineralization was not complete due to the high stability of triazine nucleus and formation of stable cyanuric acid, which, fortunately, is not toxic. For chlorinated molecules, Cl^- ion is easily released into the solution and it is the first of the ions appearing during the photocatalytic degradation (Topalov et al., 2000; Topalov et al., 2001; Konstantinou et al., 2001; Šojić et al., 2009). The herbicides containing sulfur atoms are mineralized into sulfate ions (Maurino et al., 1999; Konstantinou et al., 2001). As is known (Abramović et al., 2003; Abramović et al., 2004a, 2004b), in the course of photodegradation, the organic nitrogen is transformed mainly to ammonium ion and/or nitrite/nitrate, and sometimes to nitrogen. Ammonium ions are relatively stable and the ratio of ammonium to nitrate depends on the chemical structure of the substrate and reaction conditions (Alberici et al., 2001). It was also demonstrated that the photodegradation of heterocyclic substrates yields both ammonium and nitrate (Abramović et al., 2004a).

In the present review we showed some examples of the most important classes of herbicides and, depending on their type, the corresponding degradation pathways (Table 4).

The study of the photocatalytic degradation of **s-triazines** herbicides showed that all the herbicides were rapidly degraded within a few minutes, but full mineralization was not observed. Different intermediates have been identified, mainly the hydroxylated and dealkylated derivatives: ammeline and cyanuric acid being the common final photoproducts of all herbicides (Table 4).

Sulfonylureas exhibit similar behavior in photocatalytic treatment as s-triazines since their molecular structures contain s-triazine moiety; in this case too, ammelide and cyanuric acid are also end products. The degradation of both aliphatic and aromatic groups in sulfonylureas leads to the formation of smaller and more oxidized metabolite molecules, such as short carboxylic acids (formic, acetic and oxalic) (Table 4).

Table 4. List of herbicides mineralized by applying photocatalysis in the presence of TiO_2, main reaction pathways and transformation products (TP).

Herbicide	Reaction pathway	Transformation products (TP)	Reference
s-Triazin (s-Tr)			
Atrazin Ametrin Cianazin Irgarol Propazin Prometrin Prometon Simazin	Oxidation of the alkyl chain Dealkylation Decarboxylation Deamination Hydroxylation	Amido and dealkylated-(s-Tr) Hydroxy-(s-Tr) Amelin Cyanuric acid	Muszkat et al. (1995) Konstantinou et al. (2001) Pelizzetti et al. (1990) Pelizzetti et al. (1992a) Pelizzetti et al. (1992b) Sanlaville et al. (1996) Zwiener et al. (1995) Zhanqi et al. (2007)
Sulfonylureas (Su)			
Clorosulfuron Cynosulfuron Triasulfuron Tiphenylfuron Iodosulfuron	Cleavage of sulfonylurea functional group S-triazine reaction path Hydroxylation Demethylation	Hydroxy-Su Dealkylated-Su Melamin Cyanuric acid	Vulliet et al. (2002) Maurino et al. (1999)
Thiocarbamates (TC)			
Butylate Cycloate EPTC Eptam Molinat Thiocarbazyl Thiobencarb Vernolat Vapam	Cleavage of the amino group Oxidation of the *S*-alkyl group Oxidation via formation of peroxo radicals S & O replacement by N(C)OS group	Amino & carboxy-TP Keto & dealkylated TC Sulfoxo & dealkylated TP *S*-isomers of TC	Vidal et al. (1999) Sturini et al. (1996) Konstantinou et al. (2001) Vidal (1998) Vidal and Luengo (2001)
Amides, anilides			
3,4-dichlorpro-pio-amide alachlor Propachlor Propanil Propyzamide	Dechlorination Cleavage of the amino group Oxidation with formation of peroxo radicals Hydroxylation	Hydroxy, aliphatic TP Amino TP Dechlorinated TP Dealkylated TP Cyclization TP Benzamide derivatives	Peñuela and Bunceló (1996) Konstantinou et al. (2001) Pathirana and Maithreepala (1997) Muszkat et al. (1992) Konstantinou et al. (2002)
Fenylureas (PhU)			
Chlortoluron Diuron Fenuron Isoproturon Monuron Metobromuron Chlorotoluron Linuron	Hydroxylation Oxidation Decarboxylation Demethoxylation Dechlorination Hydroxylation Demethoxylation	Hydroxy-PhU Phenyl-hydroxyurea Aniline derivatives Alkylic-oxidation derivatives Dechlorination-hydroxylation derivatives Derivatives of benzene ring hydroxylation	Para et al. (2000) Richard and Bengana (1996) Para et al. (2002) Pramauro et al. (1993) Kinkennon et al. (1995) Amine-Khodja et al. (2005) Haque et al. (2006) Rao et al. (2009)
Phenoxy acids			
MCPP MCPA 2,4-D Benzofuran Propoksur	Hydroxylation Dechlorination Decarboxylation	Hydroxy-phenoxy acids Hydroxy-phenoxy acids Hlorophenols Salicylaldehyde Chlorophenols Chinoidal structures Carboxylic acids	Topalov et al. (2000) Topalov et al. (2001) Barbeni et al. (1987) Herrmann et al. (1998) Mylonas et al. (1996) Poulios et al. (1998)

Herbicide	Reaction pathway	Transformation products (TP)	Reference
Pyridine			
Triclopyr Clopyralid Fluroxypyr	Hydroxylation Decarboxylation Dehlcrination Deamination	Hydroxychloropyridine Dichloropyridine-2-ol Hydroxypyridine-2-carboxylic acid Bipyridine-2'-carboxylic acid	Qamar et al. (2006) Šojić et al. (2009) Aramendía et al. (2005)
Imidazolinon			
Imazamox Imazapyr	Decarbonylation Decarboxylation Hydroxylation	Nicotinic acid derivatives Imidazolinone derivatives	Harir et al. (2008) Carrier et al. (2006)

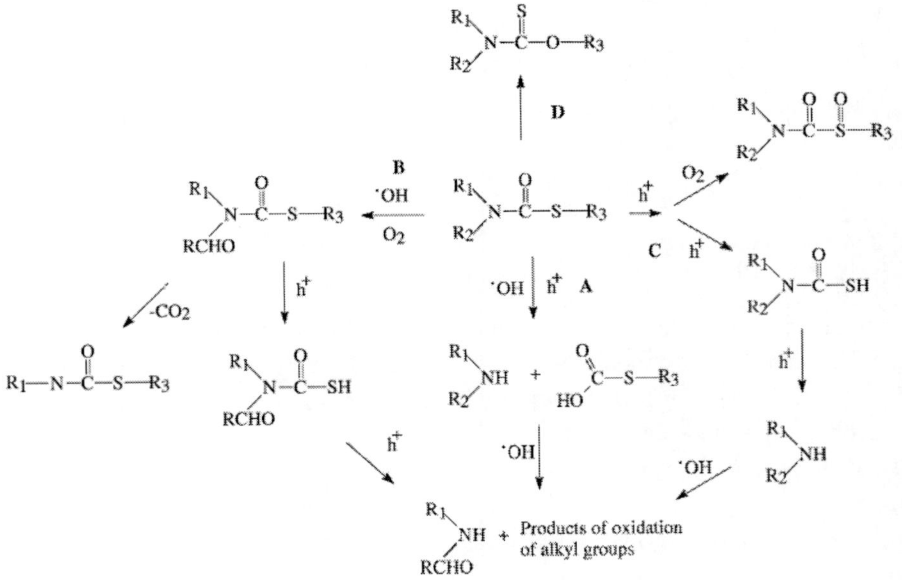

Figure 4. Photocatalytic degradation pathways and main transformation products of thiocarbamate herbicides in aqueous TiO_2 suspensions (from Konstantinou and Albanis, 2003, with permission from Elsevier)

Several **thiocarbamate** herbicides have been studied to determine the feasibility of the TiO_2 photocatalytic oxidation in water (Table 4). Figure 4 shows the reaction pathways and main transformation products of thiocarbamate herbicides. It has been observed that these herbicides can be completely mineralized to CO_2, H_2O, NO_3^-, NH_4^+, SO_4^{2-}, by irradiation in the presence of TiO_2.

When concerning the degradation of **anilide** and **amide** herbicides, their complete mineralization takes place through the formation of various aromatic derivatives and different aliphatic products. The quantitative release of Cl^- was found to proceed with the complete disappearance of the parent herbicide in a short time. The complete mineralization to Cl^-, H^+, CO_2, NO_3^- occurred after prolonged irradiation times because of the great number of nitrogen containing aromatics.

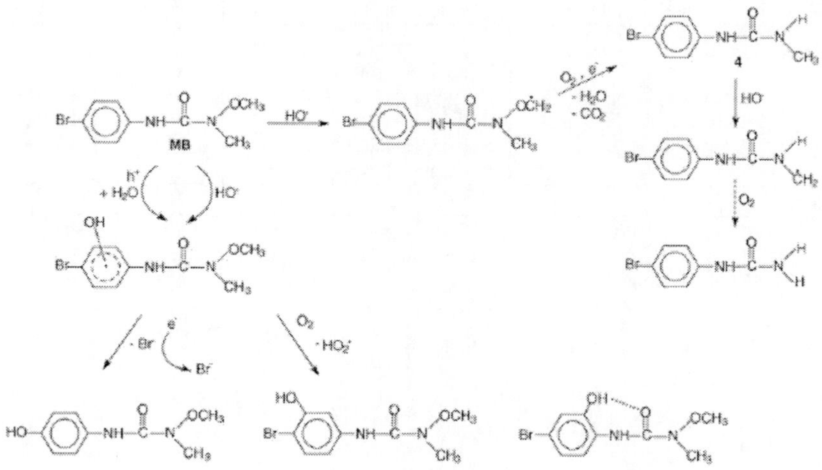

Figure 5. Reaction mechanism proposed for the photocatalytic degradation of metobromuron (from Amine-Khodja et al., 2005, with permission from Elsevier)

The photocatalytic degradation of **phenylurea** herbicides proceeds by three main pathways: i) oxidation of the methyl group of the side chain, ii) hydroxylation of the aromatic ring and iii) dechlorination. Different products of photocatalytic degradation are found depending on the polymorphic form of TiO_2 used (Canle López et al., 2005). Substituents also play an important role in the product distribution (Amine-Khodja et al., 2005). Thus, metobromuron was found to be efficiently degraded through photocatalysis with the intermediary formation of ring hydroxylated and demethoxylated products (Figure 5). Metobromuron can be mineralized at 90%, accompanied by complete debromination with bromide formation (Amine-Khodja et al., 2005).

By monitoring the process of photocatalytic degradation of **phenoxy acid** herbicides it is possible to distinguish fragmentation of the parent molecules into the aromatic moiety and the corresponding carboxylic acid (Konstantinou and Albanis, 2003). The carboxylic acids formed are mineralized by the destruction of not only the parent compounds but also all of the intermediate formed during the oxidation.

Similar behavior exhibit also **pyridine** herbicides, following the same fragmentation patterns (Figure 6). After formation of several pyridine-containing intermediates, the ring opens, and finally degrades to CO_2, H_2O, Cl^-, and NH_4^+. Triclopyr and fluroxypyr, belonging also to the class of pyridine herbicides, behave in a similar way; their complete mineralization takes place with the formation of the corresponding end products in dependence of the nature of the parent compound (Qamar et al., 2006; Aramendía et al., 2005). Thus, in the case of fluroxypyr, in addition to CO_2, Cl^- and F^-, NO_3^- is released, in contrast to clopyralid which gives NH_4^+. This phenomenon indicates that the transformation of organic nitrogen depends on the chemical structure of the substrate and reaction conditions, which has already been discussed.

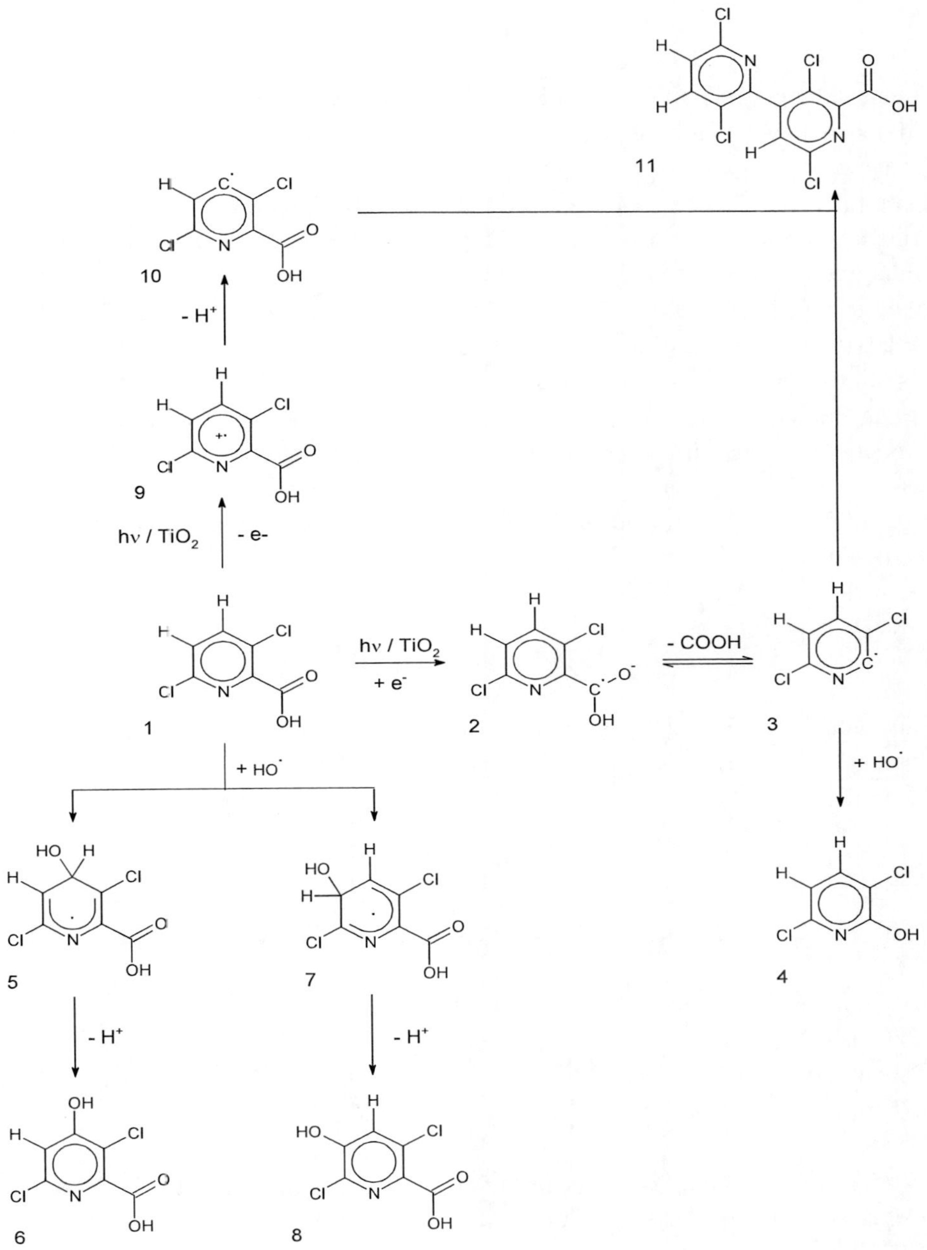

Figure 6. Tentative pathways for photocatalytic degradation of clopyralid (from Šojić et al., 2009, with permission from Elsevier)

Imidazolinone herbicides are mineralized through the formation of three transformation products resulting from dealkylation, oxidation and cyclization reactions.

As can be seen, the photocatalytic degradation on TiO_2 may be an effective way of eliminating complex organic pollutants, in this case herbicides, from water. The great number of transformation products detected shows the complexity of the photocatalytic process and suggests the existence of various degradation routes that result in complex and interconnected pathways.

4. Conclusion

Photocatalytic degradation with TiO_2 are gaining growing acceptance as an effective method for removal of herbicides from water. Various operational parameters affect the effectiveness or activities of TiO_2 photocatalysts and hence it is necessary to study their effects on the herbicide degradation rate. Some reactive herbicides are degraded at higher and some others at lower pH. Catalyst loadings, light intensity, initial concentration of the herbicide, oxidizing agents, and temperature are found to exert their individual influence on the photocatalytic degradation of any herbicide. Therefore, to study effectively the photocatalytic degradation of any herbicide, all the mentioned parameters must be given due considerations. This review presented also a survey of particular classes of herbicides, concerning the identification of products formed and type of degradation mechanism. Based on the material covered, it can be concluded that photocatalytic degradation on TiO_2 may be successfully used for the removal of herbicides as complex organic pollutants, yielding in the majority of cases a complete mineralization of both the parent compound and its intermediates.

Acknowledgment

This work was financially supported by the Ministry of Science and Technological Development of the Republic of Serbia (Project: ON142029).

References

[1] Abramović, BF; Anderluh, VB; Topalov, AS; Gaál, FF. *J. Serb. Chem. Soc.*, 2003, 68, 961-970.

[2] Abramović, BF; Anderluh, VB; Topalov, AS; Gaál, FF. *Appl. Catal. B: Environ.*, 2004a, 48, 213-221 (and references therein).

[3] Abramović, BF; Anderluh, VB; Topalov, AS; Gaál, FF. *APTEFF*, 2004b, 35, 79-86.

[4] Abramović, B; Anderluh, V; Šojić, D; Gaál, F. *J. Serb. Chem. Soc.*, 2007a, 72, 1477-1486.

[5] Abramović, BF; Šojić, DV; Anderluh, VB. *Acta Chim. Slov.*, 2007b, 54, 558-564.

[6] Alberici, RM; Canela, MC; Eberlin, MN; Jardim, WF. *Appl. Catal. B: Environ.*, 2001, 30, 389-397.

[7] Amine-Khodja, A; Boulkamh, A; Richard, C. *Appl. Catal. B: Environ.*, 2005, 59, 147-154.

[8] Aramendía, MA; Marinas, A; Marinas, JM; Moreno, JM; Urbano, FJ. *Catal. Today*, 2005, 101, 187-193.

[9] Atiqur Rahman, M; Muneer, M. *J. Environ. Sci. Health*, 2005, 40, 247-267.

[10] Barbeni, M; Morello, M; Pramauro; E. Pelizzetti, E. *Chemosphere*, 1987, 16, 1165-1179.

[11] Burns, RA; Crittenden, JC; Hand, DW; Selzer, VH; Sutter, LL; Salman, SR. *J. Environ. Eng.*, 1999, 125, 77-85.

[12] Burrows, HD; Canle LM; Santaballa, JA; Steenken, S. *J. Photochem. Photobiol. B: Biol.*, 2002, 67, 71-108.

[13] Calza, P; Medana, C; Baiocchi, C; Pelizzetti, E. *Current Anal. Chem.*, 2005, 1, 267-287.

[14] Canle López, M; Fernández, MI; Rodríguez, S; Santaballa, JA; Steenken, S; Vulliet, E. *ChemPhysChem*, 2005, 6, 2064-2074.

[15] Carrier, M; Perol, N; Herrmann, JM; Bordes, C; Horikoshi, S; Paisse, JO; Baudot, R; Guillard, C. *Appl. Catal. B: Environ.*, 2006, 65, 11-20.

[16] Chen, D; Ray, AK. *Appl. Catal. B: Environ.*, 1999, 23, 143-157.

[17] Chiron, S; Fernadez-Alba, A; Barcelo, D. *Sci. Technol.*, 1993, 27, 2352-2359.

[18] Chiron, S; Fernandez-Alba, A; Rodriguez, A; Garcia-Calvo, E. *Water Res.*, 2000, 34, 366-377.

[19] Chu, W; Wong, CC. *Water Res.*, 2004, 38, 1037-1043.

[20] Devipriya, S; Yesodharan S. *Sol. Energy Mater. Sol. Cells*, 2005, 86, 309-348.

[21] Dillert, R; Cassano, AE; Goslich, R; Bahnemann, D. *Catal. Today*, 1999, 54, 267-282.

[22] Dionysiou, DD; Suidan, MT; Bekou, E; Baudin, I; Laîné, JM. *Appl. Catal. B: Environ.*, 2000, 26, 153-171.

[23] Donald, DB; Cessna, AJ; Sverko, E; Glozier, NE. *Environ. Health Perspect.*, 2007, 115, 1183-1191.

[24] Evgenidou, E; Fytianos, K; Poulios, I. *Appl. Catal. B: Environ.*, 2005, 59, 81-89.

[25] Evgenidou, E; Bizani, E; Christophoridis, C; Fytianos K. *Chemosphere*, 2007, 68, 1877-1882.

[26] Fresno, F; Guillard, C; Coronado, JM; Chovelon, JM; Tudela, D; Soria, J; Herrmann, JM. *J. Photochem. Photobiol. A: Chem.*, 2005, 173, 13-20.

[27] Garcia, JC; Takashima, K. *J. Photochem. Photobiol. A: Chem.*, 2003, 155, 215-222.

[28] Haque, M; Muneer, M. *J. Environ. Manage.*, 2003, 69, 169-176.

[29] Gray N. Pesticides and organic micropollutants. In *Drinking Water Quality,* N. Gray (Ed.) John Wiley & Sons: Chichester, 1996, 132-148.

[30] Haque, M; Muneer, M; Bahnemann, D. *Environ. Sci. Technol.*, 2006, 40, 4765-4770.

[31] Harir, M; Gaspar, A; Kanawati, B; Fekete, A; Frommberger, M; Martens, D; Kettrup, A; El Azzouzi, M; Schmitt-Kopplin, Ph. *Appl. Catal. B: Environ.*, 2008, 84, 524-532.

[32] Herrmann, JM. *Catal. Today*, 1999, 53, 115-129.

[33] Herrmann, JM. *Top. Catal.*, 2005, 34, 49-65.

[34] Herrmann, JM; Disdier, J; Pichat, P; Malato, S; Blanco, J. *Appl. Catal. B: Environ.*, 1998, 17, 15-23.

[35] Herrmann, JM; Guillard, C; Arguello, M; Agüera, A; Tejedor, A; Piedra, L; Fernández-Alba, A. *Catal. Today*, 1999, 54, 353-367.

[36] Herrmann, JM; Guillard, C; Disdier, J; Lehaut, C; Malato, S; Blanco, J. *Appl. Catal. B: Environ.*, 2002, 35, 281-294.

[37] Iliev, V; Tomova, D; Bilyarska, L; Tyuliev, G. *J. Mol. Catal. A: Chem.*, 2007, 263, 32-38.

[38] Iliev, V; Tomova, D; Bilyarska, L; Eliyas, A; Petrov, L. *Appl. Catal. B: Environ.*, 2006, 63, 266-271.

[39] Ishiki, RR; Ishiki, HM; Takashima, K. *Chemosphere*, 2005, 58, 1461-1469.

[40] Kabra, K; Chaudhary, R; Sawhney, RL. *Ind. Eng. Chem. Res.*, 2004, 43, 7683-7696.

[41] Karunakaran,C, Dhanalakshmi, R. *Solar Energy Materials & Solar Cells*, 92, 2008, 588-593.
[42] Kinkennon, AE; Green, DB; Hutchinson, B. *Chemosphere*, 1995, 31, 3663-3671.
[43] Kitano, M; Matsuoka, M; Ueshima, M; Anpo, M. *Appl. Catal. A: Gen.*, 2007, 325, 1-14.
[44] Konstantinou, IK; Sakellarides, TM; Sakkas, VA; Albanis, TA. *Environ. Sci. Technol.*, 2001, 35, 398-405.
[45] Konstantinou, IK; Sakkas, VA; Albanis, TA. *Water Res.*, 2002, 36, 2733-2742.
[46] Konstantinou, IK; Sakkas, VA; Albanis, TA. *Appl. Catal. B: Environ.*, 2001, 34, 227-239.
[47] Konstantinou, IK; Albanis, TA. *Appl. Catal. B: Environ.*, 2003, 42, 319-335.
[48] Matthews, GA. *Pesticides: Health, Safety and the Environment*, Blackwell Publishing: Ascot, Berkshire, 2006, 3.
[49] Maurino, V; Minero, C; Pelizzetti, E; Vincenti, M. *Colloids and Surfaces A: Physicochem. Eng. Aspects*, 1999, 151, 329-338.
[50] Mills, A; Morris, S. *J. Photochem. Photobiol. A: Chem.*, 1993, 71, 75-83.
[51] Mills, A; Le Hunte, S. *J. Photochem. Photobiol. A: Chem.*, 1997, 108, 1-35.
[52] Moctezuma, E; Zamarripa, H; Leyva, E. *Rev. Int. Contam. Ambient.*, 2003, 19, 117-125.
[53] Muneer, M; Bahnemann, D. *Appl. Catal. B: Environ.*, 2002, 36, 95-111.
[54] Muneer, M; Theurich, J; Bahnemann, D. *Res. Chem. Intermed.*, 1999, 25, 667-683.
[55] Muszkat, L; Bir, L; Feigelson, L. *J. Photochem. Photobiol. A: Chem.*, 1995, 87, 85-88.
[56] Muszkat, L; Halmann, M; Raucher, D; Bir, L. *J. Photochem. Photobiol. A: Chem.*, 1992, 65, 409-417.
[57] Mylonas, A; Hiskia, A; Papakonstantinou, E. *J. Mol. Catal. A: Chem.*, 1996, 114, 191-200.
[58] Ollis, DF; Pelizzetti, E; Serpone, E. *Environ. Sci. Technol.*, 1991, 25, 1522-1529.
[59] Osajima, JA; Ishiki, HM; Takashima, K. *Monatsh. Chem.*, 2008, 139, 7-11.
[60] Parra, S; Olivero, J; Pulgarin, C. *Appl. Catal. B: Environ.*, 2002, 36, 75-85.
[61] Parra, S; Sarria, V; Malato, S; Péringer, P; Pulgarin, C. *Appl. Catal. B: Environ.*, 2000, 27, 153-168.
[62] Pathirana, HMKK; Maithreepala, RA. *J. Photochem. Photobiol. A: Chem.*, 1997, 102, 273-277.
[63] Pelizzetti, E; Minero, C; Carlin, V; Vincenti, M; Pramauro, E. *Chemosphere*, 1992a, 24, 891-910.
[64] Pelizzetti, E; Carlin, V; Minero, C; Pramauro, E; Vincenti, M. *Sci. Total Environ.*, 1992b, 123/124, 161-169.
[65] Pelizzetti, E; Maurino, V; Minero, C; Carlin, V; Pramauro, E; Zerbinati, O; Tosato, M. L. *Environ. Sci. Technol.*, 1990, 24, 1559-1565.
[66] Peñuela, GA; Barceló, D. *J. Chromatogr. A*, 1996, 754, 187-195.
[67] Pizarro, P; Guillard, C; Perol, N; Herrmann, JM. *Catal. Today*, 2005, 101, 211-218.
[68] Poulios, I; Kositzi, M; Kouras, A. *J. Photochem. Photobiol. A: Chem.*, 1998, 115, 175-183.
[69] Poulios, I; Avranas, A; Rekliti, E; Zouboulis, A. *J. Chem. Technol. Biotechnol.*, 2000, 75, 205-212.

[70] Pramauro, E; Vincenti, M; Augugliaro, V; Palmisano, L. *Environ. Sci. Technol.*, 1993, 27, 1790-1795.

[71] Prousek, J; *Chemicke Listy*, 1996, 90, 229-237.

[72] Qamar, M; Muneer, M; Bahnemann, D. *J. Environ. Manage.*, 2006, 80, 99-106.

[73] Rachel, A; Sarakha, M; Subrahmanyam, M; Boule, P. *Appl. Catal. B: Environ.*, 2002, 37, 293-300.

[74] Rao, YF; Chu, W. *Environ. Sci. Technol.*, 2009, 43, 6183-6189.

[75] Richard, C; Bengana, S. *Chemosphere*, 1996, 33, 635-641.

[76] Sanlaville, Y; Guittonneau, S; Mansour, M; Feicht, EA; Meallier, P; Kettrup, A. *Chemosphere*, 1996, 33, 353-362.

[77] Sene, JJ; Zeltner, WA; Anderson, MA. *J. Phys. Chem. B*, 2003, 107, 1597-1603.

[78] Singh, HK; Muneer, M; Bahnemann, D. *Photochem. Photobiol. Sci.*, 2003, 2, 151-156.

[79] Singh, HK; Saquib, M; Haque, MM; Muneer, M; Bahnemann, DW. *J. Mol. Catal. A: Chem.*, 2007, 264, 66-72.

[80] Sleiman, M; Conchon, P; Ferronato, C; Chovelon, JM. *Appl. Catal. B: Environ.*, 2007, 71, 279-290.

[81] Sturini, M; Fasani, E; Prandi, C; Casaschi, A; Albini, A. *J. Photochem. Photobiol. A: Chem.*, 1996, 101, 251-255.

[82] Šojić, DV; Anderluh, VB; Orčić, DZ; Abramović, BF. *J. Hazard. Mater.*, 2009, 168, 94-101.

[83] Tamaki, Y; Furube, A; Murai, M; Hara, K; Katoh, R; Tachiya, M. *J. Am. Chem. Soc.*, 2006, 128, 416-417.

[84] Teoh, W; Amal, YR; Mädler, L; Pratsinis, SE. *Catal. Today*, 2007, 120, 203-213.

[85] Thompson, TL; Yates, JT., Jr. *Chem. Rev.*, 2006, 106, 4428-4453.

[86] Topalov, A; Molnár-Gábor, D; Kosanić, M; Abramović, B. *Water Res.*, 2000, 34, 1473-1478.

[87] Topalov, A; Abramović, B; Molnár-Gábor, D; Csanádi, J; Arcson, O. *J. Photochem. Photobiol. A: Chem.*, 2001, 140, 249-253.

[88] Topalov, AS; Šojić, DV; Molnár-Gábor, DA; Abramović, BF; Čomor, MI. *Appl. Catal. B: Environ.*, 2004, 54, 125-133.

[89] Turchi, C; Ollis, D. *J. Catal.*, 1990, 122, 178-192.

[90] Vidal, A; Martin Luengo, MA. *Appl. Catal. B: Environ.*, 2001, 32, 1-9.

[91] Vidal, A. *Chemosphere*, 1998, 36, 2593-2606.

[92] Vidal, A; Dinya, Z; Mogyorodi, F. Jr; Mogyorodi, F. *Appl. Catal. B: Environ.*, 1999, 21, 259-267.

[93] Vione, D; Minero, C; Maurino, V; Carlotti, ME; Picatonotto, T; Pelizzetti, E. *Appl. Catal. B: Environ.*, 2005, 58, 79-88.

[94] Vulliet, E; Emmelin, C; Chovelon, JM; Guillard, C; Herrmann, JM. *Appl. Catal. B: Environ.*, 2002, 38, 127-137.

[95] Vulliet, E; Chovelon, JM; Guillard, C; Herrmann, JM. *J. Photochem. Photobiol. A*, 2003, 159, 71-79.

[96] Wang, KH; Hsieh, YH; Chou, MY; Chang, C. *Appl. Catal. B: Environ.*, 1999, 21, 1-8.

[97] Wang, KH; Hsieh, YH; Wu, CH; Chang, CY. *Chemosphere*, 2000, 40, 389-394.

[98] Wong, CC; Chu, W. *Chemosphere*, 2003, 50, 981-987.

[99] Yano, J; Matsuura, J; Ohura, H; Yamasaki, S. *Ultrason. Sonochem.*, 2005, 12, 197-203.

[100] Zhanqi, G; Shaogui, Y; Na, T; Cheng, S. *J. Hazard. Mater.*, 2007, 145, 424-430.

[101] Zwiener, C; Weil, L; Niessner, R. *Intern. J. Environ. Anal. Chem.*, 1995, 58, 247-264.

In: Desalination: Methods, Cost and Technology
Editor: Irena A. Urbonienė pp. 143-174

ISBN: 978-1-61668-909-4

Chapter 4

MODELING OF A MEMBRANE DISTILLATION UNIT FOR DESALINATION

***Jamel Orfi*[1*] *and Nizar Loussif*[2]**

[1] Department of Mechanical Engineering, College of Engineering, King Saud University, P.O. Box 800, 11421,Riyadh, KSA

[2] LESTE, Ecole Nationale d'Ingénieurs de Monastir, Université de Monastir, 5000, Monastir, Tunisia

ABSTRACT

We present in this chapter an overview and a state of the art on the Membrane Distillation (MD) process. Such a process is a new, rapidly increasing membrane technology, which can be used for desalination and also for recycling. It is a separation process that involves transport of vapor through porous hydrophobic membranes. The principle, the main applications and the benefits of this separation technique are outlined. Approaches to improve the process performance are also discussed.

A detailed theoretical model for the transport phenomena in an Air Gap Membrane Distillation used for desalination was developed. This model is based on the conservation equations for the mass, momentum, energy and species within the feed water solution as well as on the mass and energy balances on the membrane sides.

In order to enhance the MD process performance, a composite hydrophilic-hydrophobic membrane is proposed in an Air Gap Membrane Distillation (AGMD) unit. The configuration considered here is a vertical tubular one with two concentric tubes. Hot saline water flows inside the inner tube which has a porous hydrophilic-hydrophobic wall. The external tube maintained at a constant and low temperature is the condenser.

The results analyzed the effect of the operating parameters such as the velocity and the temperature of the hot feed water at the inlet of the unit, the properties of the membranes, the cooling plate temperature and air gap thickness on the production rate of the vapor and the efficiency of the process. These results show that this desalination process is very efficient. More than 90% of the total heat flux coming from the hot side of the unit (feed water side) is converted into latent heat for the generation of the permeate

[*] Corresponding author: Email: orfij@ksu.edu.sa.

(water vapor). The process performance is very sensitive to several parameters in particular to the porosity of the hydrophobic membrane. An increase of this parameter induces a fast growth of the mass and temperature boundary layers of the feed solution. A significant improvement of the permeate flux and the thermal efficiency are obtained by the incorporation of the composite hydrophilic-hydrophobic membrane.

Keywords: Desalination, Membrane Distillation, hydrophilic-hydrophobic membrane, permeate flux, thermal efficiency.

1. INTRODUCTION

An increasing number of areas on our planet including the Kingdom of Saudi Arabia suffer from water scarcity and many more will suffer in the near future because of the rapid depletion of ground water and surface water. Low cost desalination of brackish and sea water is a way among others to tackle this problem. Current commercially available desalination technologies can be divided in thermal and membrane processes. The membrane processes, particularly reverse osmosis (RO), continue to take market share from thermal desalination with 59 % of the total new built capacity being membrane based (Meindersma et al. [1]). Membrane distillation (MD) is a new, rapidly increasing membrane technology, which can be used for desalination and also for recycling. It is a separation process that involves transport of water vapor through porous hydrophobic membranes.

1.1. The Principle of MD

The driving force for membrane processes may have different character. For reverse osmosis, this driving force is a total pressure difference that induces the mass transport across the membrane. In other cases, it may be a concentration gradient or an electrical potential gradient. Membrane distillation differs from other membrane technologies in that the driving force for desalination is the difference in vapor pressure of water caused by an existing temperature difference across the membrane rather than total pressure difference such as for RO. Thus, vapor molecules are transported from the high vapor pressure (high temperature) side to the low vapor pressure (low temperature) side of the membrane. This trans-membrane vapor pressure difference may be maintained with one of the four following possibilities applied on the permeate side (Rommel et al.[2] and El-Bourawi et al. [3]):

An aqueous solution colder than the feed solution maintained in the direct contact with the permeate side; this configuration is known as Direct Contact Membrane Distillation (DCMD).
A cold inert gas sweeps the permeate side carrying the water vapor molecules outside the membrane module where the condensation takes place; this configuration is termed Sweeping Gas Membrane Distillation (SGMD).
An air gap is placed between the membrane and a condensation surface; the water vapor molecules cross the membrane and the stagnant air and condense on the internal side of a cooling plate; this configuration is known as Air Gap Membrane Distillation (AGMD).

A vacuum pump can be used to reduce the pressure in the permeate side; the condensation occurs outside of the membrane module; this configuration is termed Vacuum Membrane Distillation (VMD).

Figure 1 describes schematically the operation principle of each MD configuration. Exhaustive descriptions of the various physical mechanisms occurring in the MD devices can be found in [1-7].

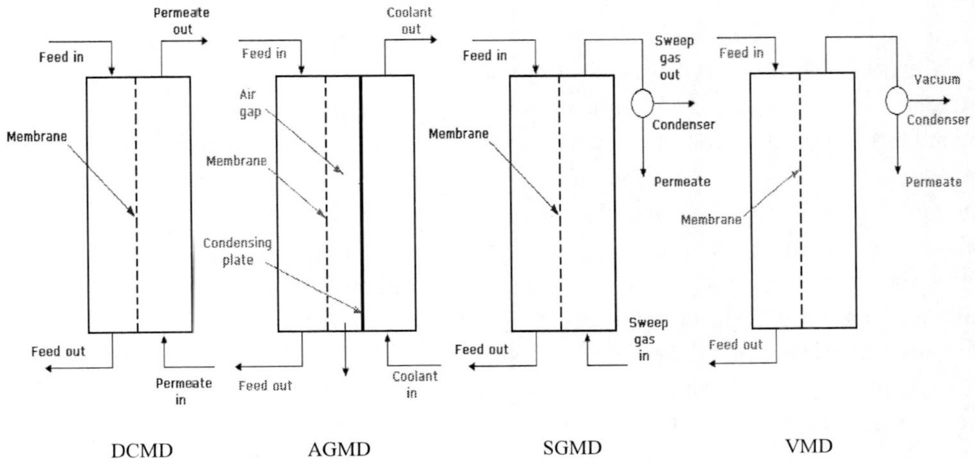

Figure 1. Membrane Distillation configurations

Table. 1 Examples of areas where MD process was successfully applied on laboratory scale (adapted from El-Bourawi et al. [3])

Application area	MD configuration			
	DCMD	AGMD	SGMD	VMD
Desalination and pure water production from brackish water	♦	♦	♦	♦
Nuclear industry (concentration of radioactive solutions and wastewater treatments, pure water production)	♦			
Chemical industry (concentration of non volatile acids, separation of azeotropic aqueous mixtures such as alcohol/water mixtures and crystallization)	♦	♦	♦	♦
Pharmaceutical and biomedical industries (such as removal of water from blood and protein solutions and wastewater treatment)	♦			
Food industry (such as concentration of juices and milk processing)	♦	♦		♦

1.2. Examples of Applications of MD

MD has been used for a wide range of applications such as water desalination, food processing and pharmaceutical and biomedical applications. Table 1 gives a summary of the areas where the MD process was applied on laboratory scale. This table shows in particular that the DCMD configuration was the most used application. As reported by Pharraranawick and Jiraratananon [8], this configuration is convenient to set up, it consumes relatively low energy and produces high water permeate flux. Besides, DCMD seems to be the best configuration for applications where the water is the major feed component. Lagana et al. [9] applied DCMD to produce a highly concentrated apple juice using fiber modules. Tomaszewska [10] presented several examples of MD applications in water and wastewater treatment as well as in food industry. He analyzed some data obtained from the MD application for concentration of diluted non volatile acids such as sulphuric or phosphoric. An other example of applications concerns the recovery of volatile components such as ethanol from aqueous solutions. Traditionally, ethanol is produced by fermentation of biomass in the batch fermenter. However, the integration of the conventional bioconversion with the MD process is more profitable and increases the fermentation efficiency [10]. Membrane distillation was also applied for processing liquid radioactive wastes because of its deep purification of water and its capability to concentrate all the impurities in the same flow [11]. The experimental tests conducted by Macedonio and Drioli [12] proved that MD is a promising technology for the total boron and arsenic removal from water.

1.3. Membrane Characteristics

Membrane used in MD have to meet several requirements. They must be hydrophobic. Thus, the liquid transport across the membrane is prevented. The Liquid Entry Pressure LEP which should be high determines the pressure value that should not be exceeded to respect the non wetting condition. This LEP is given by the Laplace equation:

$$\Delta P_{entry} = -2\frac{\gamma_l}{r_p}\cos\theta \tag{1}$$

Where ΔP_{entry} is the entry pressure difference, γ_l is the surface tension of the solution and θ is the angle of contact between the solution and the membrane surface.

Khayet et al. [13] used in their experiments LEP of water of 137.8 kPa.

Using the above equation, one can say that the non wetting conditions are met with:

a small pore size r_p. A maximum pore radius of 0.6 μm is recommended [10].
a high surface tension γ_l;
a contact angle measured on the hydrophobic surface higher than 90°;
a higher membrane porosity is required for a higher permeate flux. Membrane porosity higher than 0.7 is recommended by several authors [10].
the thickness of the membrane should be very low. Thicknesses ranging from 0.076 to 0.178 mm were used in different previous studies [14].

Polymers such as polytetrafluoroethylene (PTFE), polyvinylidenefluoride (PVDF), polyethylene (PE) and polypropylene (PP) characterized by a low surface energy and good chemical resistance and thermal stability have been used by many authors [14]. Hydrophobic ceramic membranes with pore sizes of 5 and 50 nm were recently used by Cerneaux et al. [15]. Ceramic membranes of different nature, zirconia, alumina and alumino-silicate, with pore diameters of 50 nm, 200 and 400 nm, and 800 nm, respectively, were chemically modified and tested by Gazagnes et al.[16]. On the other hand, Galvez et al. [17] added other membrane properties such as high resistance to heat conduction, low moisture adsorption and a commercially long life.

Khayet et al. [18] designed and prepared novel hydrophobic/hydrophilic composite membranes. The performance of this modified DCMD membrane was compared to that of the frequently used commercial PTFE membranes. The results expressing in particular the variation of the permeate flux with several operating parameters show higher permeate values.

1.4. Benefits of MD

The main benefits of membrane distillation over conventional distillation processes are found in its simplicity and the need for only small temperature gradients to operate. They are found also in its lower operating temperatures and pressures that reduce the equipment surface and costs, the compact modules and the possibility of overcoming corrosion problems by using plastic equipment (Rommel et al. [2]). In addition and since the process operates at liquid – vapor equilibrium, theoretically 100% of ions, macromolecules and other non volatile components can be rejected [19]. This high rejection rate and the high quality of the permeate observed by different authors are almost independent on the feed water salinity. Thus, the MD process allows obtaining fresh water from high concentrated aqueous solutions. Recently, Mohammadi [19] conducted an experimental analysis on desalination using VMD of high salinity water with concentrations ranging from 100 to 300 g/L.

The MD systems are simpler than those for RO plants. Thus, the operational behavior of the MD process is superior to RO systems especially for remote areas.

Besides, the feed water can be heated by any cheap source of energy such as solar energy, geothermal energy or waste energy from diesel engines or from cogeneration plants. The utilization of such low grade energy sources makes the MD process promising separation technique. In particular, the integration of solar energy to drive MD units has been considered by several authors. Fath et al. [20] summarized some advantages of MD systems powered by solar energy:

the MD process operates at low temperatures (60 - 90°C). At this temperature level, the efficiency of the solar thermal collectors is high.
Chemical pretreatment of the feed water is not necessary. This is contrary to RO systems.
Intermittent operation of the MD units is possible.

Koschikowski et al. [21] conducted experimental investigations on solar driven stand-alone membrane distillation to supply the needs of remote areas. Two sets of small capacity systems (between 100 and 500 L/day) and larger ones (up to 10 m^3/day) were designed and tested. Long term maintenance tests in different weather conditions (different countries) have shown that operation is possible with low maintenance. The results of such solar desalination systems obtained in Egypt (Alexandria) [20] show that for a clear and a cloudy day the productivity reaches 11.2 L/m^2.day. The electric conductivity of the feed water was reduced from 670 to 3 μS/cm for the permeate. The MD process efficiency was about 90% whereas the solar thermal collector efficiency was 50 %.

On the other side, it is also of importance to note that there are some obstacles impeding the implementations of MD processes. According to Macedonio and Drioli [12], the main obstacles are the relatively high energy consumption and high membrane maintenance and replacement cost due to the membrane fouling. One can also add it is necessary to guarantee the hydrophobic character of the MD module.

1.4. Methods for MD Improvements

Several approaches used in the literature to improve the MD performance and enhance the permeate flux were discussed by El Bourawi et al. [3]. These authors reported that the low MD permeate flux is attributed mainly to the temperature polarization. For the DCMD configuration, the conduction heat transfer across the membrane material reduces its thermal efficiency and hence the permeate flux.

In order to reduce the effects of the temperature polarization as well as the concentration polarization, different enhancement techniques were proposed and tested. For example, the inclusion of turbulence promoters in the flow channels and the use of ultrasonic systems were proposed [22-24]. One can also note that the design and fabrication of new MD membranes are of importance to provide higher permeability and reduce the heat lost by conduction through the membrane matrix.

Bonyadi and Chung [25] proposed an innovative approach by applying the co-extrusion for the fabrication of dual layer hydrophilic–hydrophobic hollow fibers for the DCMD process.

The results of this experimental study show that flux as high as 55 $kg/(m^2h)$ at 90 ◦C was obtained. This performance is much higher than most of the previous reports indicating that the application of dual layer hydrophilic–hydrophobic hollow fibers may be a promising approach for MD.

The integration of MD with other processes can improve the overall performance of the whole combined system. Macedonio and Drioli [12] designed and studied the performance of a Reverse Osmosis system followed by a Membrane Distillation unit. The results are encouraging due to the possibility of overcoming the limits of the single units. Criscuoli and Drioli [26] conducted an energy and exergy analysis of an integrated system coupling RO, MD and Nanofiltration (NF) modules where the MD unit operates on the RO brine while the NF unit was used for the RO feed pretreatment. They concluded that such integrated systems represent an attractive alternative to RO and to conventional thermal desalination processes.

1.5. Survey of the Previous Theoretical Studies

Based on a recent literature review on this subject, one can observe that the most of the research work on the MD areas was experimental and was done mainly on DCMD, VMD and

AGMD configurations. Furthermore, the theoretical studies concerned with the membrane distillation (MD) process are subjected to various limiting assumptions. The principal assumptions can be summarized as follows:

The heat and mass transfers are supposed generally one dimensional in the unit (Liu et al.[4], Banat and Simardtl [5], Guijt et al. [6]).
The transfer coefficients are generally evaluated using correlations obtained for impermeable interfaces and not for permeable ones (De Pinho et al. [7]).
The most theoretical studies on MD have been focused on the vapour transfer mechanism through the micro-porous membrane (Liu et al. [4]). The effects of convection have received less attention.

As a consequence and because of the performance of such units depends both on the membrane characteristics and on the transport phenomena occurring at the fluid phases adjacent to the membrane, it is important to pay more attention to the characteristics of the fluid behavior in the unit. On the other hand, heat and mass transfers with evaporation and condensation are very important phenomena encountered in various systems used in the desalination area. The accurate evaluation of the heat and mass transfer coefficients before the optimization study of desalination devices is, as reported by Shawaqfeh and Farid [27], an important task. Goosen et al. [28], in their overview of solar desalination systems, concluded that because the design of many desalination devices is largely semi-empirical in nature, there is a need to increase the fundamental understanding of the thermodynamics involved in such systems. The conclusions of Goosen et al. [28] are also valid for other desalination processes in particular membrane desalination processes.

Several simplified models for the heat and mass transports in the membrane distillation process were developed and the obtained results were compared with experimental data. Liu et al. [4] presented theoretical and experimental studies on air gap membrane distillation; they obtained relationships of permeate flux, total heating load and thermal efficiency of AGMD with respect to the membrane distillation temperature difference. El Bourawi et al. [3] reported a detailed review on Membrane Distillation process. They cited and classified numerous experimental and the theoretical MD studies. They concluded in particular that the heat transfer aspect in MD is far from being fully understood and much works in this field are needed to check the applicability of the heat transfer empirical correlations. Bouguecha et al. [29] studied the effect of the natural convective heat and mass transfer in an Air Gap Membrane Distillation (AGMD) system. The steady, laminar, two dimensional equations for the continuity, vorticity, energy and mass were solved using the control volume method and the successive over relaxation method with Chebyshev acceleration. The mass transfer was described using a combined Knudsen and Maxwel-Stefan model. The results show that for the configuration considered in their work when the air gap thickness reaches 5 mm, the transition from conduction to convection occurs and the permeate flux decreases. Alklaibi and Lior [14] observed that the most common approach to model MD, as found in the literature, is by assuming the process as one dimensional and applying empirical heat and mass transfer coefficients. They proposed a transport analysis in an air gap membrane distillation unit based on a two-dimensional conjugate model. They conducted also an analysis of the sensitivity of the production rate to the major operating parameters for a cartesian configuration.

The present work concerns a computational investigation of the fluid flow and the heat and mass transfer in a Membrane Distillation unit. The work is divided into two parts. The first part concerns the basic AGMD configuration while in the second part, a modified AGMD configuration is proposed and studied.

2. Mathematical Model

2.1. Process Description

The configuration of an Air Gap Membrane Distillation is a vertical tubular one. It consists of two concentric tubes. Hot saline water flows inside the inner tube (the evaporator one). The wall of this tube consists of a micro porous hydrophobic (non wettable) membrane through which only water vapour can diffuse and the liquid water is retained. The vapour is condensed on the cold surface of the outer tube (the condenser one). This tube is impermeable and can be cooled by the feed sea water. An air gap is interposed between the membrane and the condensation surface separating the evaporator and the condenser. The temperature difference between the inner and the outer tubes creates a partial pressure gradient forcing the vapour to diffuse through the membrane and the air gap. Figure 2(a) shows a description of the physical model of an AGMD configuration with a hydrophobic membrane. Figure 2(b) represents the same configuration under the same operating conditions with the adoption of a hydrophilic-hydrophobic membrane. For the two membrane distillation units, the calculation domain is limited to the flow, heat and mass transfers in the hot saline water region confined in the micro porous tube with internal radius R_1.

Figure 2(a) shows an Air Gap Membrane Distillation unit containing 5 domains:

I : Hot saline solution; **II :** Hydrophobic membrane ; **III :** Air gap; **IV :** Condensate water film; **V :** Cooling plate.

Figure 2(b) shows the same process configuration as presented in figure 2(a) with a hydrophilic-hydrophobic membrane (the two membranes adopted have the same total thickness). This hydrophilic-hydrophobic AGMD configuration contains 6 domains :

I : Hot saline solution; **II :** Hydrophobic part of the membrane; **II' :** Hydrophilic part of the membrane; **III :** Air gap; **IV :** Condensate water film; **V :** Cooling plate

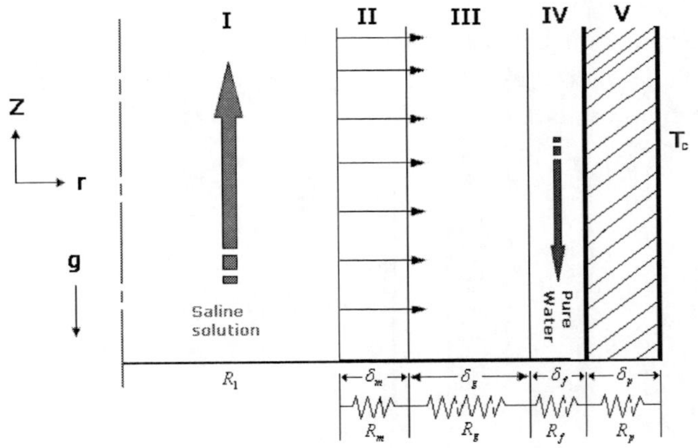

a)

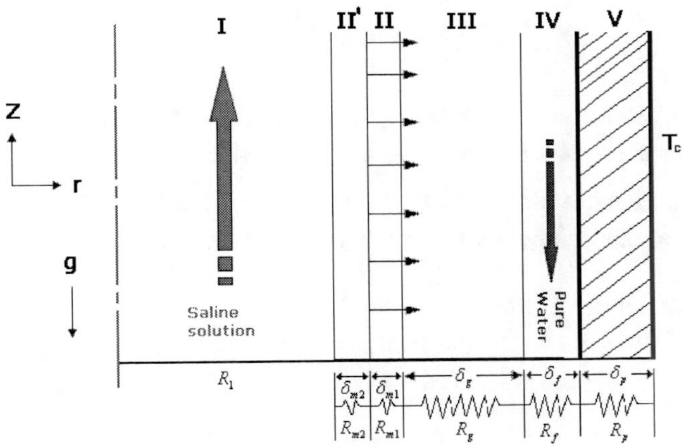

b)

Figures 2. Schematic diagram of an Air Gap Membrane Distillation unit (AGMD): a) hydrophobic membrane; b) hydrophilic-hydrophobic membrane (modified AGMD)

2.2. Model Development

The axisymmetric flow field is modeled using the two dimensional steady-state partial differential equations expressing conservation of mass, energy, momentum and species using constant fluid properties. Viscous dissipation, thermal radiation and Soret and Dufour effects are neglected. The natural convection effects in the feed solution are supposed negligible. Besides, the convective effect in the air gap is also ignored. The condensation temperature, T_c, is supposed to be uniform. Finally, the effect of salt on the vapor saturation pressure is ignored.

2.2.1. Governing equations

The partial differential equations governing the flow, heat and mass transfer within the hot feed water are those of conservation of mass, momentum in r and z directions, energy and species. These equations can be expressed in a compact form as [30]:

$$\mathrm{div}\vec{\mathrm{V}} = 0 \tag{2}$$

$$\rho \frac{D\vec{V}}{Dt} = -\overrightarrow{grad}P + \mu \nabla^2 \vec{V} \tag{3}$$

$$\rho Cp \frac{DT}{Dt} = k\nabla^2 T \tag{4}$$

$$\frac{DW_s}{Dt} = D_s \nabla^2 W_s \tag{5}$$

$\vec{\mathrm{V}}$ refers to the velocity vector of the hot saline water with V_z and V_r as respectively axial and radial components. D/Dt represents the 'material derivative' operator encountered frequently in convective heat and mass transfer. ∇^2 is the Laplacian operator.

The above equations can be normalized using the following dimensionless variables:

$$\bar{r} = \frac{r}{R_1}\ ,\ \bar{z} = \frac{z}{R_1}\ ,\ \bar{V}_r = \frac{V_r}{V_e}\ ,\ \bar{V}_z = \frac{V_z}{V_e}\ ,\ \bar{P} = \frac{P}{\rho V_e^2}\ ,\ \bar{T} = \frac{T - T_0}{T_e - T_0}\ ,\ \overline{W_s} = \frac{W_s - W_0}{W_{se} - W_0} \tag{6}$$

Where R_1 is the inner radius of the membrane, V_e and T_e are respectively the inlet velocity and the inlet temperature of the saline solution, W_{se} is the mass fraction of salt of the solution at the entrance of the membrane module. T_0 and W_0 are respectively the temperature and the salinity at a reference state.

The dimensionless equations expressed in cylindrical coordinates are:
The continuity equation:

$$\frac{1}{\bar{r}}\frac{\partial}{\partial \bar{r}}\left(\bar{r}\bar{V}_r\right) + \frac{\partial}{\partial \bar{z}}\left(\bar{V}_z\right) = 0 \tag{7}$$

The radial momentum equation:

$$\left(\overline{V_r}\frac{\partial \overline{V_r}}{\partial \bar{r}} + \overline{V_z}\frac{\partial \overline{V_r}}{\partial \bar{z}}\right) = -\frac{\partial \bar{P}}{\partial \bar{r}} + \frac{1}{\mathrm{Re}_{R_1}}\left[\frac{\partial}{\partial \bar{r}}\left(\frac{1}{\bar{r}}\frac{\partial}{\partial \bar{r}}\left(\bar{r}\overline{V_r}\right)\right) + \frac{\partial^2 \overline{V_r}}{\partial \bar{z}^2}\right] \tag{8}$$

The axial momentum equation:

$$\left(\overline{V_r}\frac{\partial\overline{V_z}}{\partial\overline{r}}+\overline{V_z}\frac{\partial\overline{V_z}}{\partial\overline{z}}\right)=-\frac{\partial\overline{P}}{\partial\overline{z}}+\frac{1}{Re_{R_1}}\left[\frac{1}{\overline{r}}\frac{\partial}{\partial\overline{r}}\left(\overline{r}\frac{\partial\overline{V_z}}{\partial\overline{r}}\right)+\frac{\partial^2\overline{V_z}}{\partial\overline{z}^2}\right] \tag{9}$$

The energy equation:

$$\left(\overline{V_r}\frac{\partial\overline{T}}{\partial\overline{r}}+\overline{V_z}\frac{\partial\overline{T}}{\partial\overline{z}}\right)=\frac{1}{Re_{R_1}\Pr}\left[\frac{1}{\overline{r}}\frac{\partial}{\partial\overline{r}}\left(\overline{r}\frac{\partial\overline{T}}{\partial\overline{r}}\right)+\frac{\partial^2\overline{T}}{\partial\overline{z}^2}\right] \tag{10}$$

The species equation:

$$\left(\overline{V_r}\frac{\partial\overline{W_s}}{\partial\overline{r}}+\overline{V_z}\frac{\partial\overline{W_s}}{\partial\overline{z}}\right)=\frac{1}{Re_{R_1}Sc}\left[\frac{1}{\overline{r}}\frac{\partial}{\partial\overline{r}}\left(\overline{r}\frac{\partial\overline{W_s}}{\partial\overline{r}}\right)+\frac{\partial^2\overline{W_s}}{\partial\overline{z}^2}\right] \tag{11}$$

Where the Reynolds, Prandtl and Schmidt numbers of the saline solution are defined respectively as follows:

$$Re_{R_1}=\frac{\rho V_e R_1}{\mu},\ \Pr=\frac{\mu C_P}{\lambda},\ Sc=\frac{\upsilon}{D_s}, \tag{11}$$

2.2.2. Heat and mass transfer in the membrane

Modeling the transport phenomenon in the membrane is quite difficult because of the nature and the variety of the physical phenomena occurring through this micro porous material. Examples of such phenomena are the capillarity, the mass diffusion and the heat transfer by conduction and convection.

Many authors have adopted empirical approaches to describe the mass transfer across the membrane. Stephan's law is used to give the general mass flux form [13,14,31]:

$$J_v=K\Delta P_v \tag{12}$$

Where:

J_v : Vapor flux generated by the membrane;

K: Permeability of the membrane;

ΔP_v : Water vapor pressure difference between the membrane sides;

The vapor pressure P_v can be calculated using the Antoine's equation [16]:

$$P_v=\exp\left(23.1964-\frac{3816.44}{T-46.13}\right) \tag{13}$$

The pore sizes of the membranes may be of the order of the mean free molecular path of saturated water vapor crossing the membrane. Therefore, the knowledge of the mechanisms of mass transfer through the membranes is very important. The established theory of gas diffusion in porous media considers three independent mass diffusion mechanisms [32]:

Poiseuille diffusion,
Knudsen diffusion, and
molecular diffusion.

Viscous Poiseuille flow is caused by total pressure gradients. In this case, the gas behaves as a continuous fluid where the molecule – molecule collisions dominate over molecule-wall collisions. However, in the Knudsen diffusion, the transport is controlled by molecule–wall collisions, so that the molecules move independently from each other. In this diffusion mode, the pore size is small or the gas density is so low that the collisions of molecules can be ignored. The third mechanism concerns the molecular diffusion where the driving force for the motion of the different species of a mixture is the concentration gradients. The molecule–molecule collisions dominate over the wall-molecule collisions.

The membrane permeability K is defined for each mechanism of vapor transport as [4,32]:

$$\text{Knuden diffusion: } K_K = \frac{2r_p\varepsilon}{3\chi\delta_m}\left(\frac{8M_v}{\pi R_u T_{moy,m}}\right)^{1/2} \tag{14a}$$

$$\text{Poiseuille diffusion: } K_P = \frac{r_p^2 \varepsilon M_v P_m}{8\chi\delta_m \mu R_u T_{moy,m}} \tag{14b}$$

$$\text{Molecular diffusion: } K_M = \frac{\varepsilon D_{v/a} M_v P_T}{\chi\delta_m P_{a,moy} R_u T_{moy,m}} \tag{14c}$$

The above relations (14a-14c) show that the permeability is not only affected by the membrane's characteristics but also by other parameters such as its average temperature and by the vapor molar mass.

Many combinations of the three models are presented in the literature [32]. Figure 3 describes four different combinations.

Let us consider now the heat transfer mechanism in the AGMD process corresponding to the basic AGMD configuration (figure 2a). The total heat is first transferred by heat convection from the feed solution to the membrane surface where evaporation takes place. Thus, the total heat involved in such a process can be divided in two parts: the latent heat and the sensible one. The latent one is associated with the evaporation of the liquid water at the hot membrane side. While, the total sensible heat transfer (Φ_{sens}) is transferred from the hot surface of the membrane to the condensation surface by:

heat conduction across the membranes and the air gap, (Φ_c);
the mass transfer of the vapor, (Φ_v):

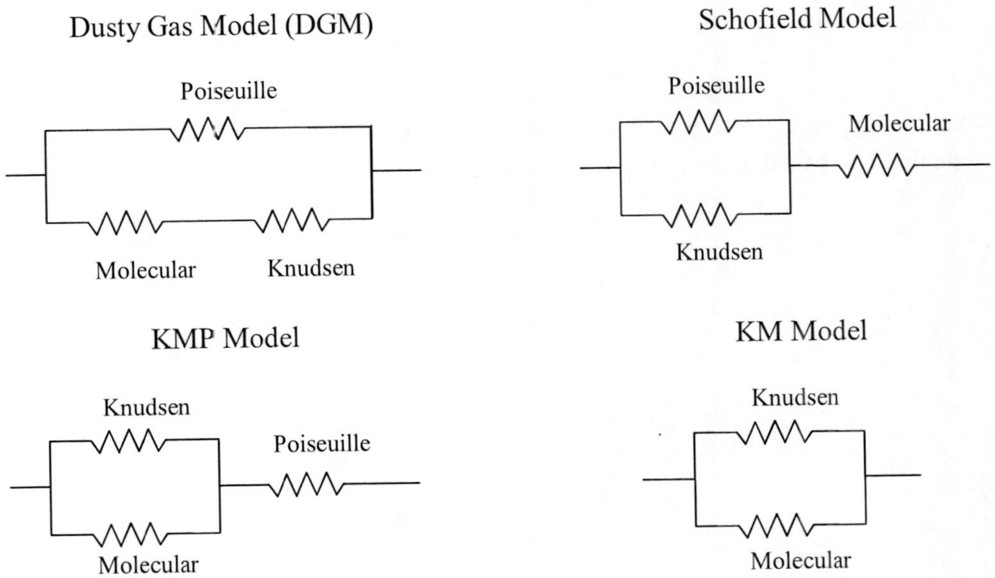

Figure 3. Some models adopted for the prediction of the mass transfer across the hydrophobic membrane

Thus,

$$\Phi_{sens} = \Phi_c + \Phi_v = \frac{T_1 - T_c}{R_T} \tag{15}$$

T_1 is the temperature at the hot side of the membrane.
R_T is the total heat transfer resistance defined by:

$$R_T = R_m + R_g + R_f + R_P \tag{16}$$

R_m, R_g, R_f and R_p are the thermal resistances respectively of the membrane, the humid air gap, the condensate liquid film and of the cooling plate.

For the modified configuration (figure 2b), the heat transfer mechanism is characterized by a conductive resistance in addition to the above mentioned heat transfer resistances.

Thus, the thermal resistances can be defined as:

Heat transfer resistance of the membrane R_m

$$R_m = R_{m1} + R_{m2} \tag{17a}$$

The total thermal resistance of the membrane is the sum of the thermal resistance of the hydrophobic layer R_{m1} and the hydrophilic one R_{m2} :

$$R_{m1} = \frac{R_v R_{mc}}{R_v + R_{mc}}, \quad R_{m2} = \frac{\ln\left((R_1 + \delta_{m2})/R_1\right)}{2\pi L k_{m2}} \tag{17b}$$

Where R_{mc} and R_{m2} are the heat transfer resistances of the solid part respectively for the hydrophobic and hydrophilic layers while R_v refers to the heat transfer resistance of the vapor flux through the hydrophobic membrane pores:

$$R_v = \frac{1}{J_v C_p S} \tag{17c}$$

$$R_{mc} = \frac{\ln\left(R_2/(R_1 + \delta_{m2})\right)}{2\pi L k_{m1}} \tag{17d}$$

$$k_{m1} = \varepsilon_1 k_a + (1 - \varepsilon_1) k_{ma1} \tag{17e}$$

$$k_{m2} = \varepsilon_2 k_e + (1 - \varepsilon_2) k_{ma2} \tag{17f}$$

Heat transfer resistance of the air gap R_g

$$R_g = \frac{R_v R_a}{R_v + R_a} \tag{18a}$$

$$R_a = \frac{\ln(R_3/R_2)}{2\pi L k_a} \tag{18b}$$

Where R_a represents the resistance to the conductive heat transfer of the air.

Heat transfer resistance of the condensate film R_f

$$R_f = \frac{\ln(R_4/R_3)}{2\pi L k_f} \tag{19a}$$

Heat transfer resistance of the cooling plate R_p

$$R_P = \frac{\ln(R_5/R_4)}{2\pi L k_P} \tag{20}$$

The temperature T_2 which is the temperature of the membrane-air gap interface can be expressed in dimensionless form as:

$$\bar{T}_2 = \frac{R_T - R_m}{R_T}\bar{T}_1 + \frac{R_m}{R_T}\bar{T}_c \tag{21}$$

2.2.3. Boundary conditions

We consider in this study that:

The flow, temperature, mass fields are symmetrical with respect to the tube axis (r = 0).
The saline hot water enters the tube (at z=0) with uniform axial velocity V_e, uniform temperature T_e, and uniform mass fraction of salt W_{se}.
At the solution tube exit, the convective terms are much larger than the diffusive terms in the governing equations;

By introducing the non-dimensional variables defined in equation (6), the boundary conditions in dimensionless form are:

Entrance of the saline solution (z=0)

$$\overline{V_z} = 1\ ,\ \overline{V_r} = 0\ ,\ \overline{T} = 1\ ,\ \overline{W_s} = 1 \tag{22a}$$

Symmetry conditions (r=0)

$$\frac{\partial \overline{V_z}}{\partial \overline{r}} = 0\ ,\ \frac{\partial \overline{T}}{\partial \overline{r}} = 0\ ,\ \frac{\partial \overline{W_s}}{\partial \overline{r}} = 0\ ,\ \overline{V_r} = 0 \tag{22b}$$

Outlet of the inner side of membrane (z=L)

$$\frac{\partial \overline{V_z}}{\partial \overline{z}} = 0\ ,\ \frac{\partial \overline{V_r}}{\partial \overline{z}} = 0\ ,\ \frac{\partial \overline{T}}{\partial \overline{z}} = 0\ ,\ \frac{\partial \overline{W_s}}{\partial \overline{z}} = 0 \tag{22c}$$

Feed water - membrane interface

$$\overline{V_z} = 0\ ,\ \overline{V_r} = \frac{J_v}{V_e \rho}\ ,\ \frac{d\overline{T}}{d\overline{r}} = \frac{R_1(Q_c + Q_L)}{k_s(T_e - T_0)}\ ,\ \frac{d\overline{W_s}}{d\overline{r}} = \frac{J_v R_1}{\rho D_s W_{se}} \tag{22d}$$

Where $Q_L = J_v h_{fg}$ represents the latent heat flux.

The vapor flux generated by the membrane will condensate on the internal side of the cooling plate, and for a thin film, the condensate film thickness δ_f can be calculated as given by [14, 30]:

$$\delta_f(z) = \left(\frac{3\mu_f \int_0^z J_v(z)dz}{g\rho_f(\rho_f - \rho_v)} \right)^{1/3} \tag{23}$$

2.2.4. Process parameters

Results of this work were expressed in terms of profiles of temperature, axial velocity and concentration as well as distributions of several process parameters; these process parameters include the average permeate flux, the conductive heat flux, the total heat transfer and the process thermal efficiency.

The averaged permeate flux is obtained by integrating equation (12) over the membrane length L:

$$J = \frac{1}{L}\int_0^L J_V(z)dz \tag{24a}$$

The averaged conduction heat flux is defined as:

$$\bar{Q}_C = \frac{1}{L}\int_0^L Q_C(z)dz \tag{24b}$$

The averaged total latent heat flux is:

$$\bar{Q}_L = \frac{1}{L}\int_0^L Q_L(z)dz \tag{24c}$$

The total heat transfer is:

$$\bar{Q}_T = \frac{1}{L}\int_0^L Q_T(z)dz \tag{24d}$$

Therefore, the process thermal efficiency can be defined as:

$$\eta_t = \frac{\bar{Q}_L}{\bar{Q}_T} \tag{25}$$

Besides, the temperature and concentration polarization coefficients are defined as:

$$I_T = \frac{T_{hm} - T_{gf}}{T_F - T_c} \tag{26a}$$

T_{gf} is the temperature at the interface of the air gap and condensate while T_F refers to the bulk temperature of the feed solution.

$$I_C = \frac{W_{Shm}}{W_{SF}} \tag{26b}$$

W_{Shm} is the concentration at the internal side of the membrane while W_{SF} refers to the bulk concentration of the saline feed water.

3. Numerical Method and Validation

The above five non linear, coupled and second order partial differential equations (7-11) together with the boundary conditions (22a-22d) were solved numerically using the control volume method and the SIMPLER algorithm for handling the pressure-velocity linking [33]. A grid dependence analysis of the solution method was performed. The number of nodes in the axial and radial directions was chosen to be 250X40 because it was found that further refinement of the mesh produced only very small differences in the values of the permeate flux and the thermal efficiency.

The proposed theoretical and numerical model was validated by following several steps including:

Comparing the obtained results of the thermal and flow fields for the forced convection case in isothermal impermeable tubes with the previous results from the literature.

Comparing our results giving the generated water vapor flux with the experimental results of Guijt et al.[6]. Figure 4 shows a good agreement between our computed results and those of Guijt et al.[6].

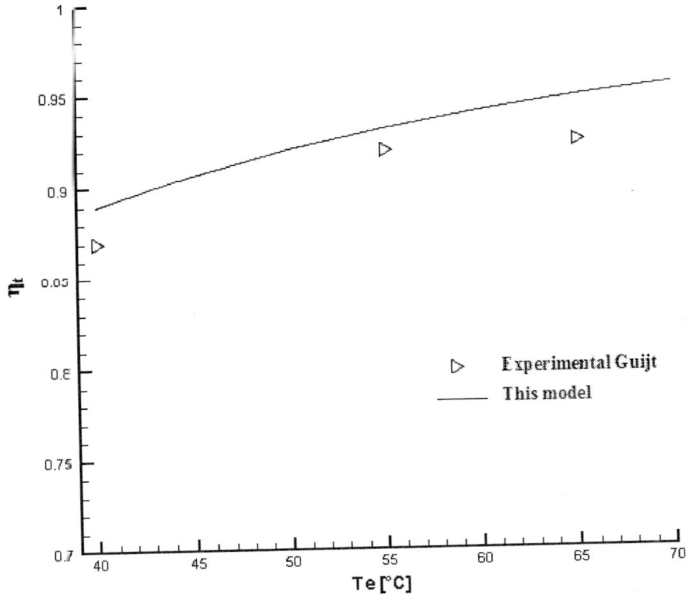

Figure 4. The effect of the saline solution temperature on the permeate flux and comparison with the experimental results of Guijt et al. [6]

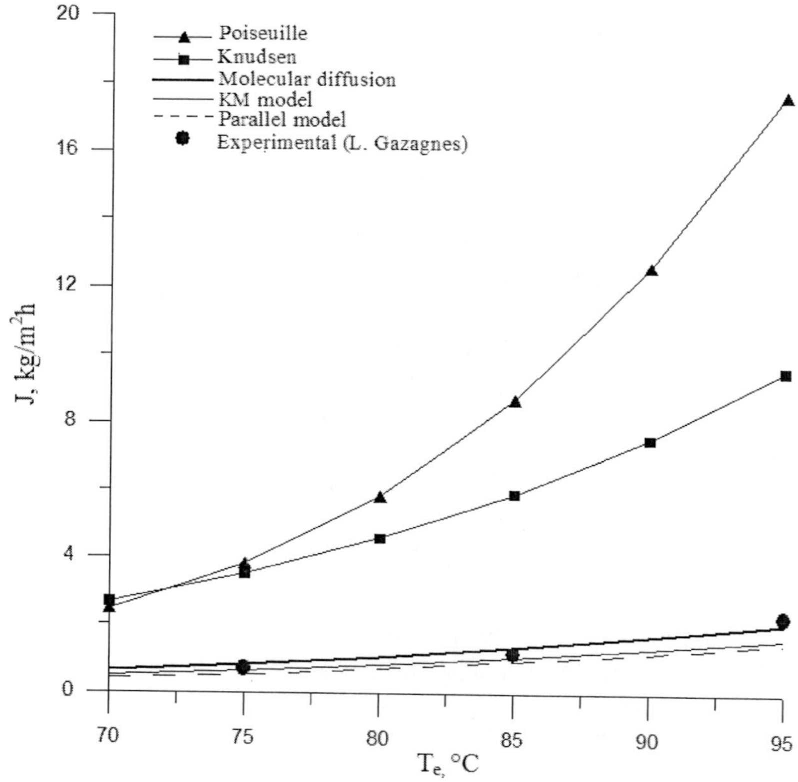

Figure 5. Evolution of the vapor flux for different mecanisms of mass transport for a pore diameter of the membrae equal to 800 nm

Comparing our results expressing the generated water vapor flux with other experimental results (for example the experimental results of Gazagnes et al.[16], see figure 5).

Figure 5 illustrates the variation of the generated vapor flux with the feed inlet temperature for pore size of 800 nm. The results of Gazagnes et al. [16] obtained for ceramic membranes for the same pore diameter are also shown. One can see that the results for the permeate flux given by Poiseuille and Knudsen diffusion models are higher than the experimental ones while the molecular diffusion model seems to be more accurate. The parallel and KM models are also very accurate for the prediction of the permeate flux across the membrane.

4. Results and Discussion

The above described model for the transport phenomena in a Membrane Distillate unit was used to investigate the effects of several variables such as the salinity and the temperature of the feed water on the flow, heat and mass characteristics and on the performance of this desalination system. The molecular diffusion model was used to perform the following simulations. The results are divided in two sections. Section one concerns the basic AGMD configuration (figure 2a) while the second section contains results for the modified configuration (figure 2b).

4.1. Case of the Basic Configuration (Hydrophobic Membrane)

For all these calculations, the following general conditions were considered: R_1= 5 mm, Re_{R1}= 200, Pr = 4.894, Sc = 521.4, L = 50 cm, δ_g = 1-5 mm, W_{se}=0.035 ; T_C=5-40°C, χ =1.5 ; ε = 0.2, 0.7, 0.8, 0.9 ; T_e = 40-80°C.

The temperature and the concentration distributions are given in figures 6 and 7. One can say first that the obtained results (not given here) show that the membrane porosity has no significant effect on the axial velocity profile meaning that the influence of the radial flux across the membrane on the axial velocity is negligible. However, this is not the case for the temperature and salinity profiles as shown in figures 6 and 7. One can see that the mass and the temperature boundary layers grow at a faster rate when the porosity increases. Figure 6 shows that increasing the membrane porosity reduce the non-dimensional temperature in the vicinity of the membrane. In this thin layer, the salinity can be more 4 times the inlet feed water salinity when the material porosity is high (ε = 0.9). This is due to the water evaporation and the water vapor diffusion across the membrane.

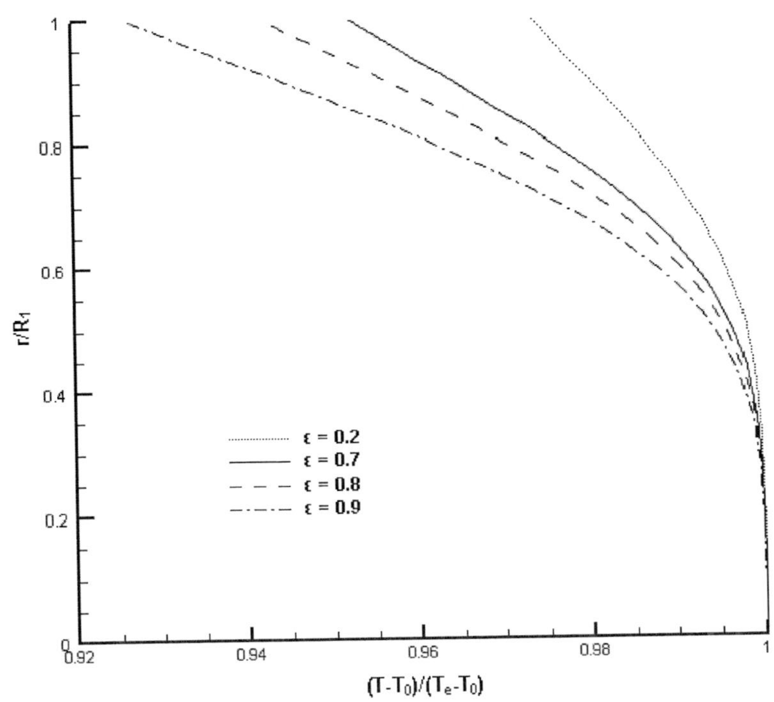

Figure 6. Effect of the porosity on the temperature profile at z = $40R_1$

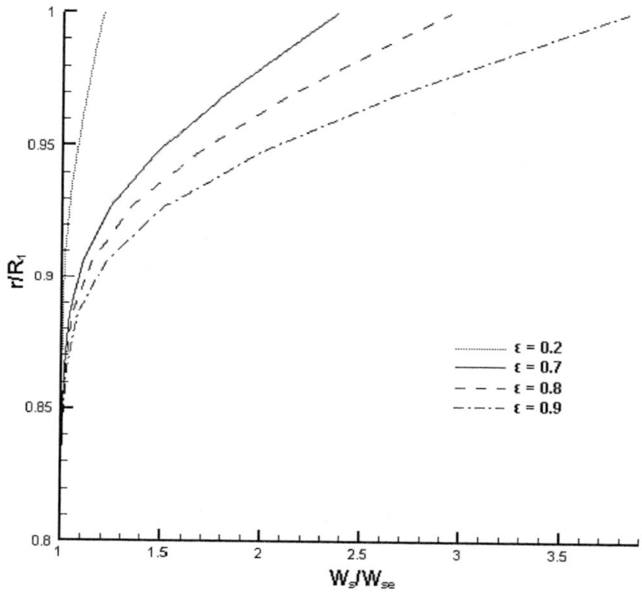

Figure 7. Effect of the porosity on the concentration profile at $z = 40R_1$

The performance of the Air Gap Membrane Distillation configuration depends also on the air-vapor gap width. Figures 8 and 9 show the permeate flux and the thermal efficiency as a function of the air gap width δ_g. The permeate flux increases 4, 3.5 and 3 fold for respectively $\varepsilon = 0.7$, 0.8 and 0.9 as δ_g is reduced from 5 to 1 mm. the increase of the permeate flux can be attributed to the fact that the air gap thermal conductivity is very low relative to the other membrane domains between the hot and the cold membrane sides, which creates a high vapor pressure drop across it.

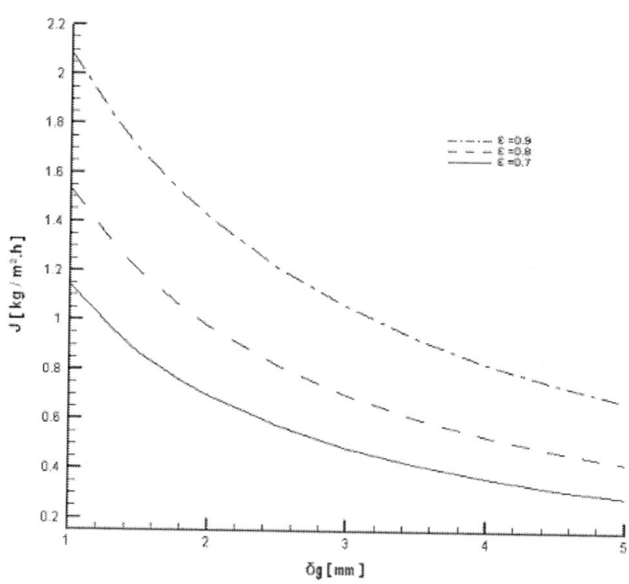

Figure 8. Effect of the air-vapor gap width on the permeate flux

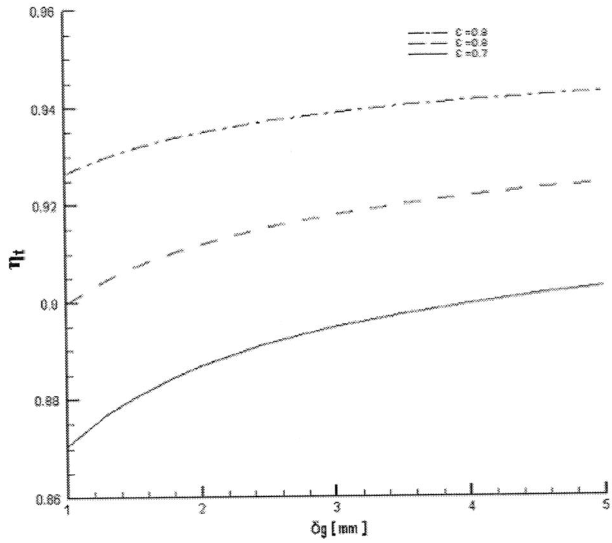

Figure 9. Effect of the air-vapor gap width on the process thermal efficiency

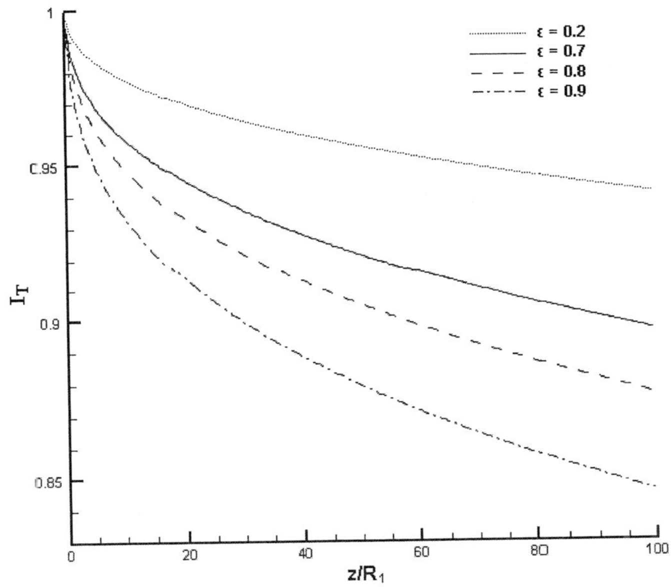

Figure 10. Variation of the temperature polarization index with for the membrane porosity

Analyzing figure 9 reveals that as the air gap width is reduced from 5 to 1 mm, the improvement of the process thermal efficiency remains very small for a fixed porosity. For example, when ε = 0.9, the thermal efficiency is already quite high (about 93 % for δ_g = 1mm) and only a slight improvement is obtained when δ_g increases. Consequently and as concluded by Al Khaibi and Lior [14] narrower air gaps are preferred since they lead to more compact and less expensive system. Figure 10 represents the variation of the temperature polarization index. This index decreases from the inlet to the outlet of the tubular membrane; its effect is more significant for a higher porosity. The value of this index falls to about 0.85

when $\varepsilon = 0.9$. The temperature polarization induces an important reduction of the driving force.

Figure 11 illustrates the concentration polarization index's evolution for different porosities. An increase in the concentration polarization occurs from the inlet to the exit of the tube; this increase is important for higher porosities. Concentration polarization induces a decrease of the vapor flux across the membrane due to salt accumulation on the membrane surface which reduces, mainly at high porosities, the performance of the separation process.

4.2. Case of the Modified Configuration (Hydrophilic-Hydrophobic Membrane)

We study in this section the effect of introducing a hydrophilic layer between the feed water solution and the hydrohobic membrane on the performance of the AGMD process. The total thickness of the composite membrane δ_m is kept constant.

Figure 12 deals with the influence of the hydrophilic membrane's thickness on the permeate flux. As shown, an increase of the hydrophilic membrane thickness induces an increase of the permeate flux. The permeate flux enhancement is important and reaches about 26%.

The minimal value of the flux occurred when δ_{m2}=0 (hydrophobic membrane).

One can also observe that for δ_{m2}/δ_m higher than 0.8, the process performance enhancement in term of permeate flux is very small (within than 3%). In addition and in order to get the maximum value of the product, we consider that:

thikness of the hydrophobic layer of the membrane $\delta_{m1}=1/6\ \delta_m$
thikness of the hydrophilic layer of the membrane $\delta_{m2}=5/6\ \delta_m$

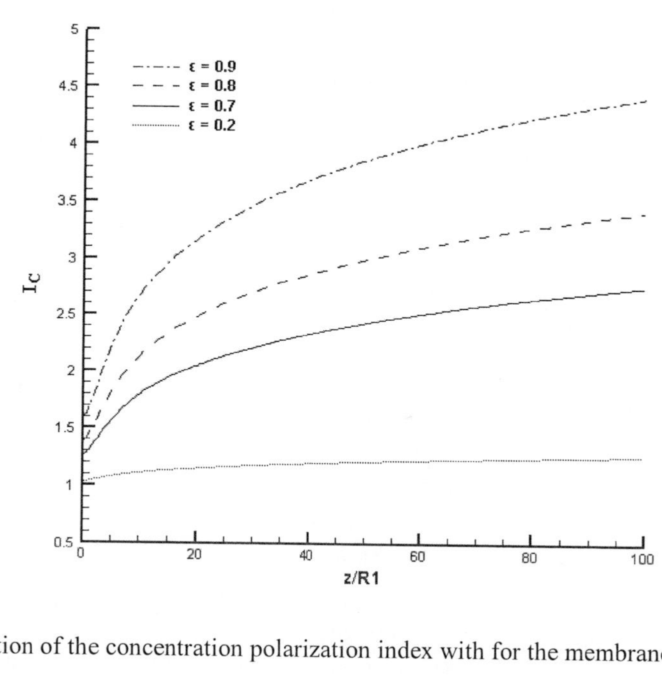

Figure 11. Variation of the concentration polarization index with for the membrane porosity

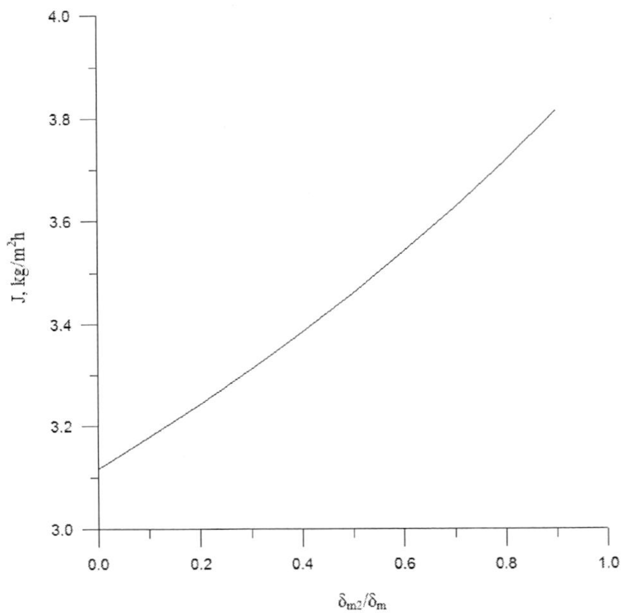

Figure 12. Hydrophilic membrane thickness effect on the permeate flux

For all the calculations, the following general conditions were considered:

R_1=510^{-3}m, δ_m= 510^{-4}m, δ_{m1}=1/6 δ_m, δ_{m2}=5/6 δ_m, ε_1=0.9, ε_2=0.9, χ=1.5, V_e=0.1m/s, T_e=70°C, L=0.5m, R_1=510^{-3}m, T_c=20°C, W_e=0.035, δ_g= 210^{-3}m, δ_p= 210^{-3}m, k_{ma1}=k_{ma2}=0.2 W/mK, k_P=60 W/mK.

Figure 13 presents the effect of the cold plate temperature T_c and that of the inlet temperature of the feed solution T_e on the permeate flux for the two configurations: the composite (hydrophilic-hydrophobic) membrane and the hydrophobic one. One can see that increasing T_e or decreasing T_c has a major positive effect on the vapor flux variation. Such an enhancement of the process performance is more important when a composite membrane is used. In fact, incorporating a hydrophobic-hydrophilic membrane induces a permeate flux enhancement reaching 23% in comparison with the permeate flux produced by the usual hydrophobic membrane. A better enhancement reaching 24.5% may occur when we decrease the cooling temperature T_c.

Figure 14 illustrates the permeate flux J as a function of both the inlet feed water velocity V_e and the air gap thickness δ_g. Here again, a clear improvement of the process performance is observed with the hydrophilic-hydrophobic membrane. The variation of the mass flux crossing the hydrophobic membrane with the inlet velocity of the feed solution seems to be negligible in particular at higher values of V_e. On the other hand and as observed in the previous section, it is not recommended to have a large value of δ_g.

The heat transferred by conduction as a function of the feed water inlet temperature is shown in figure 15. The heat flux Q_c is greatly reduced (from about 1050 to 300 kJ/m^{2}h) if T_e decreases from 80 °C to 40 °C. Concerning the thermal efficiency of the process, one can note that in all cases η_t is higher than 89% which indicates that about 90% of the total heat transferred from the hot side (feed water) to the membrane is used to evaporate the water. The remaining part (around maximum 12%) is lost by conduction. Here, using the hydrophobic-hydrophilic membrane induces only a slight enhancement of the thermal efficiency.

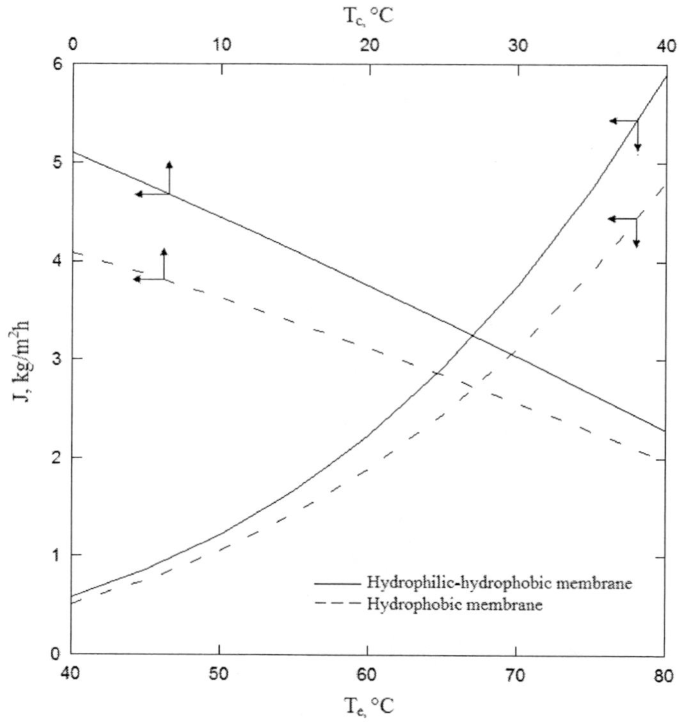

Figure 13. Effect of T_e and T_c on the permeate flux for both types of membranes

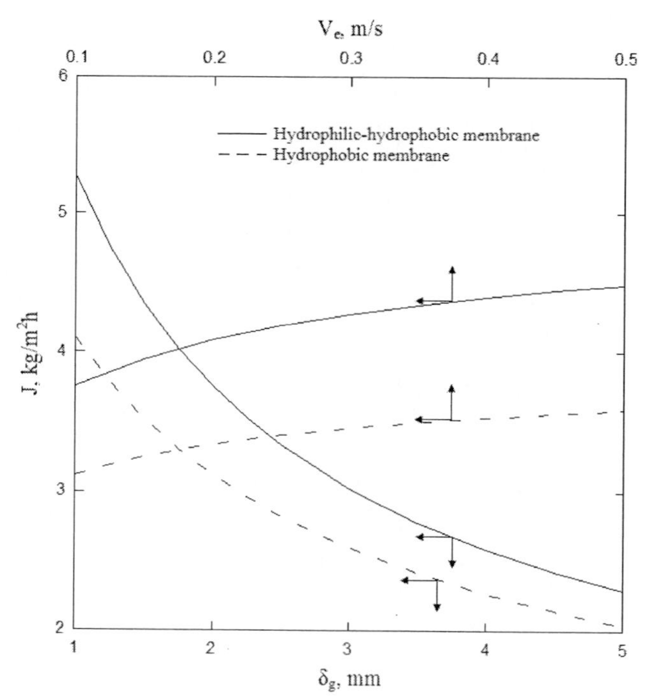

Figure 14. The permeate flux as function of inlet velocity of the hot solution and the air gap thickness for the two types of membranes

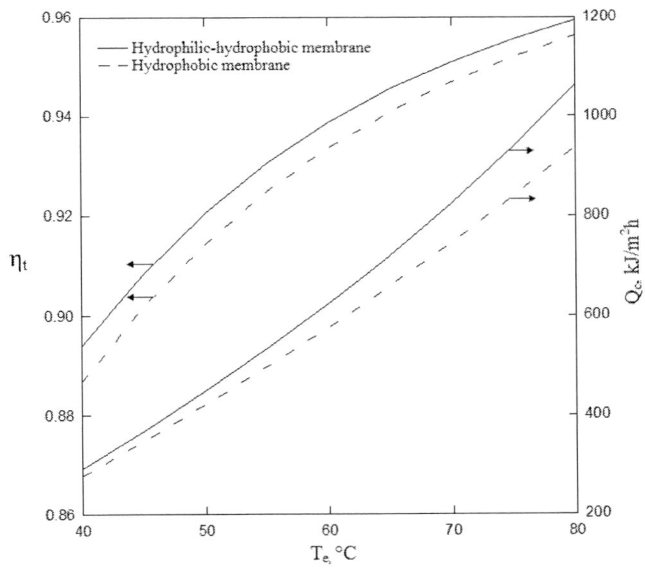

Figure 15. Evolution of thermal efficiency and the conductive heat flux as a function of the saline inlet temperature

Figure 16 illustrates the influence of the cooling plate temperature T_c on both the thermal efficiency and the conductive heat flux. Lowering T_c has a great effect on the increase of the conductive heat flux due to the increase of the temperature difference across the membrane. On the other side, the thermal efficiency which is higher than 94 % for the presented cases is slightly affected. Besides, a significant enhancement of η_t is observed when the composite membrane is used.

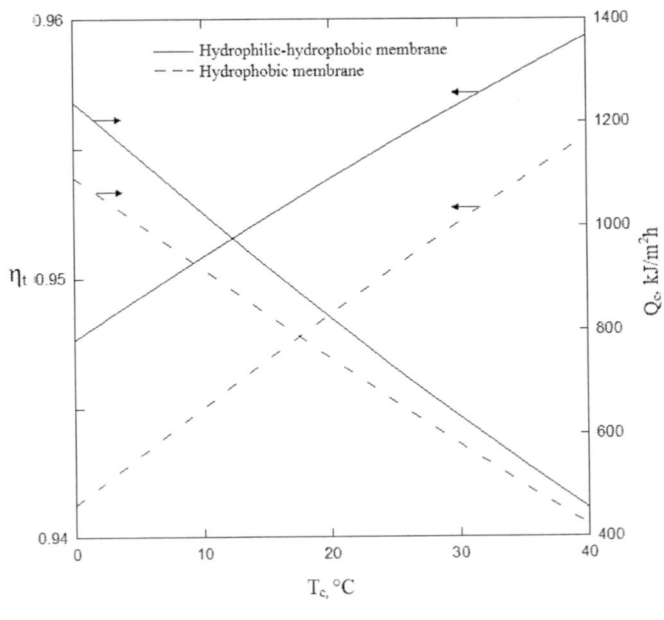

Figure 16. Cooling temperature effects on the thermal efficiency and on the conductive heat flux for the two types of membranes

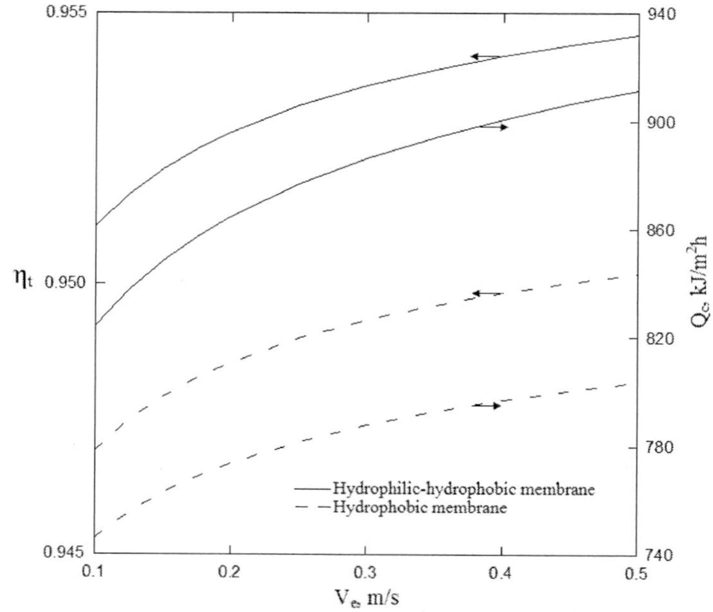

Figure 17. The feed water inlet velocity effects on the thermal efficiency and the conductive heat flux

In figure 17, the thermal efficiency and the conductive heat flux are plotted as the functions of the feed circulation velocity at the MD unit inlet. For the same velocity magnitude increase from 0.1 to 0.5 m/s, η_t changes very slightly but Q_c increases from 840 to about 910 kJ/m^{2}h.

With a turbulent flow, the concentration and the temperature polarization phenomenon may be reduced leading to the enhancement of thermal performance of the system.

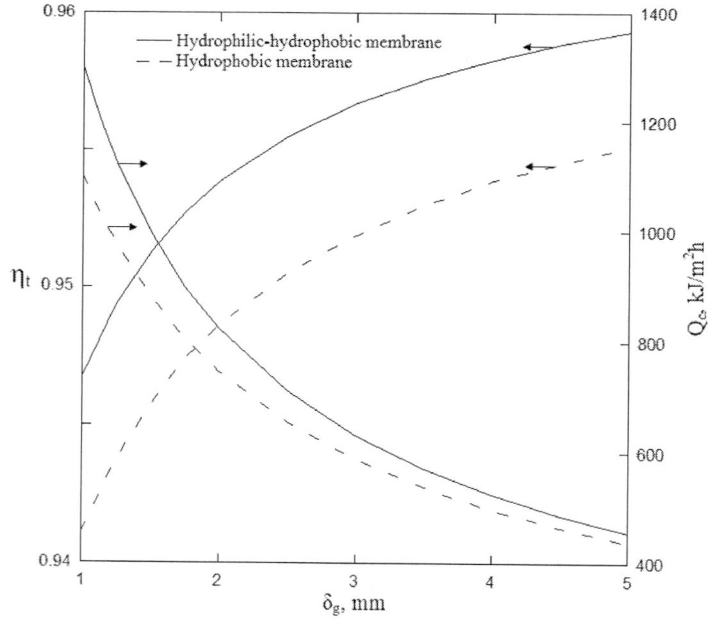

Figure 18. Air gap width effects on the thermal efficiency and the conductive heat flux

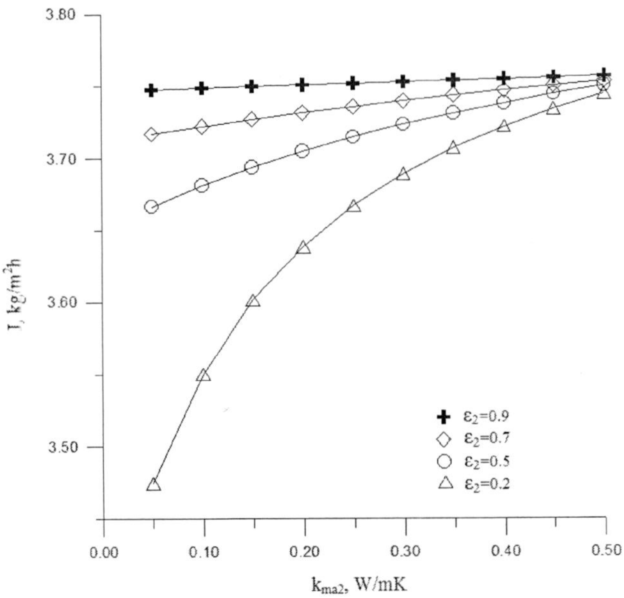

Figure 19. Effects of the thermal conductivity of the hydrophilic material on the permeate flux

Analyzing figure 18 reveals that as the air gap width is reduced from 5 to 1mm, the heat transferred by conduction increases 2.8 fold for the hydrophilic-hydrophobic membrane and 2.5 fold for the hydrophobic one. A slight improvement of the thermal efficiency is obtained when adopting a hydrophilic-hydrophobic membrane.

It is of interest here to mention that the general tendency of the results found here was also observed by Al Klaibi and Lior [14] who studied the transport phenomena in an AGMD unit using cartesian geometry.

In the last figure (figure 19), our attention is turned to analyze the effect of the hydrophilic membrane properties namely its thermal conductivity and its porosity. Figure 19 illustrates the effect of the thermal conductivity of the hydrophilic material on the permeate flux for different porosity of the hydrophilic matrix (the porosity of the hydrophobic layer is fixed equal to 0.9). It is clearly seen that a higher porosity induces a higher vapor generation rate. The worst case is obtained with a low thermal conductivity and a low porosity. Nevertheless and in all cases, the permeate flux variation is observed to be very small (approximately between 3.50 and 3.75 kg/m^2h).

5. Conclusion

We presented in this chapter an overview and a state of the art on the Membrane Distillation process. We performed a theoretical analysis on the performance of a desalination system based on the AGMD configuration. The governing equations expressing the conservation of mass, momentum, energy and species in the feed saline water as well as the energy and mass balances on the both sides of the separation membrane were developed and solved numerically using the finite volume method. The mass transfer mechanisms across the membrane were discussed.

The theoretical and numerical models were validated and applied to simulate the fluid flow with heat and mass transfer for two different configurations. The first one corresponds to the basic AGMD process. The second one concerns a modified AGMD unit characterized by a composite hydrophilic-hydrophobic membrane. The results analyzed the effect of the operating parameters such as the velocity and the temperature of the hot feed water at the inlet of the unit, the properties of the membrane and the cooling plate temperature and air gap thickness on the production rate of the vapor and the efficiency of the process.

The main results can be summarized as follows:

This desalination process is very efficient. More than 90% of the total heat flux coming from the hot side of the unit (feed water side) is converted into latent heat for the generation of the permeate (water vapor).
the process performance is very sensitive to several parameters in particular to the porosity of the hydrophobic membrane. An increase of this parameter induces a fast growth of the mass and temperature boundary layers of the feed solution.
the permeate flux as well as the thermal efficiency of the process are higher for a thinner air gap.
A significant improvement of the permeate flux and the thermal efficiency are obtained by the incorporation of the composite hydrophilic-hydrophobic membrane.

ACKNOWLEDGMENTS

The first author is grateful for the support provided by the Research Center of the College of Engineering, King Saud University (Ref 70/428) to carry out the present work.

NOMENCLATURE

Symbols

C_p	Specific heat [$Jkg^{-1}K^{-1}$]
D_s	Diffusion coefficient of NaCl [m^2/s]
$D_{v/a}$	Coefficient of vapor-air mass diffusion [m2/s]
g	Acceleration of gravity [m/s^2]
h_{fg}	Latent heat of evaporation [J/kg]
J	length-averaged permeate flux at the hot side of the membrane [kg/m^2h]
J_v	local permeate flux at the hot side of membrane, in vapor phase [kg/m^2s]
K	permeability of the membrane
k	Thermal conductivity [W/mK]
L	Membrane length [m]
M_v	Molar mass of water vapor [kg/kmol]
P	pressure [Pa]
Pr	Prandtl number
Q_C	Conductive heat flux [kJ/m^2h]

Q_L	Latent heat flux [kJ/m^{2}h]
Q_{Sens}	Sensible heat transfert [kJ/m^{2}h]
Q_T	Total flux [kJ/m^{2}h]
R_u	Universal gaz constant [J/kmol K]
R_1	Inner radius of the membrane [m]
R_2	Outer radius of the membrane [m]
Re	Reynolds number
R_g	Thermal resistance of the air gap
R_f	Thermal resistance of the condensate film
R_m	Thermal resistance of the membrane
R_p	Thermal resistance of the cooling plate
S	Surface area for heat transfer = $2\pi rL$
Sc	Schmidt number
T	temperature [°C]
V_r	radial velocity component [m/s]
V_z	axial velocity component [m/s]
V_e	Velocity at the inlet of the membrane [m/s]
W_s	Mass fraction of NaCl
W_{se}	Mass fraction of NaCl at the entrance
r	Coordinate normal to the solution flow [m]
r_p	Pore radius of the membrane [m]
z	Coordinate along the solution flow [m]
β_T	Coefficient of thermal expansion [K^{-1}]
β_W	Coefficient of mass expansion [kg^{-1}]
μ	Dynamic viscosity [kgm^{-1}s^{-1}]
ν	Cinematic viscosity [m^{2}s^{-1}]
ρ	Density [kgm^{-3}]
ε	Porosity
χ	Tortuosity
δ	Thickness or width [m]
δ_m	Thickness of the membrane
η_t	Process thermal efficiency
Φ	Heat flux = Q.S [kJ/h]

Subscripts

a	air
c	cooling plate
e	inlet
f	Condensate film
g	Air gap
K	Knudsen diffusion
M	Molecular diffusion
P	Poiseuille diffusion

Continued	
m	membrane
m_1	Hydrophobic membrane
m_2	Hydrophilic membrane
ma	Membrane material
moy	Average
s	Saline solution
T	total
v	vapor
0	Reference
1 saline solution-membrane interface	
2 membrane-air gap interface	
2' hydrophilic-hydrophobic membrane interface	
3 air gap-condensate water interface	
4 condensate water-cooling plate interface	
5 outer side of the cooling plate	

REFERENCES

[1] Meindersma, GW; Guijt, CM; de Haan AB. '*'Desalination and water recycling by air gap membrane distillation''*. Desalination, 2006, 187, 291-301.

[2] Rommel, M; Koschilowski, J; Wighaous M. '*'Solar Driven Desalination systems based on Membrane desalination''*, in Solar Desalination for the 21st Century, Editors Rizzuti et al. 2007, 247-257.

[3] El-Bourawi, MS; Ding, Z; Ma, R; Khayat MA. ''framework for better understanding membrane distillation separation process''. Journal of Membrane Science, 2006, 285, 4-29.

[4] Liu, GL; Zhu, G; Cheung, GS; Leung GW. ''Theoretical and experimental studies on air gap membrane distillation'', Heat and Mass Transfer, 1998, 34, 329-335.

[5] Banat, FA; Simandl J. "Desalination by membrane distillation: A parametric study". Separation Science and Technology, 1998, 33.201-226.

[6] Guijt, GM; Meindersma, GW; Reith, T; De Haan AB. "Air gap membrane distillation: 2. Model validation and hollow performance analysis". Separation and Purification Technology, 2005, 43, 245-255.

[7] De Pinho, MN; Semião, V; Geraldes, V. "Integrated modeling of transport processes in fluid /nanofiltration membrane systems". Journal of Membrane Science, 2002, 206, 189-200.

[8] Phattaranawik, J; Jiraratananon, R. *Direct contact membrane distillation: effect of mass transfer on heat transfer*, Journal of Membrane Science, 2001, Vol. 188, no 1, 137-143.

[9] Laganà, F; Barbieri, G; Drioli, E. '*Direct contact membrane distillation: modeling and concentration experiments*', Journal of Membrane Science, 2000, Vol. 166, no 1, 1-11.

[10] Tamaszewska, M. Membrane Distillation Examples of Applications in Technolgy and Environment Protection, Polish Journal of Environmental Studies, 2000, Vol. 9, no 1 27-36.

[11] Dyterskii, YI; Karlin, YV; Kropotov, BN. *Prospects for using Membrane Distillation for preprocessing liquid radioactive wastes*, Atomic Energy, 1993, Vol. 75, no 5, 345-350.

[12] Macedonio, F; Drioli, E. '*Pressure-driven membrane operations and membrane distillation technology integration for water purification*', Desalination, 2008, Vol 223, no 1-3, 396-409.

[13] Khayet, M; Godino, P; Mengual, JI. '*Theory and experiments on sweeping gas membrane distillation*', Journal of Membrane Science, 2000, Vol. 165, no 2, 261-272.

[14] Alklaibi, AM; Lior, N. "*Transport analysis of air gap membrane distillation*". Journal of Membrane Science, 2005, 255.239-253.

[15] Cerneaux, S; Strużyńska, I; Kujawski, WM; Persin, M; Larbot, A. '*Comparison of various membrane distillation methods for desalination using hydrophobic ceramic membranes*', Journal of Membrane Science, 2009, Vol. 337, no 1-2, 55-60.

[16] Gazagnes, S; Cernaux, M; Persin, E; Prouzet, A. Larbot, Desalination of sodium chloride solutions and sea water with hydrophobic ceramic membrane, Desalination, 2007, 217, 260-266.

[17] Blanco Gálvez, J; García-Rodríguez, L; Martín-Mateos, I. *Seawater desalination by an innovative solar-powered membrane distillation system: the MEDESOL project* Desalination, 2009, Vol. 246, no 1-3, 567-576.

[18] Khayet, M; Matsuura, T; Mengual, JI; Qtaishat, M. *Design of novel direct contact membrane distillation membranes*, Desalination, 2006, Vol. 192, no 1-3, 105-111.

[19] Safavi, M; Mohammadi, T. *High-salinity water desalination using VMD*, Chemical Engineering Journal, Volume 149, Issues 1-3, 1 July 2009, Pages 191-195.

[20] Fath, HES; Elsherbiny, SM; Hassan, AA; Rommel, M; Wieghaus, M; Koschikowski, J; Vatansever, M. '*PV and thermally driven small-scale, stand-alone solar desalination systems with very low maintenance needs*', Desalination, Volume 225, Issues 1-3, 1 May 2008, Pages 58-69.

[21] Koschikowski, J; Wieghaus, M; Rommel, M; Subiela Ortin, V; Peñate Suarez, B. J. Rosa Betancort Rodríguez *Experimental investigations on solar driven stand-alone membrane distillation systems for remote areas*, Desalination, Volume 248, Issues 1-3, 15 November 2009, Pages 125-131.

[22] Martínez-Díez, L; Vázquez-González, MI; Florido-Díaz, FJ. *Study of membrane distillation using channel spacers*, Journal of Membrane Science, Volume 144, Issues 1-2, 10 June 1998, Pages 45-56.

[23] Zhu, C; Liu, GL; Cheung, CS; Leung, CW; Zhu, ZC. '*Ultrasonic stimulation on enhancement of air gap membrane distillation*', Journal of Membrane Science, 1999, Volume 161, Issues 1-2, 1 August, Pages 85-93.

[24] Tzahi, Y; Cath, V; Dean Adams, Amy, E. Childress, *Experimental study of desalination using direct contact membrane distillation: a new approach to flux enhancement*, Journal of Membrane Science, Volume 228, Issue 1, 1 January 2004, Pages 5-16.

[25] Bonyadi, S; Shung Chung, T. '*Flux enhancement in membrane distillation by fabrication of dual layer hydrophilic–hydrophobic hollow fiber membranes*', Journal of Membrane Science, Volume 306, Issues 1-2, 1 December 2007, Pages 134-146.

[26] Criscuoli, A; Drioli, E. '*Energetic and exergetic analysis of an integrated membrane desalination system*', Desalination, Vol. 124, no 1-3, 243-249, 1999.

[27] Shawaqfeh, AT; Farid. MM. "New development in the theory of heat and mass transfer in solar stills", Solar Energy, 1995, 55, 527-538.

[28] Goosen, MFA; Sablani, SS; Shaya, WT; Paton, C; AlHanai, HA. *"Thermodynamic and economic considerations in solar desalination". Desalination*, 2000, 129, 63-89.

[29] Bouguecha, S; Chouikh, R; Dhahbi. M. "Numerical study of the coupled heat and mass transfer membrane distillation". Desalination, 2002, 15, 245-252.

[30] Bejan, A. *Convective Heat Transfer*, Third Edition, Wiley, 2004, 694.

[31] Alklaibi, AM; Lior, N. *"Membrane-distillation desalination: Status and potential"* Desalination, 2005, 171, 111-131.

[32] Ding, Z; Ma, Z; Fane, AG. A new model for mass transfer in direct contact membrane distillation, Desalination, 2002, 151, 217-227.

[33] Versteeg, H.K; Malalasekera, W. *"An introduction to Computational Fluid Dynamics: Finite Volume Method",* Second Edition, Pearson –Prentice Hall, London, 2007.

In: Desalination: Methods, Cost and Technology
Editor: Irena A. Urbonienė pp. 175-208

ISBN: 978-1-61668-909-4

Chapter 5

MEMBRANE DESALINATION: METHODS, COST, AND TECHNOLOGY

N. K. Saha and A. Bhattacharya
Reverse Osmosis Discipline, Central Salt and Marine Chemicals Research Institute
G. B. Marg, Bhavnagar, Gujarat, India

ABSTRACT

'Necessity drives invention' that is the basis of Desalination Technology. In this chapter, principles of different membrane desalination techniques (viz. Electrodialysis, Reverse osmosis, membrane distillation) are discussed. The emphasis of the chapter is on reverse osmosis technology. The membrane-based technology is dealt with special attention on technology and economics; also membrane preparation methods and materials got brief attention. Present scenario of the technology is brightened by comparing other technologies with it. The different energy sources are also coupled with the technology. The economics of the technology is discussed. Different factors influencing the cost of production are discussed. RO cost prediction techniques are also sketched.

INTRODUCTION

Technological advancement in the past few decades coupled with better acquisition, processing and interpretation techniques have made the Membrane desalination an important tool for exploration. Desalination is one of the technologies of water managements. As the world's population grow, access to fresh water declines. There are places where water scarcity is not a problem today; it might become a problem tomorrow. Reverse osmosis (RO), membrane distillation (MD), membrane electrodialysis (ED) are different separation techniques when applied for water desalination can be grouped under the common head - *membrane desalination techniques*. The common point in all these techniques is the use of membrane and the difference lies mainly in the nature of the driving force carrying out the separation. Considering the installed capacity, RO shares the credit of producing highest volume of water in the World (43.5%) along with *multi stage flash* process (thermal desalination) and second to the same when comes to sea water desalination particularly [1]. RO has several advantages over thermal desalination techniques:

- low energy cost
- smaller foot print
- low operating temperature
- modular design
- low water production cost

PRINCIPLES OF TECHNIQUES RELATED TO DESALINATION

Reverse Osmosis

The two processes viz. Osmosis and Reverse Osmosis are controlled by thermodynamic function i.e. 'chemical potential' of the system. It is essentially a driving force expressed as a change in free energy of the system because of the change in the composition of the system. Chemical potential of solvent in solution (higher concentration) is always lower than when it is in pure state (lower concentration).

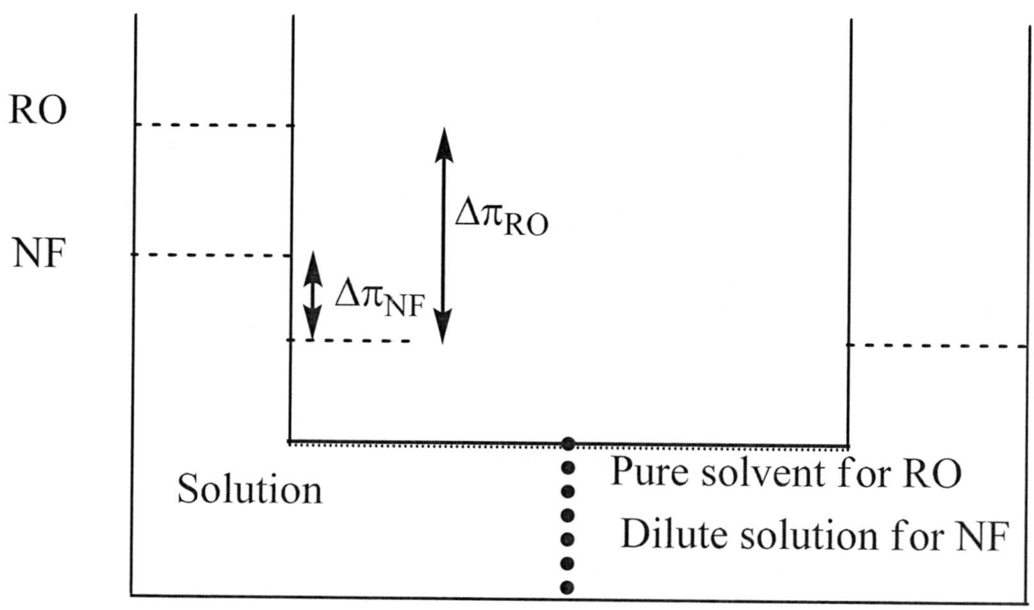

Figure 1. Osmotic pressure for RO and NF

In osmosis, solvent flows through a semi permeable membrane from high solvent concentration (dilute solution) region to low solvent concentration (concentrated solution) region (Figure 1). Important point to note that only solvent molecules movement through the barrier occurs. Osmotic pressure is equal to net hydraulic pressure require to apply on low solvent concentration side to bring the process in equilibrium i.e. a situation of no net flow (hydraulic pressure equals osmotic pressure). If the hydraulic pressure applied is greater than the equilibrium pressure or osmotic pressure, solvent flow will revert its direction corresponding to the osmotic flow i.e. solvent flow will take place along the solution concentration gradient. The resulting process is reverse osmosis that is now commonly applied in seawater desalination. In RO, hydraulic pressure is the driving force for solvent transport and since seawater of say 36000 ppm salt concentration displays typically 376 psi (26.75bar) osmotic pressure, sea water desalination by RO would require more than 376 psi hydraulic pressure. *Nanofiltration* (NF) having more permeable membrane and more concentrated permeate sometimes considered as the pretreatment of seawater RO requires less energy because of lower osmotic pressure generated across the membrane.

The main two characteristics of a membrane process are flux and rejection. If a RO membrane is considered as permeating water only, the water flux can be written as

$$J_w = A(\Delta P - \Delta\pi)$$

Where A is the permeability coefficient, and ΔP and $\Delta\pi$ are the hydraulic pressure and osmotic pressure difference across the membrane. The value of the permeability coefficient varies between 3×10^{-3} to 6×10^{-5} $m^3.m^{-2}.h^{-1}.bar^{-1}$for RO [2].

If the membrane is little permeable to solute too as in most of the cases then the solute flux can be written as in simplified form

$$J_s = B\Delta c_S$$

Where B is solute permeability coefficient and Δc_S is the solute concentration difference across the membrane.

The important transport models to explain the desalination mechanism are solution-diffusion, solution-difffusion imperfection, Kimura-Sourirajan, Irreversible thermodynamics, Extended Nernst-Planck equation. According to solution-diffusion theory, the widely accepted approach for explaining transport through RO membranes, for a given component in solution, solubility of it in membrane and its diffusibility through the membrane determines the selectivity of the process. However, diffusion and pore flow are considered in the solution-diffusion imperfection model. In Kimura- Sourirajan theory, preferential sorption of solvent at the membrane-solution interface is assumed. It then moves through the capillaries of the membrane by viscous flow in which the flux is directly proportional to the effective pressure. The solute passes through the pores of the membrane by diffusion. The membrane is treated as a 'black box' in the irreversible thermodynamics. The solvent flux is proportional to the pressure gradient across the membrane. The solute moves through the membrane by convection and diffusion. In the Extended Nernst Planck Model convection, diffusion and electrical potential gradient control the movement of ions through the membrane pores. The ions show selectivity by Donnan partitioning and steric effect.

Electrodialysis

In electrodialysis, electrical potential difference is applied as the driving force to bring about the separation of charged species from uncharged one through an electrically charged membrane. Figure 2 depicts the principle of electrodialysis. In this process a set of anion exchange membrane allowing through pass of anions only (restricting the cations) and cation exchange membrane allowing through pass of cations only (restricting the anions) are placed alternatively between two electrodes– cathode and anode in a cell containing ionic solution. When electrical potential field is applied, all the cations move towards the cathode and anions move towards the anode. In this process, cations pass through the cation exchange membrane but are restricted by the anion exchange membrane and vice versa. Thus, the alternative chambers in the cell become either concentrated with the charged species or dilute and separation is achieved.

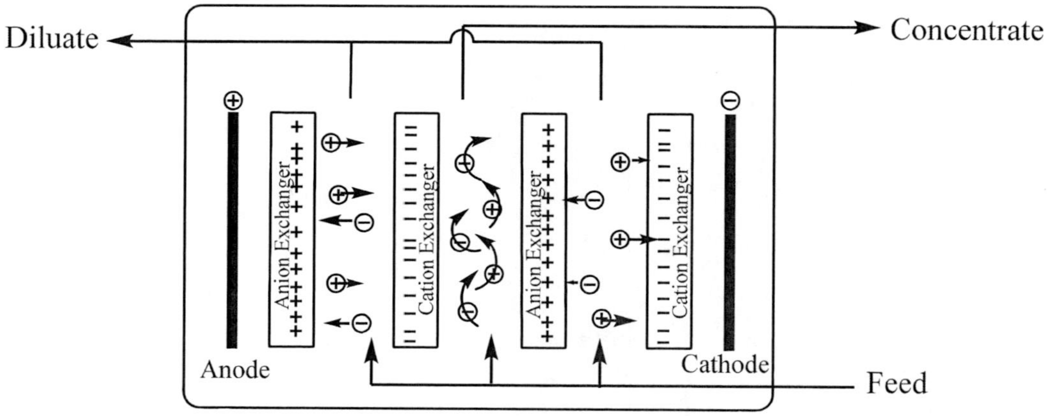

Figure 2. Schematic presentation of electrodialysis

Membrane Distillation

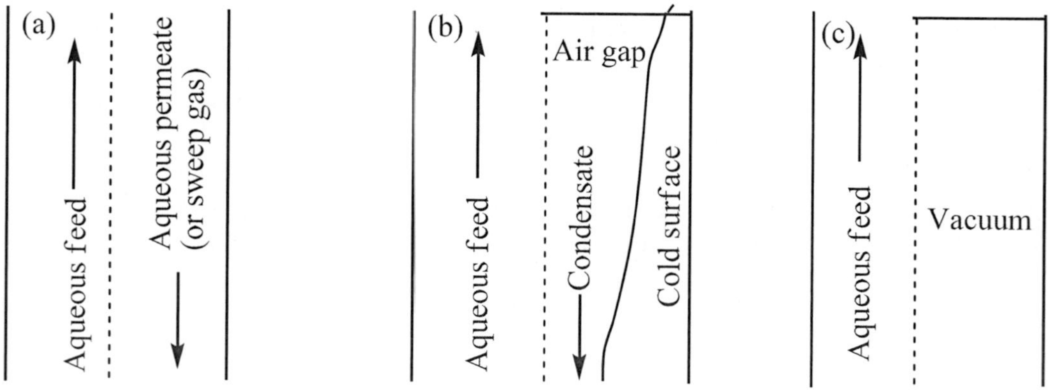

Figure 3. Membrane distillation principle: (a) Direct contact (or Sweep gas) type, (b) Air gap type, (c) Vacuum type (…… membrane) (modified from [3])

In membrane distillation, an aqueous feed or solution is brought into contact with a microporous hydrophobic membrane. The liquid does not penetrate or wet the membrane due to membrane hydrophobicity. The membrane provides support to the liquid-vapour interphase. Difference in vapour pressure of the moving species across the membrane is the driving force for separation. Different MD configurations are depicted in figure 3[3].

MEMBRANES

Properties and Preparation

The choice of membrane material and structure varies over applications depending upon the principle of separation involved. For example, microfiltration /ultrafiltration require porous membranes where the permselectivity is determined by the pore sizes; membrane material plays important role in determining fouling and cleaning properties. In pervaporation and gas separation, non-porous membranes are required since diffusion is the transport mechanism involved. Here, membrane material determines permselectivity. Reverse osmosis and nanofiltration can be considered as borderline case between the porous and non-porous dense membranes. Electrodialysis membranes require high charge density and for membrane distillation, it requires to be non-wetting hydrophobic if applied for aqueous solution. Other than these above-mentioned criteria, membrane material must show good mechanical, chemical, and thermal stability. For aqueous applications, hydrophilic materials are always preferred over the hydrophobic one because of their fewer tendencies towards adsorption, which reflects their less membrane-fouling tendency.

Cellulosic polymers like cellulose acetate, cellulose triacetate form a very good class of hydrophilic materials for RO membrane preparation. Membrane prepared from them show high water permeability and low salt diffusivity. The main disadvantages associated with these materials are their low chemical stability, low temperature resistance, and highly prone to biological attack. Other materials like polyimides, polybenzimidazole, polybenzimidazolones, polyamidehydrazide and especially polyamides are used for RO membrane preparation. Table 1 compares hydrophilicity of different membrane materials. The contact angle value is influenced by several factors and thus may vary in literatures but these values will give reader an idea about hydrophilicity of membranes prepared from different polymers.

There are different techniques for polymeric membrane preparation that includes sintering, stretching, track-etching, template leaching, phase inversion and coating. Phase inversion is the most commonly used technique for integrally skinned synthetic polymeric membrane preparation and thin film coating for composite membrane.

Table 1. Hydrophilicity/hydrophobicity of membrane materials

Material	Structure	Value	Ref.
Cellulose		55.9°*	[4]
Polyacrylonitrile (PAN)		49°	[5]
Polyethersulfone (PES)		65-76°	[6]
Polysulfone (PS)		67°	[7]
Polyamide (PA)		65-74°	[8,9]
Polyetherimide (PEI)		73-80°	[10]
Polyvinylidenefluoride (PVDF)		~115°	[11]

* Contact angle value of cellulose acetate with ~55% acetyl content

Phase separation technique involves transfer of phase of a polymer solution from liquid to solid in a controlled fashion. The main requirement is to choose the solvent (or mixture) and non-solvent of the polymer (for immersion precipitation). Membrane is formed by precipitation of polymer in solid film form. Precipitation can be induced by different methods like solvent evaporation, precipitation from vapor phase, controlled evaporation, thermal, and immersion precipitation. Immersion precipitation is the mostly followed method. In this method, polymer solution is cast on a porous (e.g., nonwoven fabric) or nonporous (e.g., glass) media using a casting knife or roller maintaining the required as cast thickness and immersed in a bath containing another liquid which is a nonsolvent with respect to the polymer but having good affinity for the solvent. In the bath, solvent-nonsolvent exchange takes place, and since the polymer is insoluble in nonsolvent, with the increase in nonsolvent concentration in polymer solution phase, polymer precipitates out and the asymmetric membrane is formed. The asymmetric membrane is of thin-skinned layer morphology on the top and the nodules are in the depth of the membrane. The nascent membrane formed can be used directly or after post treatment. Membrane morphology can be varied by varying polymer, solvent and nonsolvent compositions and preparation conditions. Concentration of polymer, choice of solvent and nonsolvent, presence of additives and/or nonsolvent in casting solution, presence of solvent and other additives in gelation bath, polymer solution viscosity and temperature, casting temperature, humidity, cast thickness, evaporation time, gelation time, gelation bath temperature are different important parameters determining the morphology of membrane formed.

Latter properties are determined by the choice of membrane materials. In integrally asymmetric membranes, the top and the sublayer are made from the same materials and the preparation conditions are maintained to yield this sort of asymmetric structure. In composite membrane, the top layer is coated separately on the sublayer that is usually a porous ultrafiltration membrane and can be prepared by immersion precipitation method. The advantage of composite membrane over the integrally skinned asymmetric membranes is that both the layers- top and the sub can be customized separately to obtain best performance. Membrane coating can be done by several methods including dip coating, spray coating, spin coating, interfacial coating, in-situ polymerization, plasma polymerization, grafting. Interfacial polymerization process is mostly followed for composite membrane preparation. In this technique, thin film is produced on the sublayer by reacting two monomers (viz. diamines and aryl halides), each dissolved in a solvent, which are mutually immiscible, and the polymerization reaction-taking place at the interface. The thin film composite membrane so formed is given heat treatment (curing) before final washing. The chemical nature of the thin film in most of the cases is polyamide (PA) though other chemicals like sulfonated polysulfone, polyphenylene oxide, polyvinyl alcohol with different crosslinkers, sulfonated poly (phthalazinone ether sulfone ketone), or chitosen, derivatives, polyamide-urethane, have also been explored as available alternatives. Examples of interfacial polymerization are shown in Figure 4. Figure 5 shows some of the SEM images of porous sublayer and interfacially polymerized thin film coated membranes.

Ion-exchange membranes used in electrodialysis require to be highly permselective and electrically less resistant, obviously mechanically and chemically stable [14]. These membranes are broadly catagorized into two classes, viz., homogeneous and heterogeneous and depending upon their degree of heterogeneity they are further classified as

- homogeneous ion-exchange membranes
- interpolymer membranes
- microheterogeneous graft- and block- polymer membranes

- snake-in-the-cage ion-exchange membranes
- heterogeneous ion-exchange membranes

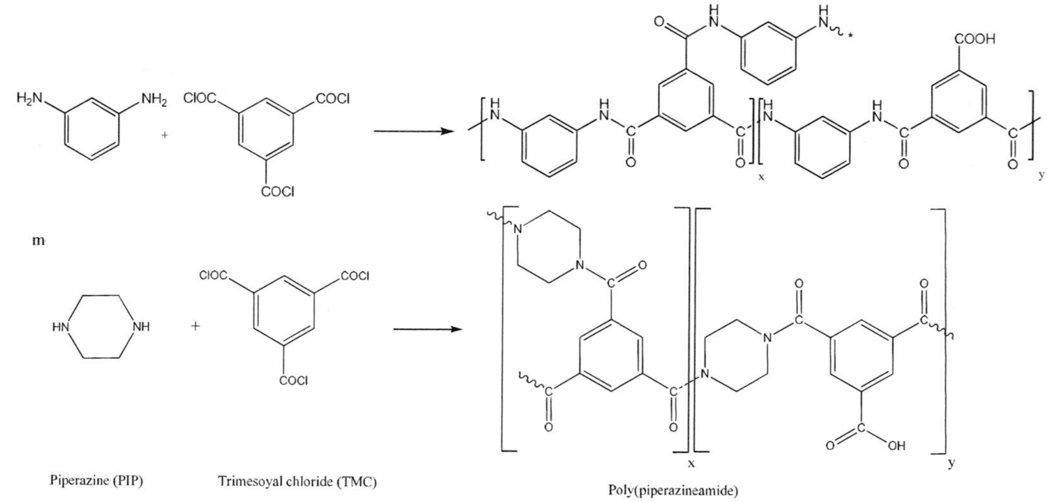

Figure 4. Interfacial polymerization reactions

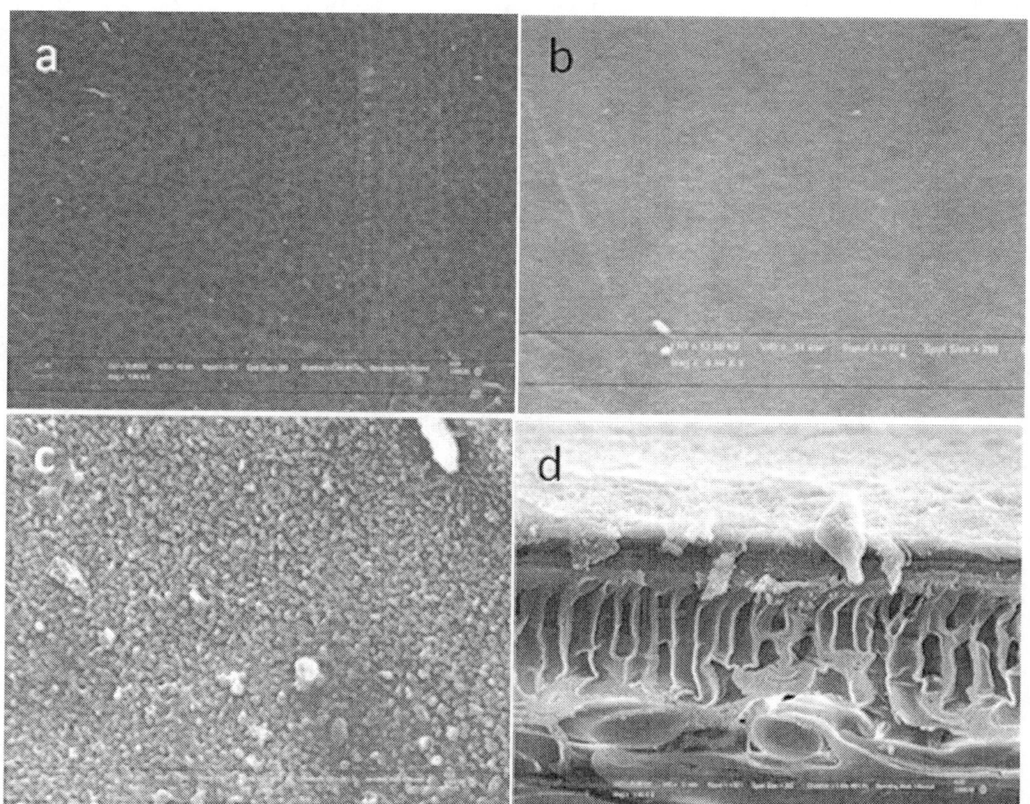

Figure 5. SEM images of sublayer (a), surface (b, c) and cross-section (d) of composite membrane

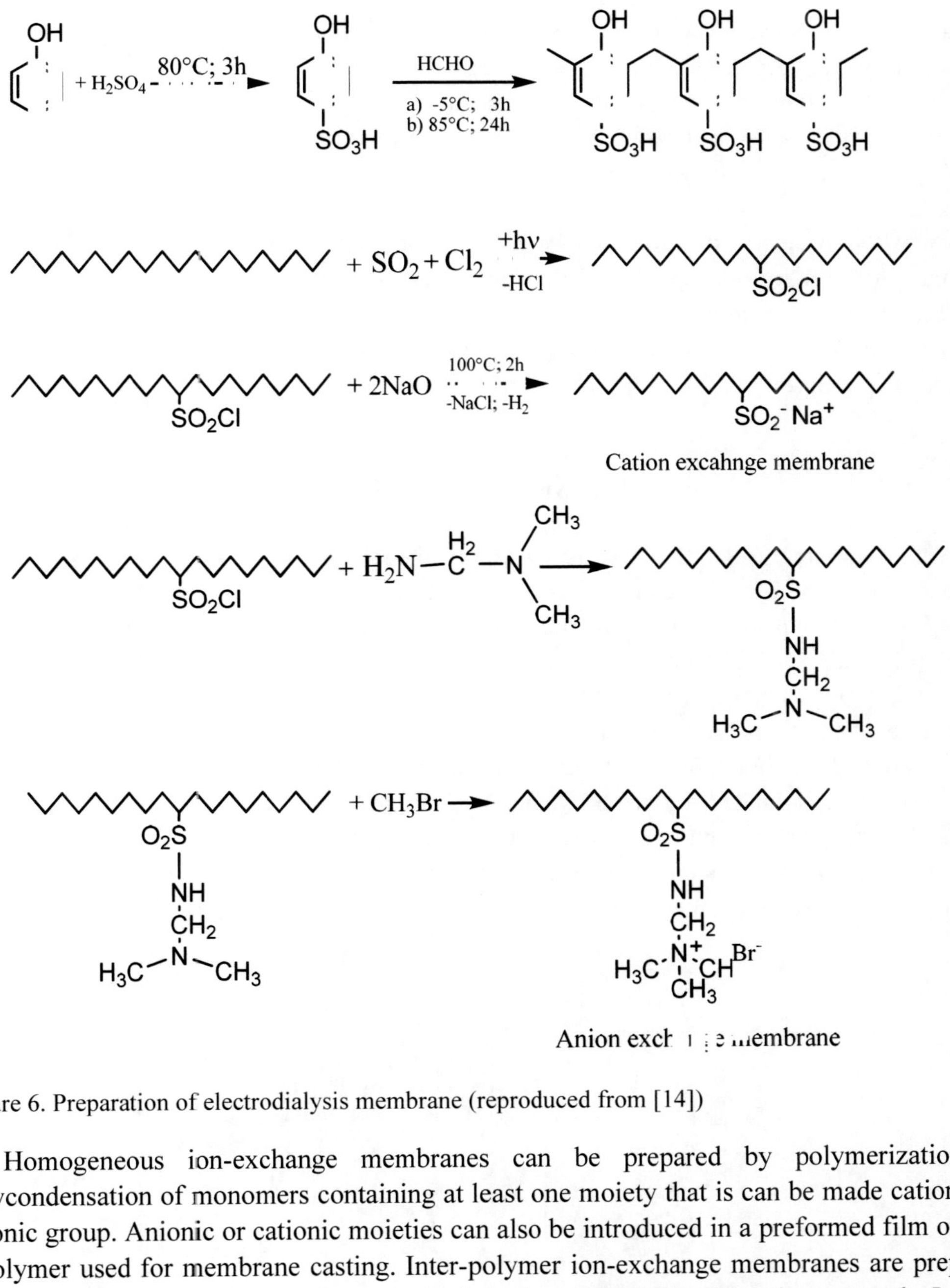

Figure 6. Preparation of electrodialysis membrane (reproduced from [14])

Homogeneous ion-exchange membranes can be prepared by polymerization or polycondensation of monomers containing at least one moiety that is can be made cationic or anionic group. Anionic or cationic moieties can also be introduced in a preformed film or into a polymer used for membrane casting. Inter-polymer ion-exchange membranes are prepared from blend of two polymers that are cross-linked to yield interpenetrating network. It adds good mechanical strength to the polymer. Different polymer blends tried to prepare such polymers are polyvinyl alcohol (PVA) with polyelectrolytes like poly(sodium styrenesulfonate) (PSSNa), poly(styrenesulfonic acid) (PSSH), poly(acrylic acid) (PAA) for cation exchange membranes and polyethyleneimine (PEI), poly(diallyldimethylammonium chloride) (PDDMACl) and poly(1,1-dimethyl-3,5-dimethylenepiperidinium chloride) (PDDPCI) for anion exchange membranes. Ion-exchange membrane preparation by grafting gives the advantage of controlling membrane composition and sometimes with two highly incompatible polymeric combinations; if grafting is done on polymeric film, it overcomes the shaping problem also. Membrane can be converted into cation- or anion-exchange one by varying the post treatment, e.g., sulfochlorination for cation exchange and amination for anion exchange membrane. Snake-in-the-cage ion-exchange membranes can be prepared by saturation of a suitable ion-exchange resin with an organic counter ion that can be polymerized to form a linear chain of oppositely charged polymer. The polymer chain remains captive in the host resin due to strong electrostatic interaction and the restricted mobility of the linear polymer chains. Heterogeneous ion-exchange membranes are prepared by mechanical incorporation of powdered ion-exchange resin into sheets of rubber, PVC, acrylonitrile copolymers or some other extrudable or mouldable matrix. Homogeneous ion-exchange membranes are usually good in their electrochemical properties but are poor in mechanical strength whereas heterogeneous membranes display good mechanical strength but inferior electrochemical properties. Figure 6 presents examples of some preparation schemes for these membranes.

PRESENT SCENARIO

Water scarcity results, as there is widening disparity between fresh water supply and demand. This is because of scanty natural resources (water) and/or increasing population as well as urbanization, uplift in living standard, climatic change and pollution. "Water stress index" is an indication of fresh water scarcity and is defined as the ratio of average amount of water withdrawal to the amount of long term available fresh water resources. A value of 40% indicates acute water scarcity and a value less than 10% is considered as the lower limit of water scarcity [15]. For most of the European countries this value is less than 40% and in some cases like Ireland, Sweden, Slovenia, Netherlands it's even less than 10%. Figure 7 shows classification of countries as water rich and poor. As can be seen from the figure most of the water poor countries come under either arid or semiarid region or in the Mediterranean region. Desalination technology is most explicitly explored in these regions.

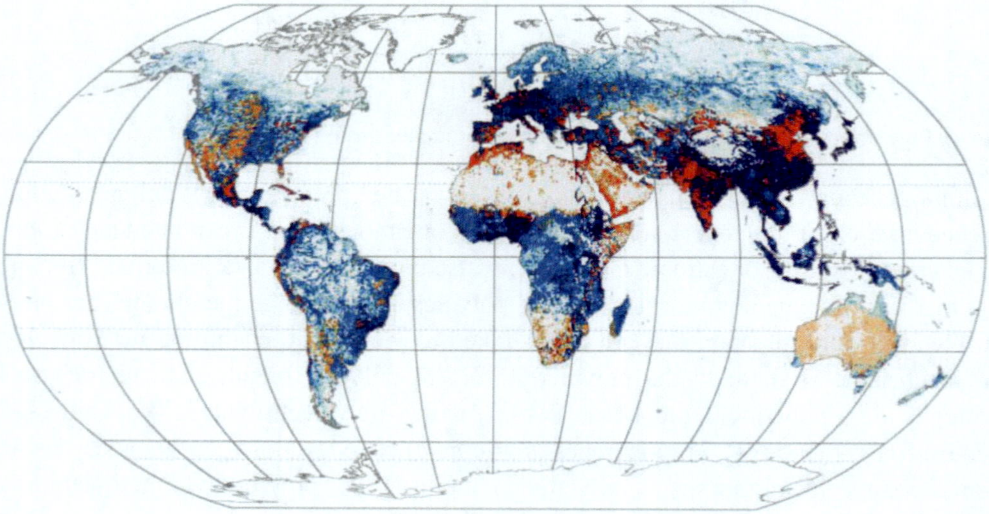

Figure 7. Water stressed regions: Red shades indicate stressed and blue shades indicate unstressed regions; darker the shades more is the stress (figure prepared by Water Systems Analysis Group, University of New Hampshire, UK)

Depending upon the nature of source water, cost effectiveness and technical feasibilities and natural resources (energy source) available different desalination technologies are in use. For example, Middle East accounts for 50% of World's desalination capacity and primarily uses thermal techniques (MSF distillation and MED) [16]. The choice of thermal method over the membrane is driven by the cheap availability of fossil fuel, and because of the poor quality of sea water having potential membrane fouling threat; another advantage of thermal desalination over RO is the advantage of being used in electricity cogeneration. Outside Middle East RO technology flourished rapidly. In the year 2001, RO accounted for 51% of the total World's desalination capacity and in 2003, 75% of desalinated water production capacity goes to RO's credit. Cost of RO is less than other thermal desalination technologies in general [17-18]. Cost of RO deceased over the period since mid 80s due to improvement in technology and innovations, introduction of energy recovery, improvement in pretreatment options, and probably due to increase in fossil fuel cost, cost of MSF increased. Figure 8 compares MSF and RO cost in 80s and 90s. Table2 compares RO capacities in different countries. Figure 9 shows the projected growth in RO around the World. There are some ED plants in operation mostly located in Libya, Canary islands (Spain), also earlier in Egypt and are operated in small capacities. For example, report in the year 2001 says that desalinated water produced by ED, MSF and RO individually accounts for 11.72%, 11.45% and 65.76% respectively of total desalinated water produced in Canary islands [19], and 0.09%, 66.45% and 21.54% respectively in Libya [20]. The technology is mostly in use for brackish water treatment, water supplying in urban and tourist areas, and in wastewater treatment.

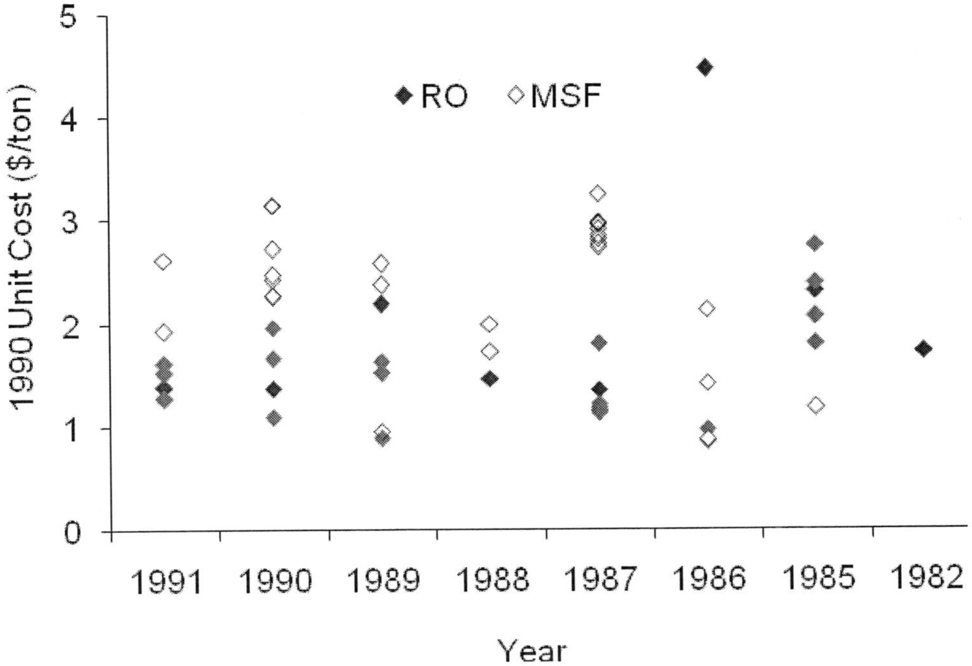

Figure 8. Comparison of desalination cost over the decades (modified and reproduced from [18])

Table 2. Countrywise RO desalination capacity in Europe[15]

Country	Capacity	Country	Capacity	Country	Capacity
Spain	2326.3	Malta	145.2	Austria	18.3
Italy	608.6	Cyprus	102.1	Portugal	11.0
Germany	229.8	Greece	52.6	Belgium	6.4
U.K.	229.0	Czech Republik	33.4	Sweden	1.5
Netherlands	224.7	Poland	25.8	Slovenia	1.5
France	195.6	Denmark	25.0		

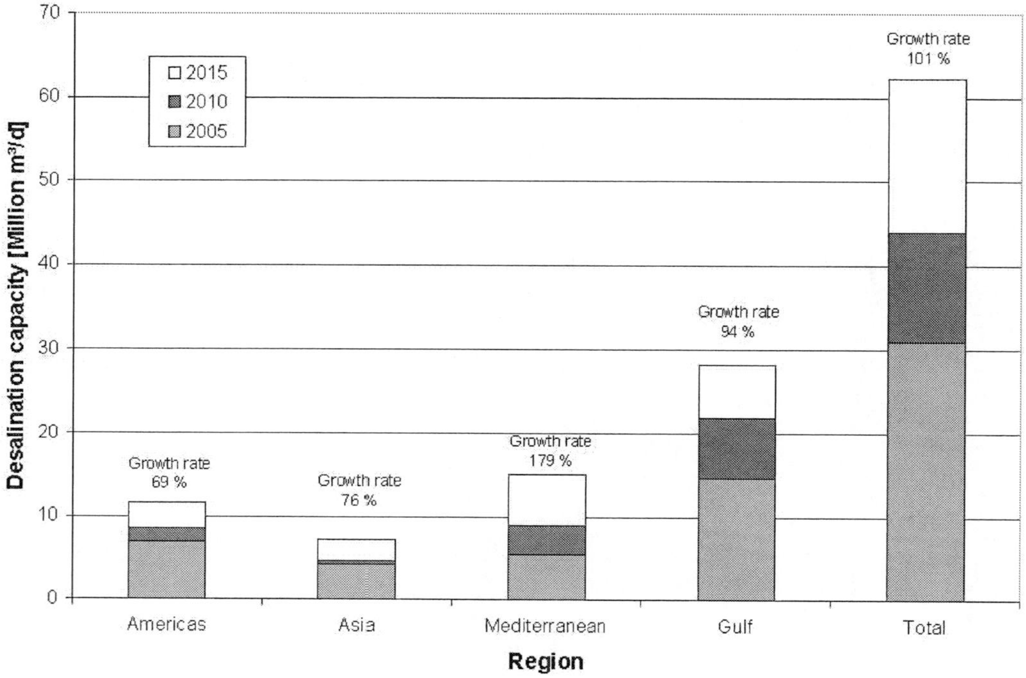

Figure 9. Worldwide projected growth of RO [15]

TECHNOLOGY

Reverse osmosis technology includes membrane and plant engineering technology which are vast in their respective areas. Therefore, detail discussion of these two is out of scope of this chapter. A brief idea about membrane preparation has been discussed in earlier section. This section deals with basic RO technology and modification made in it to meet the area specific requirements. The choice of RO desalination technology is influenced by the feed water quality and required product water quality as well as some site-specific factors such as labour cost, available area, energy cost, and local demand for electricity. As feedwater quality is one of the basic technologies design criteria, sea and brackish water RO technologies will be discussed separately.

Seawater RO

Figure 10 shows the basic RO design, is a single pass RO system [16]. In multi-pass system, permeate of first membrane system is passed through subsequent membrane system to further polish product water. The choice of single or multi pass RO system depends on certain factors including energy cost, feed water characteristics (temperature, composition, and total dissolved solute (TDS) concentration), desired recovery, and product water quality [16]. All these factors are inter-related. The energy required to pump feed water depends on TDS concentration of seawater. Higher the TDS concentration more is the osmotic pressure on the feed side of the membrane and therefore higher hydraulic pressure is required to carry out separation. Table 3 shows seawater quality of different regions. TDS concentration in feed side increases with increasing recovery consequently increasing energy demand for feed pumping. However, increased recovery also reduces the feed flow rate required to maintain a particular permeate flux and this reduces the energy requirement for pumping. Therefore, product recovery in RO system plays role in both ways and can be optimized keeping the overall energy consumption minimum. The energy minimum has been observed to exist typically between 50-55% recoveries for TDS concentration of 34,000 mg/L. This can vary with feed water quality. As the number of passes increases, energy requirement increases since the pressure drops increasingly in subsequent passes.

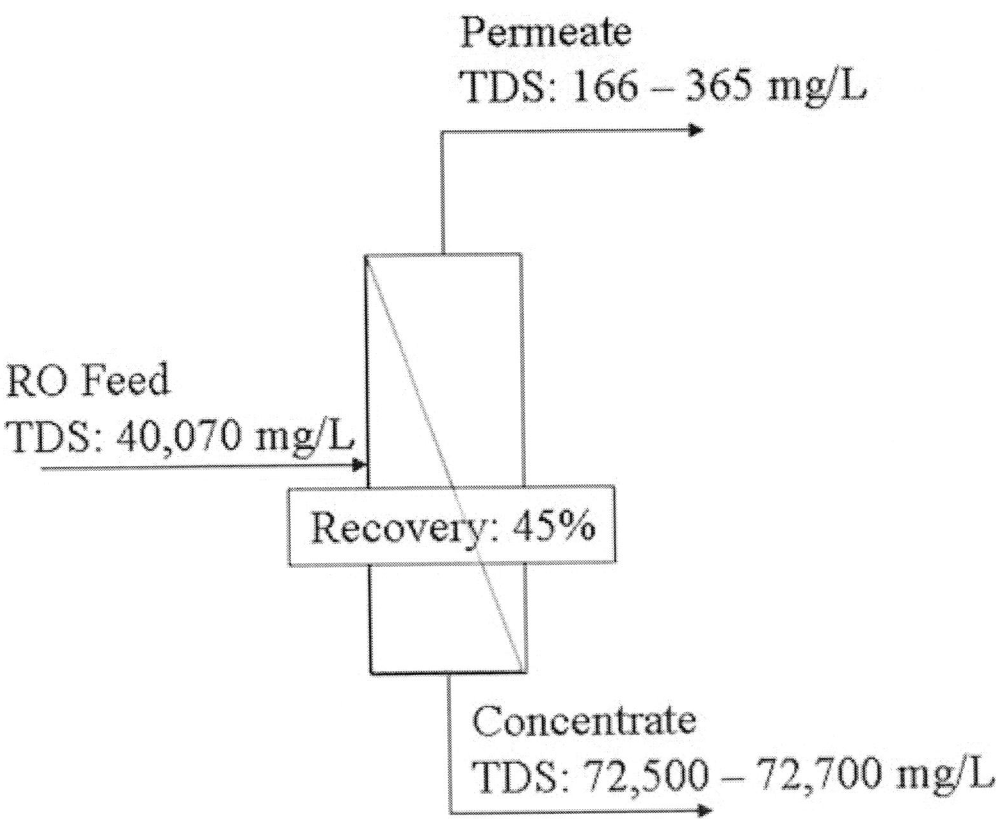

Figure 10. Single pass SWRO system (figure represents the Eni Gela Plant in Sicily) [16]

Table 3. Sea water quality of different regions

Quality parameters	Plant location (region and city)				
	Gulf of Oman (Indian Ocean)	Mediterranean Sea	Red Sea		Indian Ocean
	Fujairah	Ashkelon	Yanbu	Jeddah	Kwinana
TDS, mg/l	40000	40679	41300-46400	43300	35000-37000
Temperature, °C	22-35	15-30	22-33	24-32	16-24
Turbidity, NTU	0.5-1.3	NA	0.5-1		NA
pH	7.8-8.5	NA	8.2	8.2	NA
Reference	[21]	[22]	[23]	[24]	[25]

There have been developments of several energy recovery devices that recover hydraulic energy from concentrated RO reject before final disposal that can reach even upto 95%. This recovered energy can reduce the main pump load resulting in reduced overall energy consumption. The recovered energy can also be used to compensate energy losses in subsequent passes if a multi pass system is used. The overall recovery of RO system decreases with increase in number of passes since permeate of one pass is further polished in the subsequent pass though the higher passes are operated at high recovery (85-90%). Therefore, maximum recovery and minimum energy consumption can be expected in a single pass RO system. The realistic recovery of RO system is restricted by increasing osmotic pressure. For example for 34000 mg/L TDS concentration, required hydraulic pressure to maintain 13.5 $Lm^{-2}h^{-1}$ reaches above 7000 kPa at recovery above 60% (figure 11) [16]. Of course, with the advancement of membrane quality plants having even higher feed TDS concentrations reported to operate with 40% recovery at lower pressure than predicted (in figure 11).

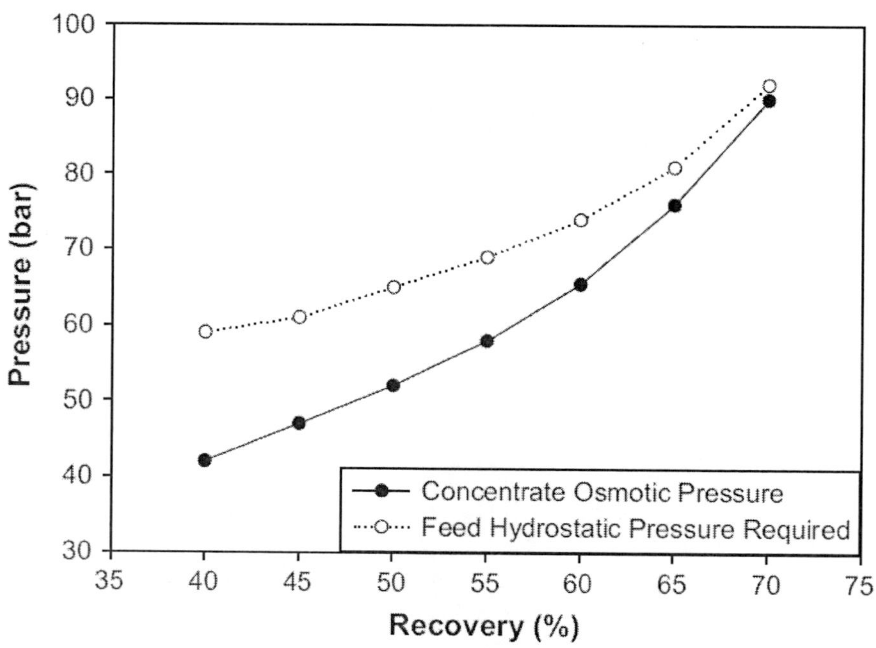

Figure 11. Recovery vs concentrate osmotic pressure for TDS value 34000 mg/l [16]

Product water quality also influences the choice of multi-pass RO system. Most of the RO plants are designed to produce 400-500 mg/L product water. With 99.7% rejection and 35% recovery, the product water TDS concentration ranges between 300-400 mg/L. If the product water TDS concentration needs to be reduced further, a multi-pass RO system is opted. Moreover, individual component present in RO feed can determine the RO process design. For example, in Ashkelon plant in Israel, uses a multi-pass RO system comprising of four stages with an aim to produce water with boron and chlorine concentration less than 0.4 and 20 mg/L respectively from feed water with TDS concentration 40700 mg/L.

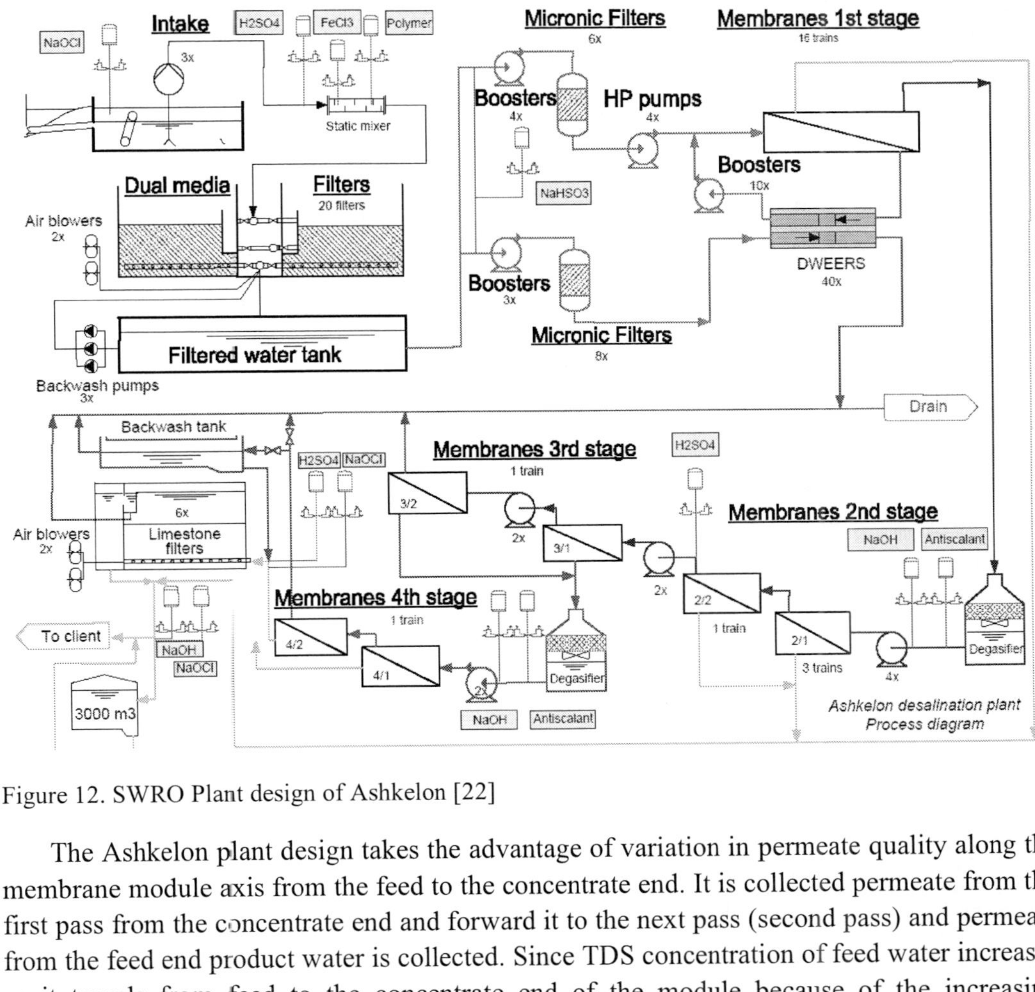

Figure 12. SWRO Plant design of Ashkelon [22]

The Ashkelon plant design takes the advantage of variation in permeate quality along the membrane module axis from the feed to the concentrate end. It is collected permeate from the first pass from the concentrate end and forward it to the next pass (second pass) and permeate from the feed end product water is collected. Since TDS concentration of feed water increases as it travels from feed to the concentrate end of the module because of the increasing concentration of rejected dissolved solid in feed water which is gradually getting converted into reject water, the permeate produced at the beginning near the feed end of the module is less concentrated than that produced near the concentrate end. Permeate from the concentrate end of the second pass is again treated in third pass and the total permeate from the third pass is treated in the fourth pass to attain the final requirement. The first pass operates at 45% recovery and rest all operate at high recovery (85-90%). The overall recovery of the system is 44% and the product water TDS concentration is less than 80 mg/L. Fujairah plant in Saudi Arab operates (Figure 13) in simpler way. The plant design consists of two passes. The first pass uses high- pressure SWRO membrane and the second pass is split into two low-pressure brackish water RO membrane systems. The concentrate of second two passes is recycled to feed of the first pass. The first pass operates at 43% and the second two passes operate at 90% recovery. The overall recovery of the plant is 41%. The feed water TDS concentration is reduced from 38000-38500 mg/L to 70-120 mg/L. With increase in feed water temperature salt and water permeation rates increase through RO membrane. Since most RO plants operate at fixed permeation rate, salt concentration in permeate will increase with increase in feed water temperature. It has been observed that for every one-degree rise in feed water temperature salt concentration in permeate increases by 3-4%. Multi-pass RO design may be useful to produce low TDS water in such cases if high feed water temperature is expected. Table 4 summarizes the data for different single and multi-pass RO plant designs.

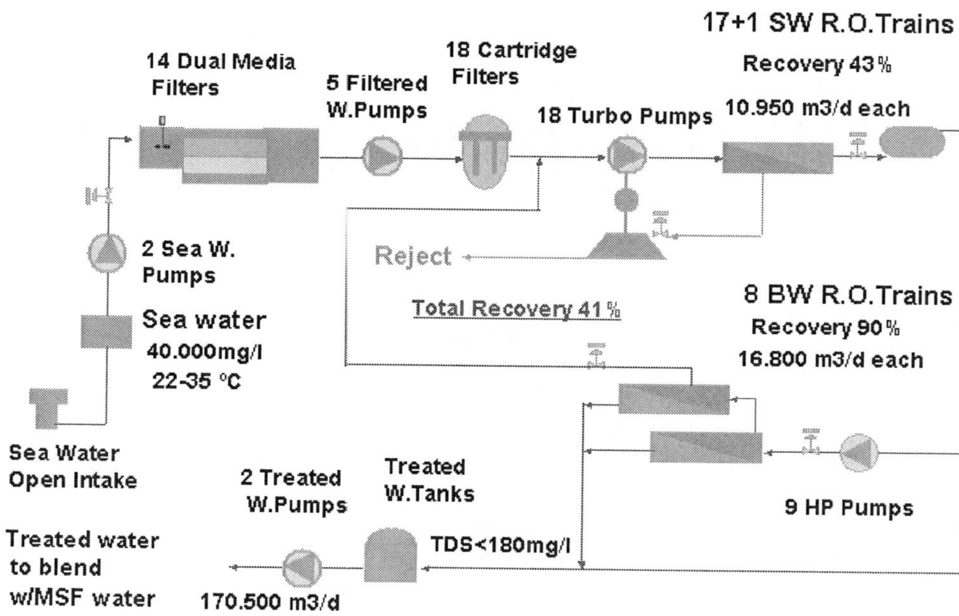

Figure 13. Design of Fujairah SWRO plant [21]

Table 4. SWRO plant data

Parameter	Plant location			
	Fujairah	Ashkelon	Kurnell	Tampa Bay
Capacity (m^3/day)	170500	330000	250000	95000
Number of passes	2	4	1	2
Membrane	SWC3, ESPA1 (Hydranautics)	FilmTec	NA	NA
Energy source	Power plant	Gas turbine	Wind	Power station
ERD	Calder Pelton turbines	Double Work Exchanger Energy Recovery (DWEER)		Energy Recovery turbine
Energy consumption, kWh/m^3	3.8	<3.9	4.2	NA
Product water price	0.87 US$/$m^3$	0.53 US$/$m^3$	1.44 AUS$/$m^3$	0.497 US$/$m^3$
Contract type	DB	BOT	DBOM	DBOOT
Reference	[21]	[22]	[25]	[26]

Combination of membrane desalination with thermal desalination technique like MSF (multistage flash) distillation or MED (Multi-effect distillation) provides an attractive option for reducing the cost of producing water with low TDS concentration. Dual hybrid technology like MSF/RO or MED /RO, or even triple hybrid technology when power generation is also associated with these have the advantages of using the common infrastructure like water intake and disposal facility, pretreatment and post treatment facilities, other site specific advantages. Moreover, blending of product water from RO with MSF or MED can allow the RO plant to operate in single pass mode producing water with relatively high TDS concentration which could be blended with MSF or MED treated water to produce the final product. Usually thermal desalination processes produce water with very low TDS concentration in the range less than 50 mg/L which is independent of feed water quality [1]. Therefore, a multi-pass arrangement can be avoided and the membrane replacement rate can be reduced. All these contribute to the cost effectiveness of the hybrid technology. For example, annual membrane replacement cost is reduced by 40% if membrane life can be expanded from three to 5 years [27]. Use of preheated seawater from the heat rejection of MSF plant or last effect of MED plant can enhance RO permeation rate, and thus increases production. Hybrid technology also gives production flexibility. For example, in Saudi Arab, MSF or thermal desalination technology was in operation for water production much before than RO. Principle of power cogeneration was adopted in those plants. The power demand in Saudi Arab remains constant except in winter when it falls by 40%, whereas the water demand remains constant through out the year. Therefore, adoption of tri hybrid SWRO/MSF/power plants gives the advantage of maintaining the water demand throughout the year. Even if the power demand drops, the reduced water production by MSF unit can be compensated by the SWRO plant. Seawater desalination plants in Jeddah, Yanbu, Al-Jubail, Fujairah in UAE, all adopted the tri hybrid technology.

Brackish Water RO

The feed water in brackish water RO is much less concentrated than that of seawater and could be well below 9000 mg/L. Therefore, brackish water RO systems are operated at higher recoveries around 50-60% than SWRO system. The basic difference in brackish water and seawater reverse osmosis technology lies in the fact that in multi-pass seawater RO system the permeate is polished in the second pass and in brackish water RO system the concentrate of the first membrane is passed through the second pass. Therefore, in brackish water RO system, recovery increases with increase in number of passes and this is just opposite to what happens in SWRO system. Membrane selection and concentrate disposal are important part of brackish water RO technology. With increasing period of operation, if the concentrate is disposed back to the feed water source or nearby area, feed TDS level may get increased and at some higher TDS level SWRO membranes are required to be used instead of brackish water one.

Electrodialysis reversal (EDR) technology is often in use in treating brackish water for example in Libya. In EDR the polarity of the electrodes are reversed in certain interval that results in disassociation of foulants, scalants and thus increases the effectiveness of the system.

Pre-/ Post-Treatment Technology

Feed pretreatment is vital for sea or brackish water RO system to avoid membrane fouling and thus to maintain the performance steady. Membrane fouling not only hampers the production it also shortens the life of the membranes. Therefore, cleaning is required. During membrane cleaning, production is shut down that incurs economic loss. Moreover, membrane cleaning may require chemical treatment and may result in membrane degradation with excessive use and ultimately resulting in membrane replacement. Therefore, pretreatment mainly aims at lowering membrane foulant concentration in feed water. Surface water (sea or brackish water) often contain high biological pollutants and turbidity and therefore more extensive pretreatment than groundwater is required. Since RO membrane is a depth filter, much of the suspended solid will be retained by it [28] and therefore the suspended solid content should be as low as possible. 'Modified fouling index' or $MFI_{0.45}$ replaced the 'Silt density index' or SDI, which was earlier, used to check feed water quality before RO treatment. SDI considers the ratio of two flow measurements, one at the beginning, and the other at the end by passing feed water through a 0.45μm filter paper in dead end mode at constant pressure. In case of $MFI_{0.45}$, it is the rate of change in flow rate is measured. The MFI value is calculated from the gradient of the linear position obtained from the plot of t/V vs. V of the following equation

$$\frac{t}{V} = \frac{\eta R_m}{A\Delta P} + \frac{\eta I}{2A^2\Delta P}V$$

Where t is filtration time, V is filtrate volume, ΔP is applied trans-membrane pressure (standard value is 2 bar), η is water viscosity (at 20°C), A is filtration area (1.38×10^{-5} m^2), R_m is membrane resistance and I is the fouling index [29]. However, observed flux decline rate is many times much higher than predicted by the MFI value. This is because of the particles sizing lower than 0.45μm. Therefore, $MFI_{0.05}$ value using membrane of pore size 0.05μm was developed and MFI-UF data reports MFI value using ultrafiltration membrane of 13000 Da having average pore size 9 nm. Conventional pretreatment processes include acid addition, coagulant/ flocculant additions, disinfection, media and cartridge filtration. Acids are added to carry out salt precipitations. Coagulations and floculations are commonly known water treatment process. In media filtration, water is treated by passing through granular media like pumice, anthracite, gravels etc that can be used in combination. The last treatment before RO is cartridge filter that is used to retain particles in the size range 1-10μm. Anti-scalants are used either before or after the cartridge filter. They are used to delay or prevent early precipitation of salts. Antiscalants disrupts the crystallization process and does not allow salt to precipitate up to a certain concentration level. Usually, they are used at low concentration mostly below 10mg/L. When disinfectants like chlorine is used, precaution should be taken so that before membrane treatment water become chlorine free or within the prescribed tolerance level for the membrane because most of the membrane materials used for RO membrane available in the market are susceptible to chlorine attack. Table 5 compares different chemical pretreatments and their purpose

Table 5. Chemicals in SWRO pretreatment [15]

Pretreatment	Purpose	Chemicals added	Fate of chemicals
pH adjustment	Lower carbonate concentration Protect membrane from hydrolysis	Acid (H_2SO_4)	Sulfate stays in concentrate and pH decreases
Antiscalants	Prevent formation of membrane scaling	Sequestering agent dispersants	Complexes formed stay in concentrate
Coagulation-filtration	Prevent membrane fouling and clogging	Coagulants-flocculants	Flocs settle, removal by filtration
Disinfection	Prevent biological fouling	Chlorine (or UV)	Forms hypochlorite, (chlorination by products by UV)
Dechlorination	Protect chlorine sensitive membranes	Sodium bisulfate	Sulfate and chloride generated stay in concentrate

In brackish water RO plant similar pretreatment scheme is followed with exceptions where water is contaminated with specific entity and special pretreatment scheme and membranes are required to produce water of desired quality.

Now-a-days MF, UF, and even NF are also considered as possible pretreatment options while treating water of specific quality. Micron filters are extensively in use in many commercial SWRO systems just before RO filtration. UF and NF are considered as pretreatment option particularly wastewater reuse facilities. The conventional pretreatment scheme involves some additional steps and usually consists of flocculation, lime treatment, settling, re-carbonation by CO_2, media filtration.

Most of the plants still use SDI data for monitoring feed water quality before RO; MFI data is less used or published. Table 6 compares various pretreatment methods and MFI for seawater collected from SWRO plant in Chang-won in Korea.

Table 6. Comparison of various pretreatment methods [30]

Pretreatment	$MFI_{0.45}$ (S/L^2)	Turbidity (NTU)
SW	13.2	1.2
SW + Coagulation	55.0	2.0
SW + MF	4.0	0.3
SW + MF + NF	3.4	0.1
SW + Coagulation + NF	1.9	0.1

After coagulation MFI value increases because of coagulants form larger addition compounds.

Examples of pretreatment schemes in some of the large scale SWRO or brackish water plants around the World are summarized in table 7.

Table 7. Pretreatment schemes in practice

RO Plant location (capacity in m^3/d)	Pretreatment scheme	SDI value	Reference
Sea water			
Boujdour city, Morocco (800)	Chlorination Dual media filtration (Sand and gravel filtration) Acidification Antiscalant agent injection Microfiltration Sodium bisulfite addition	<1	[31]
Yanbu, Saudi Arabia (5000)	Chlorination Coagulation ($FeCl_3$/ polyelectrolyte) Acid treatment Media filtration Cartridge filtration Antiscalant dosing Sodium bisulfate addition	<4	[23]
Ashkelon, Israel (330, 000)	Chlorination Acid treatment Coagulation ($Fe_2(SO_4)_3$/organic polyelectrolyte) Dual media filtration (quartz sand and anthracite) Sodium bisulfate addition Cartridge filtration	<90%	[22, 32]
Fujairah, Saudi Arabia (170, 500)	Acid treatment Coagulation ($FeCl_3$/ polymer) Dual media filter (Sand and Pumice) Cartridge filtration (5μm)	<3.6	[21]
Brackish water			
Tan Tan Demineralization plant, Maroc (20*)	Pre-chlorination Acidification Sand filtration Antiscalants addition Sodium bisulfate addition Microfiltration	NA**	[31]
Las Palmas III, Canary Island, Spain (36000)	Hypochlorite treatment Coagulation Gravity filtration Bisulfate addition Acid treatment Antiscalants Precoat base Precoat adsorbent Precoat filtration Cartridge filtration	NA	[33]
Wastewater			
Riyadh Refinery, Saudi Arabia (20000)	Two stage lime clarification Dual media filtration	NA	[34]
Water Factory 21, California, U.S.A. (57000)	Lime clarification Air stripping Re-carbonation Pre-chlorination Filtration Activated carbon filter Post chlorination	NA	[34]
Chennai Petrochemical Corporation Ltd, India (510)	$FeCl_3$, lime and polymeric treatment Chlorination Multi-media filter Ultrafiltration	NA	[35]

* L/s, ** Not Available

RO water post-treatment involves blending of RO water with other water to obtain water of required taste. Hardness imparts natural taste to water, and lime or lime stone contactors are used to adjust hardness. Alkalinity can be adjusted by carbonates and bicarbonates and pH can be adjusted by adding caustic soda. Additional disinfection treatment can also be done for example by UV treatment. In brackish water RO permeate is sometimes blended partially with RO feed.

Disposal of RO Concentrate

Disposal of RO concentrate is another important part of technology that one cannot ignore. Disposal of SWRO concentrate back to sea is commonly in practice and therefore is not a problem, even for brackish water RO concentrate if the plant is located in coastal area [16]. Main problem related to brackish water RO concentrate disposal is often its need for treatment before disposal that may be costlier than RO treatment itself. Usually brackish water RO concentrate is disposed through irrigation in agricultural lands, through salt concentration in concentration pond and deep well injection. Water cannot be even disposed into surface water if met the surface water disposal standard. However, an irrigation or concentration pond appears to be cheap and hassle free but it has the potential to damage the ground water with time. With the increasing environment awareness, such disposal technique is less practiced or need to take extra precaution. Deep well injection is less expensive and considered as safe technique for concentrate disposal though the technique is not old enough to be sure about its adverse effect. It may result in ground water aquifer disturbance or contamination due to salt precipitation or leaching. It is the most widely practiced wastewater disposal technique through out the World. Zero liquid discharge RO plants appears to be very good options but there is no commercial operation report. For such process, recovery will have to be nearing 100% and which is not cost effective one. ED technology can be considered for higher recovery and low disposal but the commercial acceptance of technology is lagging behind RO because of high capital cost and energy cost.

Energy Sources for RO

Fossil fuel is so far the predominant source of electric energy running RO plants. RO consumes minimum energy among the others established desalination technologies but is an energy intensive process. One of the estimate reported 1,000m^3/d fresh water production requires 10, 000-ton oil per year [36] and it is often a burden for many water shortage countries. Moreover, with increasing fossil fuel scarcity there is always an accompanying possibility for price to go even higher. Additionally, there is an indirect environment pollution encouragement by consuming electricity produced by burning fossil fuel. For example, one of the estimations projected a total emission of 13-264 million ton/y CO_2, 1,350-13,10,000 ton/y SO_x, 21,100-5,40,000 ton/y NO_x and 1190-40,000 ton/y particles by a desalination plant of 20.1 million m^3/d capacity depending upon the source of energy and technology chosen (RO, MSF, MED) [37]. Therefore, with increasing fossil fuel scarcity, cost, and restriction on carbon emission, search for alternative energy source for RO desalination has gained research interest. RO operation using alternative energy sources are still at the pilot scale level in R and D status and operated in non-continuous mode. Of course, there are plants that are in regular operation but only few in number and are of small capacities. Different alternative energy sources envisaged for RO are solar thermal and photovoltaic, wind, nuclear heating, ocean thermal, ocean wave, gravitational potential, geothermal, and their success is highly region specific in nature because of their available energy density. Spain has coupled RO desalination with wind energy generation. A 5000 m^3/d SWRO desalination plant in Gran Canaria (Canary island) is powered by wind energy [16]. Due to the variable energy generation through out the year, excess power generated is either sold to the grid or RO unit purchase excess power if required from the grid. A small-scale solar PV-RO (PV: Photovoltaic) system is also running successfully in Gran Canaria that is one of the smallest SWRO systems [38]. The system produce 0.8-3 m^3/d desalinated water and is supported by 4.8kW stand-alone PV system. The solar thermal system appears to be less costly than the solar PV system [36] and there are many research programs undertaken around the World. In principle, the technology involves conversion of solar thermal energy into steam that generates energy to drives the RO pump. This is done either by electricity generation or by mechanical energy production. Some times low boiling organic liquids are also used to extract energy from steam through heat exchange. This superheated organic expands and this expansion generates mechanical energy or produces electric energy to run RO pump. Coupled with the ancient technique (bullock/camel driven) to generate the electrical energy to run the RO technique is one of the sensations in recent days [39]. This can be useful technique where electrical energy is unavailable.

Environmental Impact of RO

Water desalination technology, aimed to encounter the fresh water scarcity and to meet the increasing demand for it, must be environment friendly to be sustainable and reverse osmosis got its acceptance because of its apparent non-polluting nature among several other reasons. The technology is relatively new and therefore effect on environment is yet to be discovered. RO desalination may have impact on following possible domains and in that case, it will be more localized in nature: land uses, ground water, marine environment, noise pollution, indirect air pollution [40]. The land used for RO set-up results in industrialization and may attract other concurrent development. This may or may not be of interest to the local government and may even go against the nature and lose in natural beauty. Ground water table may get disturbed and even there may be infiltration of saline water due to the feed intake facility, pipe carrying brine, or concentrate. The marine environment may be destroyed or at least may change because of concentrate discharge that also contains additional chemicals, which are used for water treatment and other purpose. The high pressure pumps need to be taken proper care otherwise can become a source of noise pollution. Use of energy by RO system will promote more energy generation that is often share the same land as in trihybrid technology and therefore more air pollution can be expected.

Economics of Desalination

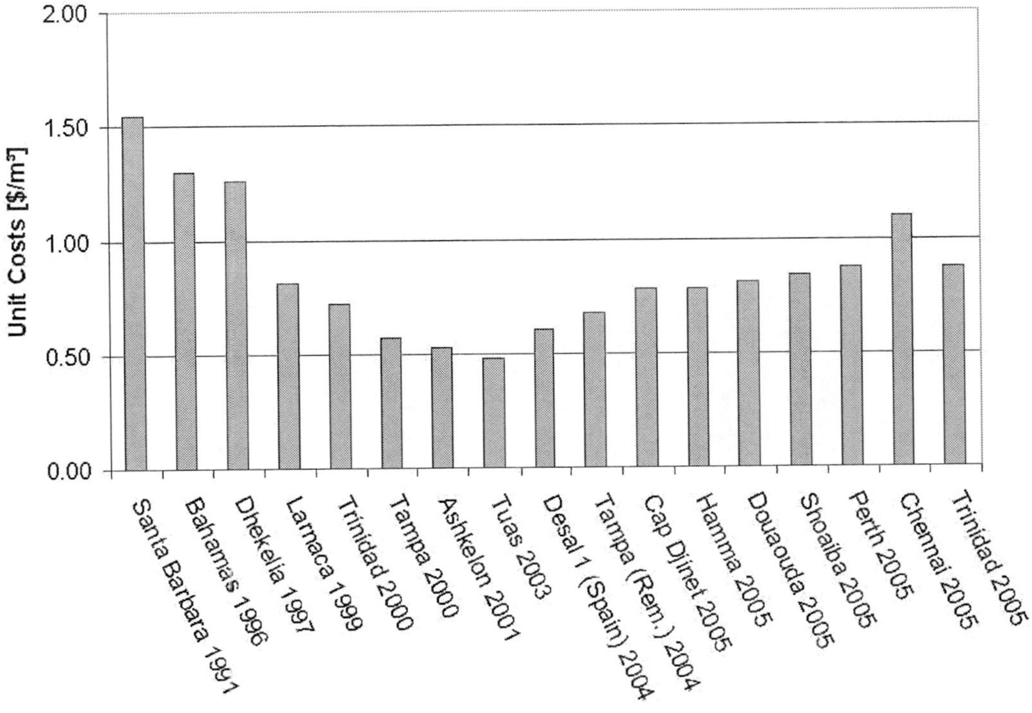

Figure 14. Specific cost of RO water in different plants around the World [15]

The economics of the water desalination is highly site specific and mostly depends on feed water quality. Precise technology selection plays a vital role in providing sustainable solution. Technology selection is based on different factors like capital cost, energy cost, reliability (continuous production with good quality), operation skill, feed water quality, plant size, space requirement, ease of operation and geographical position [1] Although there is no strict methodology, after acquiring all these information weighing and ranking methodology can be followed tc choose one out of many options. Among the different desalination technologies, RO has gained more acceptances compared to the thermal technologies probably due to the increasing fuel cost, improvement in membrane performance and enlargement in plant capacity. Thermal desalination was preferred in Middle East countries and reverse osmosis is more preferred in U.S. RO entered the Middle East market with increasing acceptance of new hybrid desalination technology. Figure 14 [15] (Fritzmann, 2007) compares the price of RO desalinated water around the World over the period 1991-2005. The major factors that influence the unit cost of production are fuel cost, chemical cost, operation and maintenance cost in addition to the capital cost. Cost decreased over the decade upto 2003 and then increased but remained cheaper than the 1997 price. This increase can be accounted by the increased energy cost and construction cost. A not too old data based on current technology around the World depicted in figure 15 [15] presents the cost structure of desalinated water production by RO and it quickly reveals the energy cost and the fixed charge (essentially the capital cost of RO equipment) as mostly contributing to the desalinated water cost. Itemized energy consumption analysis in RO desalination process indicates hydraulic high-pressure pump as the major consumer and rest of the process consumes very little compared to this (figure 16) [15]. Therefore, an attempt to reduce cost of RO water production attracted more attention in developing technology to reduce net energy cost, be it by introduction of renewable and/or nonconventional energy sources, energy recovery devices or process parameter modification.

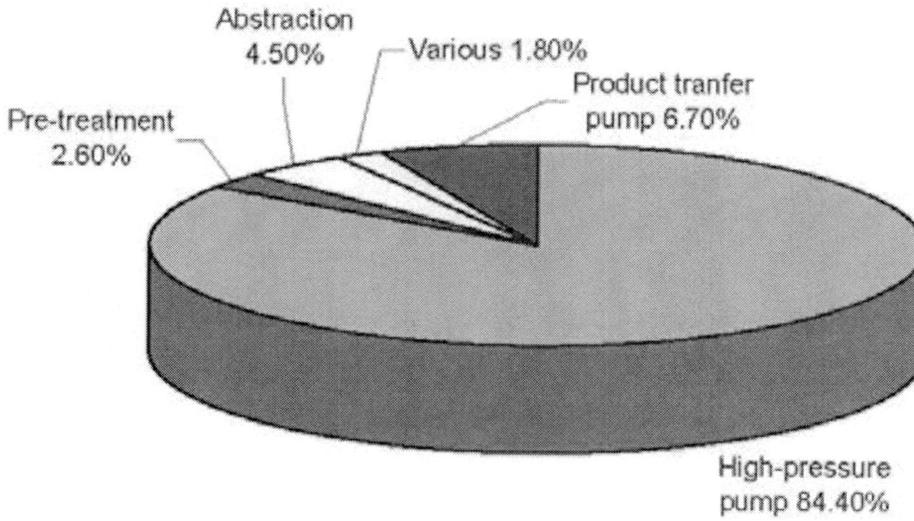

Figure 15. Cost struction of RO plant [15]

Cost Calculation for RO

While calculating RO desalinated water cost, the initial point can be capital cost and operation and maintenance (O&M) cost calculation. Individual items that are considered while calculating fixed capital cost are listed below [41]:

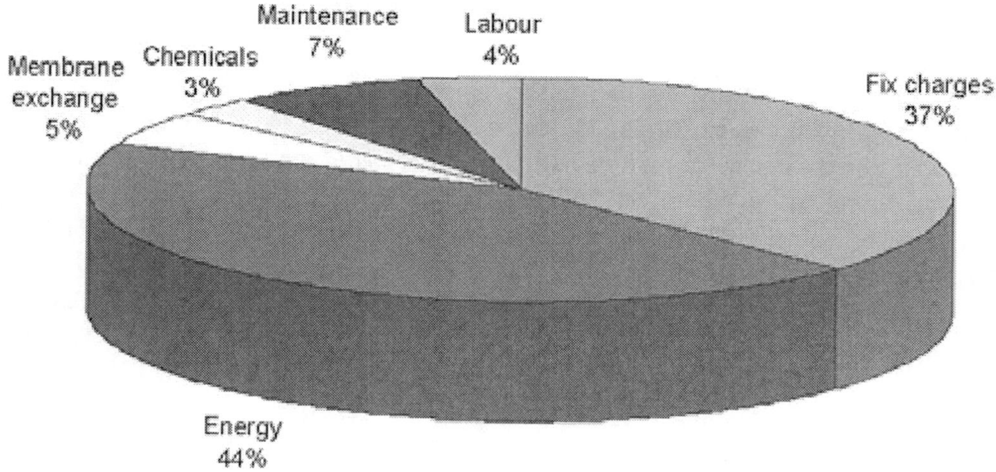

Figure 16. Energy consumption in RO plant [15]

Raw water intake system
Intake pipeline
Raw water storage
Pretreatment
Media filtration
Microfiltration
Low pressure pump
High pressure pump
Membrane system
Energy recovery system
Post treatment
Disposal pump
Disposal pipe
Fresh water storage
Disposal deep well
Land
Building
Site work
Distribution'
Laboratories
Engineering services
Permitting
Legal services
Capitalized interest
Financing
Insurance

Table 8 presents capital cost of different RO plants.

Table 8. RO plant capital cost

Year	Location	Capacity, m^3/d	Specific cost, \$/(m^3/d)	Reference
1984	Tajura (Libya)	10,000	7500.0	[42]
1985	Jeddah (Saudi Arabia)	56,781	757.3	
1989	Fujairah (UAE)	9,085	1651.0	
1991	Jubail (Saudi Arabia)	75,708	1828.0	
1995	Tripoli (Libya)	30,000	3400.0*	
2000	Tripoli (Libya)	2,50,000	2495.0*	
NA	Hurgada (Egypt)	4800	1363.0	[41]
NA	Sharm (Egypt)	3500	1685.0	

*offer price

The capital cost varies over plant size and other local factors influencing the capital cost, e.g. labour cost, building material cost, site cost, location, taxes and government policies. In general higher the plant capacity, less is the production cost and capital unit cost [41]. Operation and maintenance cost includes replacement, repair and spare parts any other cost which is a recurring expenditure. Followings are the itemized heads come under this cost:

Electrical energy
Land lease
Membrane replacement
Pretreatment chemicals
Post treatment chemicals
Repairing
Labour
Laboratory services
General services
Sell tax

While calculating the actual water production cost, the nature of the project execution is also important from the economic point. There are different types of project contract, for examples:

- Investor's property (IP)
- Operation and maintenance contract type (O&M)
- Build, Own, Operate and Transfer (BOOT)
- Build, Own, Operate (BOO)

In IP type, investor owns the whole project and pays for everything including operation and maintenance. In O&M type, the investor gives the operation and maintenance contract to others who supplied the RO plants and pays for the service. The contract also includes supply of chemicals, spares, and membrane replacement in cost plus scheme. In BOOT type, the investor provide space with or without electrical energy and the contractor carries out whole project including intake, building, installation, supply, treatment, brine disposal, operation and maintenance for a pre-agreed period (10-30years). During this period the investor is committed to buy all the fresh water produced from the contractor at some pre-agreed accelerating price schedule and at the end of the agreement period, the whole project is transferred to the investor. In BOO type project, the investor provides land with or without electricity, intake, buildings, disposal, infrastructure; the contractor or the water production company is committed to cover the investor's need of water at reduced special price for a pre-agreed quantity without any limitation on consumption. The contractor or the company adopts excess production scheme and sells the excess to others at higher price. BOO type contracts are becoming more popular than others. The investor is not loaded with the fixed consumption. In IP type plant, the annualized fresh water production cost, (or annualized cost expenses for fresh water production) is sum of fixed capital cost depreciation rate (FCCDR) per year and annual operation and maintenance cost. FCCDR per year is calculated by dividing the individual itemized capital cost or the grand total by the effective lifetime of the project. In BOO type contract, since the investor provides land, building, intake, infrastructure, brine disposal facility, and if in some cases electricity, then annualized fresh water production cost is calculated by deleting these items from the annualized capital cost. Thus, the production cost comes lower in BOO type contract than in IP type.

Tables 9-12 summarize this section taking examples of cost calculation from different SWRO plants [41]. Table 9 presents description of different SWRO plants considered under this study. Data presented in tables are recalculated and it varies in some places from the original report.

Table 9. Information on selected SWRO desalination plants

Basic information	Case 1	Case 2	Case 3	Case 4	Case 5
Location (in Egypt)	Hurgada	Safaga	El-Tor	Sharm	Hurgada
Plant Production (m^3/d)	250	500	2000	3500	4800
Capital cost (US$/$m^3$)	2166	2095	1714	1685	1363
Total capital cost (US$)	541500	1047500	3428000	5897500	6542400
Intake type	Well	Well	Well	Well	Surface
Feed water TDS (mg/l)	47000	49000	47000	43000	43000
Feed water quantity (m^3/d)	500	1750	7000	12250	16800
Number of RO unit	1	1	2	7	4
Number of Vessels /unit	10	6	14	6	20
Number of RO elements /vessel	2	7	7	7	6
RO membrane type	B-1	FilmTec	FilmTec	FilmTec	Fluid
Total number of RO elements	20	42	196	294	480
Pure water recovery (%)	45	30	30	30	30

Table 10. Itemized capital costs (in US$, 2001) of the selected SWRO plants [41]

Capital item		Case 1	Case 2	Case 3	Case 4	Case 5
Intake system cost	Intake group	180320	331010	870712	1403605	1118750
Pretreatment phase cost	Settling equipment	19633	50925	166656	286713	318065
	Dosing equipment	6052	15698	51372	88380	98044
	Media filter	9292	24103	78878	135701	150541
	Microfilter	6385	16563	54204	93251	103448
	Testing equipment	6290	16316	53395	91859	101904
	Subtotal	47652	123605	404505	695904	772002
RO desalination phase cost	Membranes and vessels	36366	64495	300978	451466	737088
	Centrifugal pumps	8772	12394	43542	87908	93726
	High pressure pump	20225	47653	120868	244021	260172
	Energy recovery units	8285	14531	51050	103065	109886
	Steel structure	14377	19873	34534	69720	74335
	SS pipes and valves	19436	29528	146459	451865	226032
	Monitoring equipment	12062	21155	47296	95486	101806
	Subsides	2315	4060	6006	12125	12928
	Subtotal	121838	213689	750733	1515656	1615973
Post-treatment cost	pH adjustment	1186	1285	1501	2583	2866
	Chlorination	1522	1648	1927	3314	3677
	Subtotal	2708	2933	3428	5897	6543
Brine disposal cost	Pump station	7011	15650	51214	88109	97743
	Disposal pipeline	4524	10229	33474	57589	63887
	Deep wells	9524	21259	69571	119690	132778
	Subtotal	21059	47138	154259	265388	294408
Infrastructure cost	land	48194	93228	305092	524878	582274
	Buildings	12996	25140	82272	141540	157018
	Site works	3249	6285	20568	35385	39254
	Distribution	126711	245115	802152	1380015	1530922
	Subtotal	191150	369768	1210084	2081818	2309468
Professional and financing cost	Engineering	13538	26188	85700	147438	163560
	Legal	5957	11523	37708	64873	71966
	Financial and interest	2708	5238	17140	29488	32712
	Insurance and sales tax	12996	25140	82272	141540	157018
	Subtotal	35199	68089	222820	383339	425256
Grand total		599926	1156232	3616541	6351607	6542400
Capital unit cost (US$/m^3)		2400	2312	1808	1815	1363

Table 11. Annualized cost expenses for production of fresh water per year (in US$) of the selected SWRO plants on the basis of IP contract type

	Case1	Case2	Case3	Case4	Case5
Plant capacity (m^3/d)	250	500	2000	3500	4800
Annual depreciation (US$)					
Intake system (10years)	18032	33101	87071	140361	111875
Pretreatment phase	4765	12361	40450	69591	77200
RO desalination phase (10 years)	12184	21369	75073	151566	161597
Post treatment (10 years)	271	293	343	590	654
Brine disposal (10years)	2112	4714	15426	26539	29441
Infrastructure (40 years)	4075	7883	25796	44379	49232
Professional and financing (10years)	2653	6809	22282	38334	42526
Subtotal	44953	84854	266441	471359	506546
Annual O&M cost					
Pretreatment chemicals	4050	7297	21734	34677	38993
RO membrane replacement	21670	39038	116274	185524	208611
Post-treatment chemicals	8303	14959	44554	71088	79935
Brine disposal	3240	5838	17387	27742	31194
Power	85666	154329	459664	733425	824696
Repair and replacement	44412	80008	238302	380226	306787
Labor	18863	33982	101214	161494	130302
Land lease	0	0	0	0	0
Insurance	16316	29394	87548	139689	112708
Annualized O&M cost	202520	364845	1086677	1733865	1733226
Annual depreciation + O&M (per year)	247473	449699	1353118	2205224	2239772
Annual depreciation + O&M (per day)	678	1232	3707	6042	6136
Production cost (US$/$m^3$$day^{-1}$)	2.71	2.46	1.85	1.73	1.28

Table 12. Annualized cost expenses for production of fresh water per year (in US$) of the selected SWRO desalination plants on the basis of BOO contract type

	Case1	Case2	Case3	Case4	Case5
Plant capacity (m^3/d)	250	500	2000	3500	4800
Annual depreciation (US$)					
Pretreatment phase	4765	12361	40450	69591	77200
RO desalination phase (10 years)	12184	21369	75073	151566	161597
Post treatment (10 years)	271	293	343	590	654
Cost of financing	26582	45445	146151	238725	263848
Subtotal	43802	79468	262017	460472	503299
Annual O&M cost					
Pretreatment chemicals	101261	170219	543338	866933	961733
RO membrane replacement	20252	34044	108668	173387	192347
Post-treatment chemicals	10126	17022	54334	86693	96173
Repair and replacement	10632	17873	57050	91028	100982
Labor	17721	29788	95084	151713	168303
Annualized O&M cost	159992	268946	858474	1369754	1351235
Annual depreciation + O&M (per year)	203794	348414	1120491	1830226	1854534
Annual depreciation + O&M (per day)	558.34	955.56	3069.84	5014.32	5081.92
Production cost (US$/$m^3$$day^{-1}$)	2.23	1.91	1.53	1.43	1.06

Predicting RO Cost

Any prediction on RO cost based on present data can be done using statistical forecasting technique like autoregressive integrated moving average model (ARIMA) [43] . This kind of model is fitted to a time series data to forecast on future values on the series. To estimate water production cost before actual construction of plant it is always preferred to gather data on plants around the World under operating conditions similar to the plant to be constructed. Once this information is gathered then factors affecting RO unit water production costing can be expressed as

$$C(\$/m^3) = C_1 + C_2 + C_3 + \cdots + C_n$$

Where the individual C_n is unit cost component for each variable n. Influence of some important cost components on unit water cost production are explained in the following.

To evaluate capital cost ($C_{captial}$) on unit water cost production the following equation can be used

$$C_{capital} = \frac{D_{capital}}{V_L}$$

Where $D_{capital}$ is the direct capital cost, L is the estimated life time of the system in hours, and V_L is the total volume of water produced throughout the life.

Pumps are the main electricity consumer in RO system and therefore to evaluate effect of electricity cost on unit water production cost total work done *(W_{pump})* by pump is calculated following the equation

$$W_{pumps}(kW) = \sum_k \frac{\Delta P_k(kPa) \times V_{pump,k}(m^3/s)}{\eta_k}$$

where $\Delta V_{pump,k}$ is the volumetric flow rate in pump k, ΔP_k is the pressure increase and is the efficiency associated with pump k.

Hourly cost of electricity production is

$$E_{pump}(\$/h) = W_{pump}\ (kW) \times electricity\ unit\ cost\ (\$/kWh)$$

The effect of electricity consumption cost on unit water production cost is

$$C_{electrictiy} = \frac{E_{pump} \times L}{V_L}$$

Maintenance cost can be considered as 2% of the capital cost per year and the effect of this cost component can be written as

$$C_{Maintenance} = \frac{D_{capital} \times L \times 0.02}{V_L}$$

Chemical cost component is (C_{chem}). Variation in membrane replacement may range between 5-20% of the yearly membrane purchase cost ($C_{membr.purchase}$). So it estimates to

$$C_{membr.replacement} = \frac{\lambda \times C_{membr.purchase} \times L}{V_L}$$

where λ varies between 0.05-0.20.

The unit cost component related to pre-treatment and post-treatment, brine disposal are $C_{pre\text{-}treatment}$, $C_{post\text{-}treatment}$, $C_{brine\ dispcsal}$ can be calculated by multiplying by lifetime of the plant and dividing by V_L the capital cost involved initially and the cost of other component related to these. Similarly, the unit manpower cost component can be calculated and assigned say $C_{man\ power}$. The cost component related to interest rate is calculated as follows

$$C_{int} = \frac{D_{capital} \times a \times years}{V_L}$$

Where a is the amortization factor and is given by the following equation

$$a = \frac{i(1+i)^{years}}{(1+i)^{years} - 1}$$

Where i is the interest rate.

Finally, the unit production cost C is calculated as follows

$$C = C_{capital} + C_{electricity} + C_{maintenance} + C_{membr.replacement} + C_{chem} + C_{pre-treatment} + C_{post-treatment} + C_{brine\ disposal} + C_{man\ power} + C_{int} + C_{others}$$

C_{others} is the cost factors not included here.

Unit production cost of water for the plant of interest can be predicted by adding the unit production cost for a reference plant (C_{ref}) and the sum of differences between the cost components corresponding to the plant of interest and reference plant (ΔC_n) using the following equation

$$C_{predict} = C_{ref} + \sum_n \Delta C_n$$

A simple approach to estimate net total profit of RO operation is exemplified below [44] . The net profit of an RO operation can be given by

$$TP_{net}=TP-TC_e-TC_m-TC_{pump}-TC_{ERD}$$

Where TP_{net} is net total profit, TP is total profit; TC_e, TC_m, TC_{pump} and TC_{ERD} are total cost corresponding to electricity, membrane replacement, pumps and energy recovery devices respectively.

$$TP = 8400 * PWP * Q_p$$

$$TC_e = 8400 * uc_e(W_{pump}-W_{ERD})$$

$$W_{ERD} = Eff_{ERD} * Q_f * \Delta P$$

Where PWP indicates product water price (per m^3), Q_p is permeate water flow rate (m^3/h), uc_e is the electricity unit cost, W_{pump} and W_{ERD} are the required energies for operating pumps and energy recovery devices. Eff_{ERD} indicates the efficiency the devices, Q_f indicates feed flow rate (m^3/h). ΔP is the trans-membrane pressure. The number 8400 represents actual operation hour in a year that has been calculated considering 350 working days per year.

$$TC_m = 0.33 * uc_m * A_m$$

$$TC_{pump} = 0.1 * uc_{pump} (\Delta P\ Q_f Eff_{pump}/3600)^{0.3434} + (8400 * UC_e\ \Delta P Q_f Eff_{pump})/3600$$

$$TC_{ERD} = 0.1 * uc_{ERD} (\Delta P\ Q_f Eff_{ERD}/3600)^{0.7132} + (8400 * UC_e\ \Delta P Q_f Eff_{ERD})/3600$$

uc_m, uc_{pump} and uc_{ERD} are the unit cost for membrane, pump and ERD respectively. Model assumed 3 years as membrane lifetime and 10 years as pump and ERD for lifetime. Details of the calculation are given in [45]. Finally table 13 summarizes the RO desalinated water cost World wide.

Table 13. Cost RO desalinated water in countries across the World [16, 46]

Year	Location	Price (US\$/m^3)	Year	Location	Price (US\$/m^3)
1989	Galder-Agaete, Spain	1.94	1999	Larnaca, Cyprus	0.80
1989	Jeddah, Saudi Arabia	1.31	2000	Trinidad	0.70
1990	Ad Dur, Bahrain	1.30	2000	Tampa, Florida	0.57
1991	Lanzarote III, Spain	1.62	2003	Eilat, Israel	0.81
1991	Agragua, Spain	1.34	2003	Tuas, Singapore	0.47
1990	Catalina Island, California	1.92	2004	Spain	0.61
1991	Santa Barbara, California	1.53	2004	Tampa, remedied	0.68
1993	Monterey, California	1.57	2005	Fujairah, UAE	0.87
1996	Bahamas	1.25	2005	Cap Djinet, Algeria	0.70
1997	Dhekelia, Cyprus	1.20	2005	El Hamma, Israel	0.70
1998	Mallorea, Spain	1.03	2005	Douaouda, Algeria	0.72
1998	Marbella, Spain	1.00	2005	Shoaiba, Saudi Arabia	0.73
1998	Eilat, Israel	0.72	2005	Perth, Australia	0.90
2001	Ashkelon I, Israel	0.51	2005	Chennai, India	1.10
2002	Ashkelon I, Israel	0.49	2005	Trinidad	0.85

Conclusion

Desalination technology for fresh water production appears to be a valuable gift to the mankind. Among the techniques available reverse osmosis fits well in today's socio-economic scenario. The technology gained more acceptances because of its cheap production cost, environment friendly nature, flexibility in operation, small footprint, high potential to meet the requirements. Of course success is determined by the demand for product and availability of alternatives. So far the technology is more successful in Mediterranean, arid and semi arid region because of the high demand for fresh water but with increasing demand for the same in other parts of the World, RO is envisaged as viable solution to meet the crisis. The price of RO desalinated water in an average comes around US$0.80/m^3 and energy is the most intensive cost contributing factor. Use of alternative energy sources particularly renewable one can bring it down further. Lowering in membrane replacement cost and therefore improvement in membrane performance is another area of attraction for scientific research.

References

[1] Al-Subaie, K. Z., Precise way to select a desalination technology, *Desalination*, 206 (2007) 29-35

[2] Mulder, M. *Basic principles of membrane technology*, Kluwer Academic Publishers, Dordrecht, The Netherlands, 2000

[3] Meindersma, G.W.; Guijt, C.M.; de Haan, A.B. Desalination and water recycling by air gap membrane distillation, *Desalination*, 187(2006)291-301

[4] Lv. C., Su. Y., Wang Y., Ma X., Sun Q., Jiang Z., Enhanced permeation performance of cellulose acetate ultrafiltration membrane by incorporation of pluronic F127, *Journal of Membrane Science*, 294 (2007) 68-74

[5] Jung B., Preparation of hydrophilic polyacrylonitrile blend membranes for ultrafiltration, *Journal of Membrane Science*, 229 (2004)129-136

[6] Susanto H., Ulbricht M., Characteristics, performance and stability of polyethersulfone ultrafiltration membranes prepared by phase separation method using different macromolecular additives, *Journal of Membrane Science*, 327(2009)125-135

[7] Kim K.S., Lee K.H., Cho K., Park C.E., Surface modification of polysulfone ultrafiltration membrane by oxygen plasma treatment, *Journal of Membrane Science*, 199(2002)135-145

[8] Hurwitz G., Guillen G. R., Hoek E.M.V., Probing polyamide membrane surface charge, zeta potential, wettability, and hydrophilicity with contact angle measurement, *Journal of Membrane science*, 2009

[9] Saha N. K., Joshi S. V., Performance evaluation of thin film composite polyamide nanofiltration membrane with variation in monomer type, *Journal of Membrane Science*, 342(2009)60-69

[10] Qtaishat M., Rana D., Khayet M., Matsuura T., *Preparation and characterization of novel hydrophobic/hydrophilic polyetherimide composite membranes for desalination by direct contact membrane distillation*, 327 (2009)264-273

[11] Boributh S., Chanachai A., Jiraratananon, Modification of PVDF membrane by chitosen solution for reducing protein fouling, *Journal of Membrane Science*, 342(2009)97-104
[12] Xi Z.-Y., Xu Y.-Y., Zhu L.-P., Zhu B.-K., Modification of polytetrafuoroethylene porous membranes by electron beam initiated surface grafting of binary monomers, *Journal of Membrane Scienec*, 339(2009)33-38
[13] Yu H.-Y., Liu L.-Q., Tang Z.-Q., Yan M.-G., Gu J.-S., wei X.-W, Surface modification of polypropylene microporous membrane to improveits antifouling characteristics in an SMBR: air plasma treatment, *Journal of Membrane Science*, 311(2008)216-224
[14] Kariduraganavar M.Y., Nagarle R.K., Kittur A.A., Kulkarni S.S., Ion-exchange membranes: preparative methods for electrodialysis and fuel cell applications, *Desalination*, 197 (2006) 225-246a
[15] Fritzmann C., Löwenberg J., Wintgens T., Melin T., State-of-the-art of reverse osmosis desalination, *Desalination*, 216(2007)1-76
[16] Greenlee L. F., Lawler D. F., Freeman B. D., Marrot B., Reverse Osmosis desalination: water sources, technology and today's challenges, *Water Research*, 43(2009)2317-2348
[17] Kamal I. An assessment of desalination technology for the Rosarito repowering project, *Desalination*, 102(1995)269-278
[18] Park M.-H., Park N.-S., Park H., Shin H.-S., Kim B.-D., An economic analysis of desalination for potential application in Korea, *Desalination*, 114(1997)209-221
[19] Veza J.M., Desalination in the Canary islands: an update, *Desalination*, 133 (2001) 259-270
[20] Abufayed A.A., El-Ghuel M.K.A., Desalination process application in Libya, *Desalination*, 138(2001)47-53
[21] Sanz M.A., Bonnélye V., Cremer G., Fujairah reverse osmosis plant: 2 years of operation, *Desalination*, 203 (2007) 91-99
[22] Sauvet-Goichon B., Ashkelon desalination plant- a successful challenge, *Desalination*, 203 (2007) 75-81
[23] Khawaji A. D., Kutubkhanah I.K., Wei J.-M., A 13.3 MGD seawater RO desalination plant for Yanbu industrial city, *Desalination*, 203 (2007) 176-188
[24] Al-Sheikh A. H. H., Seawater reverse osmosis pretreatment with an emphasis on the Jeddah plant operation experience, *Desalination*, 110 (1997) 183-192
[25] Saliby I. E., Okour Y., Shon H.K., Kandasamy J., Kim I. S., Desalination plants in Australia, review and facts, *Desalination*, 247 (2009) 1-14
[26] *http://www.water-technology.net/projects/tampa/* (accessed on 24th December, 2009)
[27] Hamed O.A., Overview of hybrid desalination systems – current status and future prospects, *Desalination*, 186 (2005) 207-214
[28] Chakravorty B., Layson A., Ideal feed pretreatment for reverse osmosis by continuous microfiltration, *Desalination*, 110 (1997) 143-150
[29] Boerlage S.F.E, Kennedy M., Aniye M.P., Schippers J.C., Applications of the MFI-UF to measure and predict particulate fouling in RO systems, *Journalof Membrane Science*, 220 (2003) 97-116
[30] Choi J.-S., Hwang T.-M., Lee S., Hong S., A systematic approach to determine the fouling index for a RO/NF membrane process, *Desalination*, 238 (2009) 117-127
[31] Hafsi M., Analysis of Boujdour desalination plant performance, *Desalination*, 134 (2004) 93-104

[32] Bar-Zeev E., Berman-Frank I., Liberman B., Rahav E., Passow U., Transparent exopolymer particles: potential agents for organic fouling and biofilm formation in desalination and water treatment plants, *Desalination and Water Treatment*, 3 (2009) 136-142

[33] Veza J.M., Desalination in the Canary islands: an update, *Desalination*, 133 (2001) 259-270

[34] Madwar K., Tarazi H., Desalination techniques for industrial wastewater reuse, *Desalination*, 152 (2002) 325-332

[35] *http://www.ciigbc.org/images/Photos/Imp49.pdf* (accessed on 24th December, 2009)

[36] Qiblawey H.M., Banat F., Solar thermal desalination technologies, *Desalination*, 220 (2008) 633-644

[37] Nisan S., Benzarti N., A comprehensive economic evaluation of integrated desalination systems using fossil fuelled and nuclear energies and including their environmental costs, *Desalination*, 229 (2008) 125-146

[38] Lindemann J.H., Wind and solar powered seawater desalination applied solutions for the Mditerranean, the Middle East and the Gulf countries, *Desalination*, 168 (2004) 73-80

[39] CSMCRI, *Oxen power desalinates water, Appropriate Technology*, 30 (2003) 23

[40] Einav R., Harussi K., Perry D., The footprint of the desalination processes on the environment, *Desalination*, 152 (2002) 141-154

[41] Hafez A., El-Manharawy S., Economics of seawater RO desalination in the Red Sea region, Egypt. Part 1. A case study, *Desalination*, 153 (2002) 335-347

[42] Ashour M.M., Ghurbal S.M., Economics of seawater desalination in Libya, *Desalination*, 165 (2004) 215-218

[43] Dore M.H.I., Forecasting the economic costs of desalination technology, *Desalination*, 172 (2005) 207-214

[44] Kim Y.M., Lee Y.S., Lee Y.G., Kim S.J., Yang D.R., Kim I.S., Kim J.H., Development of a package model for process simulation and cost estimation of seawater reverse osmosis desalination plant, *Desalination*, 247 (2009) 326-335

[45] Maskan F., Wiley D.E., Johnston L.P.M., Clements D.J., Optimal design of reverse osmosis module networks, *AIChE Journal*, 46 (2000) 946-954

[46] Hoang M., Bolto B., Haskard C., Barron O., Gray S., Leslie G., *Desalination in Australia, National Research Flagships*, February, 2009

In: Desalination: Methods, Cost and Technology
Editor: Irena A. Urbonienė pp. 209-227

ISBN: 978-1-61668-909-4

Chapter 6

RECENT ADVANCES IN SUPPORTED IONIC LIQUID MEMBRANE TECHNOLOGY

A.P. de los Ríos**, *F.J. Hernández, L.J. Lozano and C. Godínez

Department of Chemical and Environmental Engineering,
Technical University of Cartagena, Campus La Muralla,
C/ Doctor Fleming S/N, E-30202 Cartagena, Murcia, Spain

ABSTRACT

Supported liquid membranes have being recognized as an effective method for the selective separation of compounds in a wide range of processes. Nevertheless, their industrial application is still limited, mainly due to concerns about supported liquid membrane stability and long-term performance. The use of ionic liquids as a liquid membrane phase could overcome these inconveniences due to their negligible vapour pressure, their greater viscosity and the possibility of minimizing their solubility in the surrounding phases by adequate selection of the cation and anion. The possibility of design suitable ionic liquids for specific separation problems has also opened up new fields of application of supported ionic liquid membranes. In this review an overview is given of recent advances in supported membranes based on ionic liquids, including issues such as methods of preparation, fields of application, stability and process intensification using supported ionic liquid membranes.

1. OVERVIEW

Membrane separation processes are widely used in industry because they offer several advantages over other separation processes including low-energy consumption, environmental friendliness operating conditions [1]. The use of supported liquid membranes (SLMs), which consist of a thin microporous support whose pores are filled with a liquid,

* Corresponding author. Tel.: +34 968 325 548. E-mail address: antonia.perez@upct.es.

have been extensively studied during recent years [2,3,4]. The SLM technique is a combination of three simultaneously occurring processes: molecule extraction from the feed phase to the SLM, diffusion through SLM and re-extraction to the received phase. Among their numerous advantages are specificity, the minimal amount of solvent needed and the fact that extraction and stripping steps are combined into a single step [5]. Nevertheless, the application of supported liquid membranes on a large scale is still limited due to inadequate membrane stability [6,7]. This apparent instability, as shown by a progressive loss of the liquid embedded in the membrane into the surrounding, is due to liquid vaporization, dissolution into the adjacent liquid phases and/or shifting from the porous structure of the supported membrane under low-pressure gradients [8].

Recent studies have suggested that the use of ionic liquids (ILs) could overcome these inconveniences due to their inherent characteristic properties [9-11]. ILs are organic salts which are liquids at room temperature consisting in an organic cation (e.g. imidazolium, pyridinium, pyrrolidinium, phosphonium, ammonium) and a polyatomic inorganic anion (e.g. tetrafluoroborate, hexafluorophosphate, chloride) or, more and more current, an organic anion (e.g. trifluoromethylsulfonate, bis[(trifluoromethyl)sulfonyl]imide) [12]. The structures of common cations and anions of ILs are shown in Figure 1. The negligible vapour pressure of the ILs makes them interesting clean alternatives to classic organic solvents for use in a wide variety of chemical processes, such as separation and purification [13-18,19], reaction media [20-23] and chemical catalysis [24-26].

The use of these new solvents as a liquid membrane phase results in the stabilization of the SLMs due to their negligible vapour pressure, the possibility of minimizing their solubility in the surrounding phases by adequate selection of the cation and anion [27], and their high viscosity, which could reduce displacement of the liquids from the micron pores under pressure.

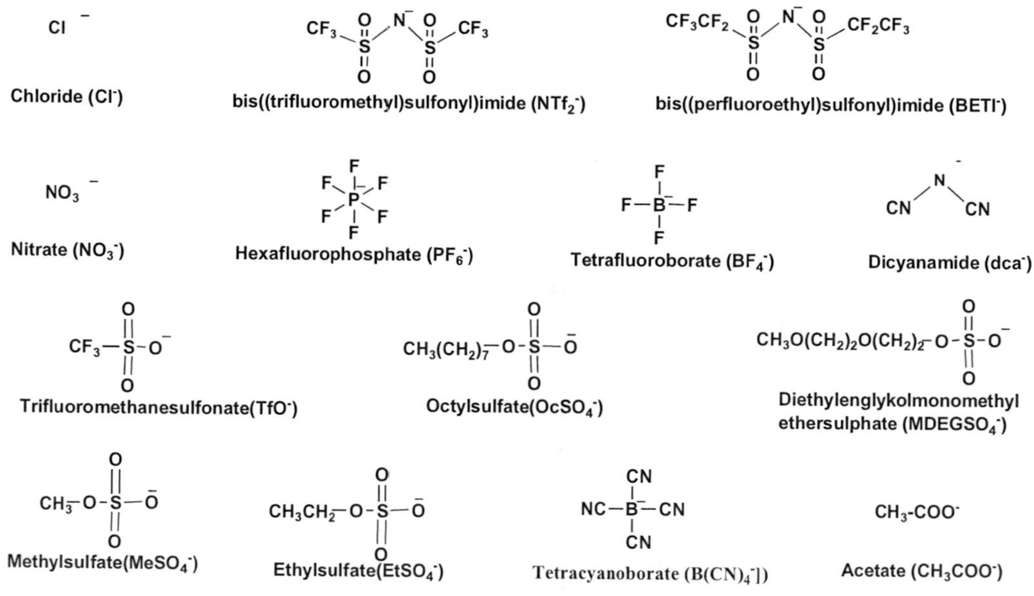

Figure 1. Example of ions involved in ILs

Other interesting properties of ILs in the context of liquid membranes are their high chemical and thermal stability, high ion conductivity [28] and high solvent power [29]. Additionally, the properties of ionic liquids (hydrophobicity, viscosity, solubility, etc.) can be varied by alternating the substitutive group on the cation or the combined anion [12]. All these mentioned properties have led ILs to being considered “green designer solvents” [30].

In the last decade some researchers have reported that SLMs based on ILs achieved selective transport organic compounds such as amines, alcohols, organic acids, ketones, ethers, and aromatic hydrocarbons [13,14,16,31,32], mixed gases [33,34] and small water-soluble ions, such as NaCl [11]. An overview of the recent advances in supported ionic liquid membrane technology is given in this monograph, including issues such as methods of preparation, fields of application, stability and process intensification using supported ionic liquid membranes.

2. Methods of Preparation of Supported Ionic Liquid Membranes

Due to the relatively high viscosity of the ionic liquids, the method of preparation of supported ionic liquid membranes can play an important role on the membrane performance. Three methods are commonly used for the preparation of supported ionic liquid membranes: direct immersion, pressure or vacuum [17,35,36]. In the former case, immobilization takes places by contacting the supporting membrane with the ionic liquid, allowing it to soap up the liquid [17]. In the pressure method, the immobilization is achieved by placing the membrane in an ultrafiltration unit, adding an amount of ionic liquid and applying nitrogen pressure to force the ionic liquid to flow into the pores of the membrane [35]. In the vacuum method, the supporting membrane is submerged in a volume of IL and vacuum is applied to release all air occluded in the pores of the membranes [36]. After all procedures, the excess ionic liquid should be removed from the membrane surface either by leaving to drip overnight or by blotting with paper tissue.

Scanning electron-microscopy (SEM) combined with Energy Dispersive X-ray (EDX) analysis was used to analyze the influence of the immobilization method on the SILMs performance [36]. This technique allows the characterization of the membrane surface morphology and the examination the global chemical composition of the membranes and the distribution of the ILs within them. Figure 2 shows examples of SEM micrographs of a plain Nylon membrane and supported liquid membranes based on [bmim$^+$][PF$_6^-$] prepared by using the pressure and the vacuum methods, respectively.

The morphological study showed that, after impregnation by either method, the IL was homogeneously distributed in the pores of the membrane, except for the largest macropores that remained partially filled. These pictures also show the presence of a small amount of excess IL located on the external membrane surface. The SEM-EDX and weight studies showed that the amount of IL immobilized was independent of the ionic liquid used when the immobilization was carried out under pressure. In contrast, following immobilization under vacuum, less ionic liquid was adsorbed into the membranes when more viscous ILs were used. This behaviour could be explained by the fact that the high viscosity of these ILs makes it difficult their penetration into the middle of the deeper pores of the membrane under vacuum. From these findings it was concluded that immobilization under vacuum can be considered suitable for low viscosity ionic liquids, with the added advantage that it is also an easier preparation method. However, for high viscosity ionic liquids, immobilization of the ionic liquid should use pressure in order to ensure that all membrane pores are filled with liquid.

The nature of the supporting membrane also plays an important role in the performance of supporting ionic liquid membranes. In this context, *De los Rios et al.* (2007) [9] studied the used of two polymeric membranes, Nylon and Mitex, as supporting membranes. Nylon membrane was a hydrophilic polyamide membrane with a pore size of 0.45μm and a thickness of 170 μm. Mitex membrane was a hydrophobic polytetrafluoroethylene membrane with a pore size of 10 μm and a thickness of 130 μm. It was observed that less ionic liquid was absorbed into the Mitex membranes, which was explained by the different textural properties and the high hydrophobic character of these membranes, which probably restrict interaction with the hydrophilic ionic liquids [37] used.

The use of nanofiltration membranes as supporting membranes have been also reported [38]. In this case, direct filtration of ionic liquids through the nanofiltration membrane was not possible at a gas pressure up to 7 bars. The ionic liquids make up of cations which contain straight or branched hydrocarbon chains were easily absorbed into the polymeric membrane allowing the nanoporous structure soaked up and saturated with the ionic liquids.

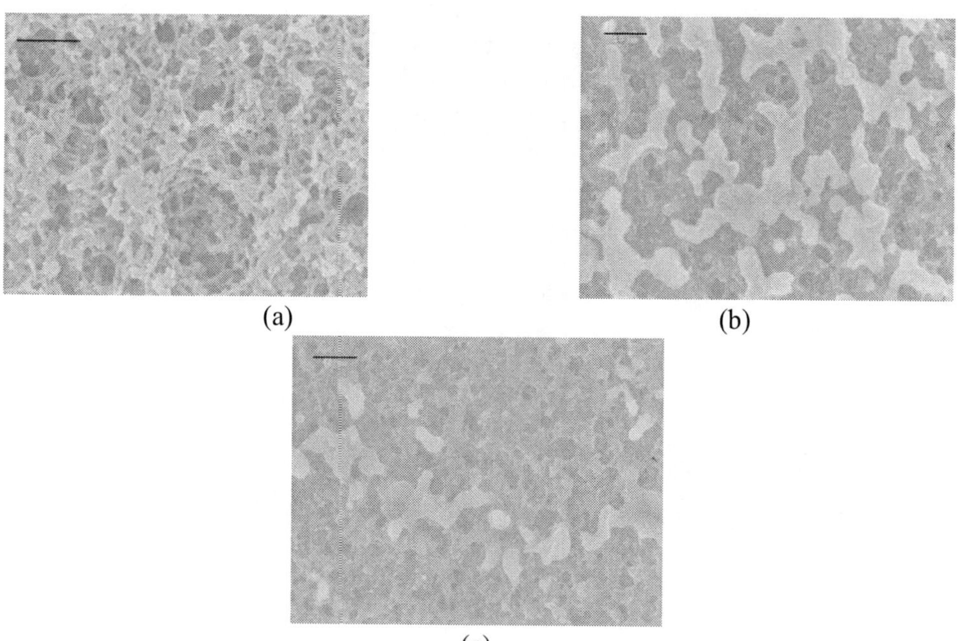

(a) (b) (c)

Figure 2. Scanning electron micrographs of the (a) Nylon membrane (3040x) (scale bar = 5 μm) and the Nylon membrane impregnated with $[bmim^+][PF_6^-]$ by pressure (b) (1010x) (scale bar = 10 μm) and by vacuum (c) (1010x) (scale bar = 10 μm)

3. Fields of Application of Supported Liquid Membranes

The use of supported ionic liquid membranes in different fields of application have received growing attention in the last decade, due to the advantages that SILMs offer, presenting the number of paper published in this field per years an exponential growth. Some representative examples of these applications will now be described.

3.1. Separation of Organic Compounds

One of the most studied applications of SILMs is the selective separation of organic compounds. The first example was reported by Branco *et al.* (2002) [13], who studied the selective transport of 1,4-dioxane, 1-propanol, 1-butanol, cyclohexanol, cyclohexanone, morpholine and methylmorpholine as a model seven-component mixture of representative organic compounds. For that, four ionic liquid based on the 1-n-alkyl-3 methylimidazolium combined with the anions hexafluorophosphate or tetrafluoroborate, immobilized in different organic polymeric membranes were used. The use of the ionic liquid $[bmim^+][PF_6^-]$ immobilized in a polyvinylipolyvinylidene fluoride membrane, allowed an extremely highly selective transport of secondary amines over tertiary amines (up to a 55:1 ratio).

Other interesting examples were reported on the use of SILM for the selective separation of the substrates and products of transesterification reactions. The biosynthesis of organic esters commonly used in the perfumery and flavour industries can be carried out by transesterification from vinyl esters and alcohols catalysed by enzymes in non-conventional media (i.e n-hexane [39] and ionic liquid [40] at low water content. Therefore, as can be observed in **Figure 3**, when the biochemical reaction reach the equilibrium, the reaction medium could consist of a mixture of alcohol, vinyl ester, organic acid and alkyl ester (flavour ester). The possibility of using SILMs for the selective separation of these reaction mixtures has been extensively analysed [32,41-44].

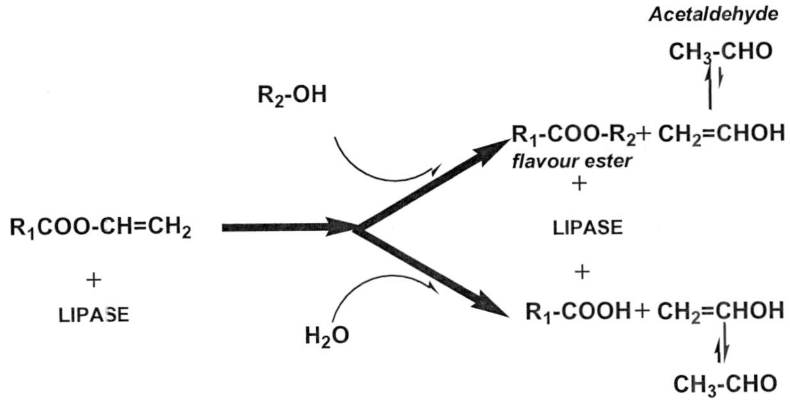

Figure 3. Stoichiometric scheme of kinetically controlled synthesis of flavour esters by transesterification from vinyl esters and alcohols catalyzed by CaLB

No appreciable permeability differences between the organic compounds were observed with the plain Nylon membranes, suggesting that such membranes can not be used for selective separation of the target compounds. The situation changed dramatically when imidazolium-based ionic liquids containing [PF_6^-], [BF_4^-] or [dca^-] anions were supported in the polymeric membranes. In these cases, substantial permeability differences between the target compounds were found on the basis of not only their functional groups but also of their alkyl chain lengths. These results were quite encouraging and suggested that these SLMs based on ILs could be use for the selective separation of the organic esters from the reaction mixture. After knowing the feasibility of using the IL as liquid phase for the selective separation of transesterification reaction mixtures, researchers have tried to establish rules for the optimal design of ILs for use in these separations [44]. For that, the permeability of four different compounds involved in a transesterification reaction (vinyl butyrate, 1-butanol, butyl butyrate and butyric acid) through thirteen supported liquid membranes based on 1-n-alkyl-3-methylimidazolium and 1-n-alkyl-3-methylpyridinium cations and different anions (bis{(trifluoromethyl)sulfonyl}imide, hexafluorophosphate, methylsulfate, 2(2-methoxyethoxy)ethylsulfate, ethylsulfate, n-octylsulfate, dicyanamide, nitrate, tetrafluoroborate and chloride) immobilized in Nylon membranes were determined. It was found that selectivity values were mostly dependent on the anion composition of the ionic liquid, being increasing for a given cation ([bmim]) in the sequence: [$OcSO_4^-$] < [PF_6^-] < [NO_3^-] < [BF_4^-] < [dca^-] < [Cl^-] < [$EtSO_4^-$] < [$MDEGSO_4^-$] < [$MeSO_4^-$]. As regards cation composition, it was observed that more effective supported liquid membranes were obtained with the use of dialkylimidazolium-based ionic liquid respect to the dialkylpyridinium ones. Furthermore, the decrease in the chain length of the alkyl substituent of the imidazolium ring from [$omim^+$] to [$bmim^+$] (i.e [$omim^+$][PF_6^-]/[$bmim^+$][PF_6^-]) and from [$bmim^+$] to [$emim^+$] (i.e [$bmim^+$][$EtSO_4^-$] /[$emim^+$][$EtSO_4^-$] pair) led to an increase in the selectivity. Therefore, it was concluded from these results that ionic liquids containing dialkylimidazolium cations with a short alkyl chain length and sulfate anions with a short alkyl substituent could allow the separation of the target compounds with high selectivity.

SILMs have also be used for the separation of aromatic hydrocarbons from aliphatic hydrocarbons. In this context, the selective separation of benzene, toluene and *p*-xylene from n-heptane was analyzed using SILMs based on [$bmim^+$][PF_6^-], [$hmim^+$][PF_6^-], [$omim^+$][PF_6^-] and [$Et_2MeMoEtN^+$][Tf_2N^-] supported in a polyvinylidene fluoride membrane [14]. It was found that aromatic hydrocarbons were successfully transported through the membrane based on these ionic liquids, the maximum selectivity to n-heptane being reached using benzene in the aromatic permeation and [$bmim^+$][PF_6^-] in the liquid membrane phase.

The concentration of organic compounds of pharmaceutical interest from diluted buffered solutions using SILMs has been also reported. One interesting example was the concentration of penicillin G using SILMs based on [$bmim^+$][PF_6^-], [$hmim^+$][PF_6^-], [$omim^+$] [PF_6^-] and [Oc_3MeN^+][Cl^-] supported in polyvinylidene fluoride membranes [45]. Uphill transport was achieved by using [Oc_3MeN^+][Cl^-] as a membrane solution and chloride concentration difference as a driving force of permeation.

Ionic liquids have been also used as extract agent diluted in an organic solvent in SILM [15]. Marták *et al.* (2008) [15] reported the effectively transport of lactic acid through a SILM based on the ionic liquid tetradecyl(trihexyl)phosphonium bis(2,4,4-trimethylpentyl) phosphinate dissolved in *n*-dodecane.

3.2. Separation of Mixed Gases

Other interesting field of application of supported ionic liquid membranes is the separation of mixed gases. Since SILMs may have the potential for industrial applications, specifically, low pressure systems such as the treatment of bio-methane from anaerobic digesters and CO_2 capture from flue gases, much effort and many resources have been expended on developing new SILMs.

Scovazzo *et al.* (2009) [46] explored the selective separation of the gas pairs CO_2/CH_4 and CO_2/N_2 using continuous flows of the mixed gases using [emim][BF_4] [$emim^+$][TfO^-] and [$emim^+$][dca^-] supported in polyvinylidene fluoride and [$emim^+$][Tf_2N^-], [$hmim^+$][Tf_2N^-] and [$emim^+$][$BETI^-$] supported in a polyethersulfone membrane. The highest CO_2/ CH_4 and CO_2/N_2 selectivities were 27 and 21.2 using the ionic liquids [$emim^+$][BF_4] and [$emim^+$][Tf_2N^-], respectively. The separation of CO_2, from N_2 and CH_4 was also successfully carried out with polymer films of ILs synthesized from monomers of ILs with polymerizable groups such as styrene and acrylate [47]. CO_2/ CH_4 separations were also reported using SILMs based on a task-specific ionic liquids such us [NH_2pmim^+][Tf_2N^-] and [NH_2pmim^+][$CF_3SO_3^-$], which have functional groups capable of chemically complexing with CO_2 [48]. These SILMs, with the amine-terminated ionic liquids, facilitated CO2 transport through the membrane, showing higher selectivities than with [$bmim^+$][Tf_2N^-] for CO_2 separation from the CO_2/CH_4 gas mixture.

Recently, the separation and enrichment of bio-hydrogen from CO_2 and N_2 has also been assayed using SILMs [33,49]. This separation is a previous step required for utilization of gaseous mixture obtained in the fermentation process (H_2, CO_2, and N_2) in fuel cells since the concentration of hydrogen in this gaseous mixture is not high enough for its direct application. A wide range of ionic liquids such as [$bmim^+$][BF_4^-], [$bmim^+$][PF_6^-] [$hmim^+$][PF_6^-], [$omim^+$][PF_6^-], [$EtOEtmim^+$][PF_6^-], [$MeOEtmim^+$][PF_6^-], [$Et(OEt)_2mim^+$][PF_6^-] and [$Et(OEt)_2mim^+$][PF_6^-] supported in PVDF were used. It was observed that the SILMs prepared have a much higher permeability to CO_2 than to N_2 and H_2. The higher selectivity values towards CO_2 (CO_2/N_2=35 and /CO_2/H_2 = 11) were reached by using [$bmim^+$][BF_4^-]. Hydrogen can also be produced from a wide variety of resources, including petroleum coke, coal, and even biomass. These resources can be gasified to produce syngas (a mixture of CO and H_2) and the resulting gas stream can be further reacted with water to produce CO_2 and more H_2. The final gas stream is a mixture of CO_2 and H_2, therefore, a separation and enrichment step is needed. SILMs have been also applied to this separation process [50]. For that, an amine functionalized ionic liquid ([NH_2pmim^+][NTf_2^-]) supported in a nylon support was used. High selectivity and permeability values were reached with this membrane, since the supporting ionic liquid facilitates the transport of CO_2.

Gan *et al.* (2006) [51] analyzed the permeation of H_2, O_2, N_2, and CO in four types of ionic liquids ([$bmim^+$][NTf_2^-], [$dmim^+$][NTf_2^-], [Oc_3Me^+][NTf_2^-], [oPy^+][NTf_2^-]) supported on nanofiltration membranes. It was found that gas permeation exhibited an exponential increase with increasing gas phase pressure in the range from 3.0 to 7.0 bars. The best H_2/CO selectivity was reached with [Oc_3Me^+][NTf_2^-], being the separation factor around 4, but offered permeation rate far less than [oPy][NTf_2] which had the best permeation performance but the worst selectivity. Other interesting gas separations using SILM have been reported such as CO_2/He [52] and H_2S/CH_4 [53] separations, in which selectivities upto 8.7 and 260, respectively, were reached.

3.3. Pervaporation and Vapour Permeation

Supported ionic liquid membranes have also been successfully applied to the separation of various liquids or vapour mixtures by pervaporation and vapor permeation [54-56]. The major problem associated with solute recovery by these techniques is the restrictive compromise between selectivity and flux: High solute selectivity involves the use of conditions that lead to relatively low mass fluxes in the membrane. This problem would be overcome, if the feed solvent could not permeate through the membrane, for example using SLMs based on ILs. In this case, solute to solvent selectivity would be unlimited and the process could be operated under conditions that provide high fluxes [54]. In this context, the use of SLMs based on the ionic liquid $[Pr_4N^+][B(CN)_4^-]$ in a nanofiltration ceramic module allowed the removal of 1,3-propanediol from aqueous solution by pervaporation, increasing the selectivity of the process by more than two orders of magnitude while the permeability was only in one order of magnitude slower [54].

The removal of acetone and butan-1-ol from aqueous solution was also carried out by pervaporation using ultrafiltration membranes impregnated by two ionic liquids ($[eneim^+][PF_6$-] and $[Pr_4N^+][B(CN)_4^-]$) and polydimethylsiloxane [55], founding a higher average enrichment factor of butan-1-ol using SLM based on ionic liquids than that obtained with polydimethylsiloxane membranes. These membranes were also applied to the pervaporation separation of water and butan-1-ol binary mixtures, the higher permeation flux and enrichment factors of butan-1-ol being reached with the supported ionic liquid membranes.

An interesting example reported on the use supported ionic liquid membranes in vapour permeation processes was the separation of benzene and cyclohexane [56]. It was found that benzene selectively permeated the SILMs and the separation factor increased with the increase in ionic liquid hydrophilicity.

3.4. Separation of Small Ions

Transport of small ions such as sodium (Na^+) and chloride (Cl^-) through SILM based on 1-n-alkyl-3-methylimidazolium cation has been studied [11]. The mechanism of water transport through these SILMs was analyzed founding that it was mainly regulated by the mobility of water microenvironments inside the IL, rather than by molecular diffusion through the bulk of the ionic liquid. Sodium and chloride, with low affinity towards the ILs used, were found to be mainly transported through the SILM, by transport through water microenvironments inside the IL.

3.5. Analytical Applications

Supported ionic liquid membranes have also found application in analytical chemistry, specifically in the trace determination of toxic. Because of the low concentration of some toxic and the complexity of the environmental samples, an enrichment step is usually needed prior to the instrumental analysis. Liquid–liquid extraction (LLE) and solid-phase extraction (SPE) are the most commonly used techniques for separation or preconcentration of toxic in environmental samples. However, these techniques often require large amounts of organic solvents, which are harmful (as VOCs) for the environment. Therefore, a variety of microextraction techniques that use no or small amounts of solvent were developed in recent years. Among them, liquid-phase microextraction (LPME) and solid-phase microextraction (SPME) are the two predominant extraction techniques for analysis of toxic [57]. Several examples of the application of ionic liquids as liquid membrane in hollow fiber supported liquid-phase microextraction have been reported [58,59]. This technique, combined with HPLC, was used for the determination of chlorophenols and sulfonamides in environmental water samples. Regarding the application of ionic liquids in solid-phase microextraction (SPME), a Nafion membrane-supported ionic liquid (IL)–solid phase microextraction coupled to gas chromatography–mass spectrometry (GC-MS) technique were developed for the simultaneous sampling and determination of ultra trace polycyclic aromatic hydrocarbons (PAHs) in aqueous samples [60]. Other example of SPME using ILs was reported by Zhao *et al.* (2008) [61], who developed polymeric imidazolium-based IL coatings for the extraction of esters. In the same context, López-Darias *et al.* (2010) [62] carried out the determination of a group of eighteen pollutants in waters, including polycyclic aromatic hydrocarbons and substituted phenols, in direct immersion solid-phase microextraction (SPME) using the polymeric ionic liquid (PIL)poly(1-vinyl-3-hexadecylimidazolium) bis[(trifluoromethyl) sulfonyl]imide as a novel coating material.

4. STABILITY OF SUPPORTED IONIC LIQUID MEMBRANES

One of the most important properties for the industrial application of SILMs is their stability. Therefore, stability studies of SILMs prepared by several methods at different operating conditions (e.g. surrounding phases, pressure and temperature) have been reported [8, 63-71].

The immobilization method was also found to have influence on the membrane stability. A comparative study of the preparation of SILMs by two different methods, under pressure and vacuum were reported by Hernández-Fernández *et al.* (2009) [63]. They use the ionic liquids $[bmim^+][Cl^-]$, $[bmim^+][BF_4^-]$, $[bmim^+][PF_6^-]$ and $[bmim^+][NTf_2^-]$ as liquid phase supported on a Nylon membrane. Small losses of ionic liquid were observed when the ionic liquid was immobilized under pressure after 7 days of operation in a diffusion cell. However, the losses of IL were higher when immobilization was carried out under vacuum, specially with the most viscous ionic liquids ($[bmim^+][PF_6^-]$ and $[bmim^+][Cl^-]$). This behaviour was explained by the higher capillary force associated with the higher viscous ILs which makes it difficult to penetrate into the middle of the deeper pores of the membrane and the ionic liquid was mainly immobilized on the most external layer of the membrane and, consequently, the immobilized ionic liquid is more easily removed during operation.

The influence of the ionic liquid composition on the stability of the resulting SILM in organic environments has also been analyzed. In this context, de los Rios. *et al.* (2008) [8] analyzed the stability of SILMs based on $[bmim^+][PF_6^-]$, $[bmim^+][BF_4^-]$ and $[bmim^+][NTf_2^-]$ using Nylon membranes as support in n-hexane/n-hexane. Stability tests were performed by keeping the respective impregnated membranes immersed for a week in a diffusion cell including two independent compartments and using n-hexane/n-hexane as the respective feed and receiving phases. Figure 4 shows the microstructure of the Nylon membrane impregnated with $[bmim^+][PF_6^-]$ after cell operation, which could be compared with the appearance of the SILM before cell operation (Figure 2b).

It was concluded that most of the ionic liquid from the external surface of the membrane disappeared during cell operation. However, comparison of the SEM-EDX spectra taken from membranes before (Figure 5a) and after (Figure 5b) immersion in the n-hexane/ n-hexane solution showed that they were very similar. The EDX spectra are taken from a sample of up to a few micrometers thick/depth and shows that the contribution of ionic liquid within the membrane pores is more important than the accumulated liquid found on the surface. Consequently, from the SEM study it was deduced that only the ionic liquid deposited on the external membrane surface has been stripped off during operation. The amount of ionic liquid retained in the membrane pores, however, was apparently kept constant and, consequently, the membrane was stable against the possible solvent action of n-hexane.

From the SEM-EDX study, it was concluded that, in all three cases studied of ILs supported on Nylon membranes, no substantial losses of ionic liquid occurred after a week of operation in an n-hexane/n-hexane diffusion cell. Only minimal losses of the ionic liquid accumulated on the external surface of SLMs were detected qualitatively by the SEM pictures.

The stability of a SILM based on $[bmim^+][BF_4^-]$ supported in a Nylon membrane has been also been analyzed in other organic solvents, such as *n*-hexane/*tert*-butyl methyl ether and *n*-hexane/dimethyl sulfoxide [64]. The SEM–EDX study of the membranes after continuous operation showed that the stability of the supported liquid membrane increase with the decrease of the polarity of the solvent used. The relationship established between polarity and stability allows the design of more stable SILMs for use in organic environments.

The stability of SILMs based on 1-n-alkyl-3-methylimidazolium hexafluorophosphate (n=4,8,10) supported in PVDF membrane towards contacting aqueous phase has been also analysed [10,35,65], resulting to be stable at the assayed conditions. These authors highlighted the importance of the consideration of two main possible effects on the performance and stability of SILMs in water mediums: (a) the loss of ionic liquids organic phase from the supporting membrane to the adjacent aqueous solutions by dissolution/emulsification; and (b) the formation of water microenvironments inside the ionic liquids organic phase, which constitute new, non-selective environments for solute transport, leading to a deterioration of the SLM performance and selectivity.

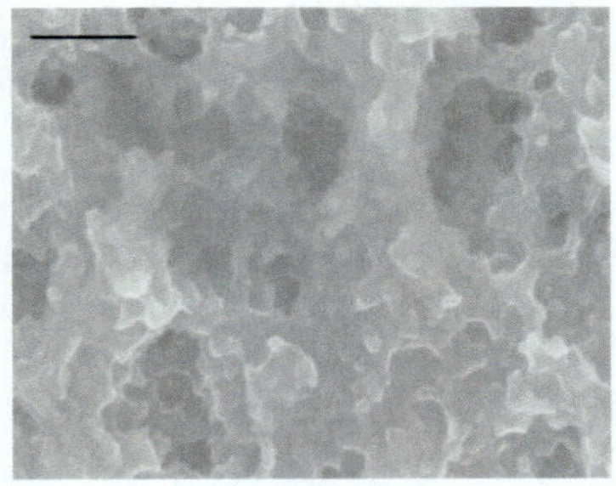

Figure 4. Scanning electron micrograph of the Nylon membrane impregnated with [bmim$^+$][PF_6^-] after seven days´ operation (3040x) (scale bar = 5 μm) in n-hexane/n-hexane

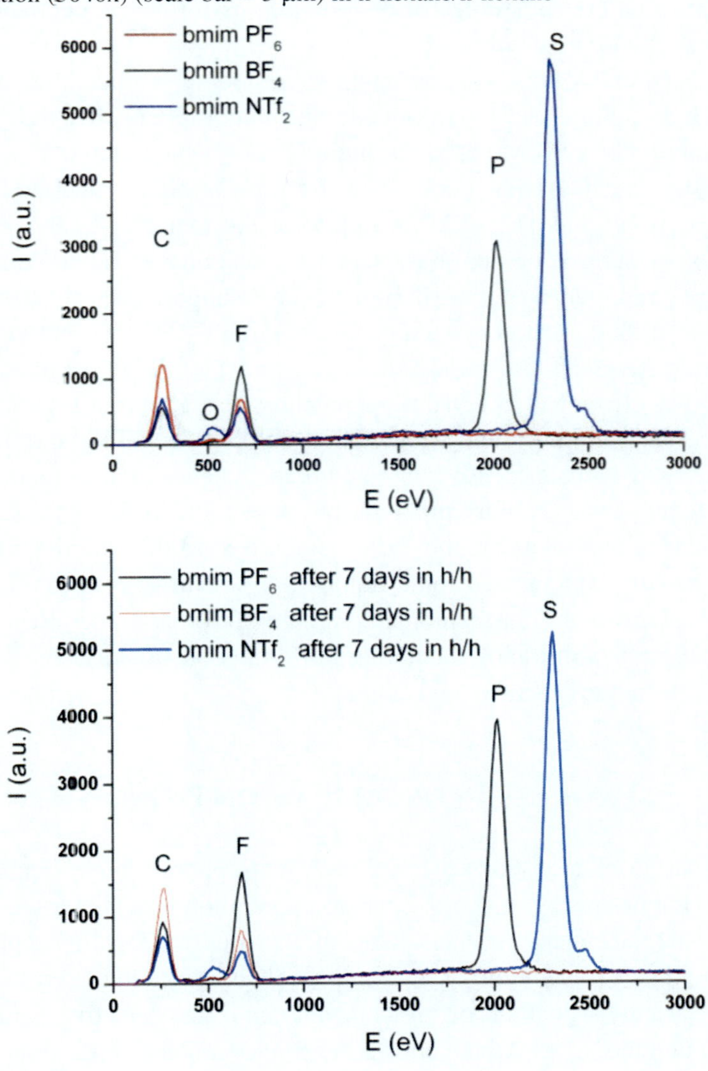

Figure 5. EDX spectra of the Nylon membranes impregnated with [bmim$^+$][PF_6^-], [bmim$^+$][BF_4^-] and [bmim$^+$][NTf_2^-] (a) before and (b) after seven days´ operation in n-hexane/ n-hexane

Stability studies of SILM have been also carried with gases as contacting phases [50,66-67]. The stability of SILMs based on the ionic liquid [$Et(OEt)_2^+$][PF_6^-] supported in Durapore hydrophobic PVDF was studied at 25°C with a pressure difference of 1.2 bar [66]. Experiments concerning stability showed that after eight cycles the permeability of H_2 slightly decreased while the CO_2 /H_2 ideal selectivity improved. Other interesting example was reported by Hanioka *et al.* (2008) [67], who examined the long term stability of a SLM based on a functionalized ionic liquid ([NH_2pmim$^+$][NTf_2^-]) in the separation of the CO_2/CH_4 gas mixture at atmosferical pressure. These authors did not found observable change in permeability or selectivity after 260 days operation. Scovazzo *et al.* (2009) [50] analyzed the stability of SILMs used in the separation of gas pairs CO_2/CH_4 and CO_2/N_2 using continuous flows of the mixed gases. They achieved long-term stability in continuous operation, from 24 to 106 days, by using imidazolium ionic liquids supported in hydrophilic-PVDF or hydrophilic-PES membranes, respectively, without performance degradation under CO_2-partial pressures of at least 207 kPa.

Temperature stability has a great interest for some applications of SILMs in gas separation, such as capture of CO_2 from coal gasification plants. Ilcovich *et al.* (2007) [68] analyzed the stability of a SILM based on [hmim$^+$][NTf_2^-] supported in a polysulfone organic membranes in the selective separation of CO_2 from He at high temperature. This membrane was found to be stable upto 125 ºC, the failure of the membranes above that temperature being attributable to support failure rather than any effect on the ionic liquid. Recently, Myers *et al.* (2008) [69] reported operating [hmim$^+$][NTf_2^-] supported in Nylon membranes up to 300 ◦C. It was also found that permeability in this [hmim$^+$][NTf_2^-]-membrane increased with temperature while the selectivity decreased.

All the above mentioned SILMs were prepared using microfiltration membranes and operated in constant volume system with low varying pressure. To avoid the pitfall of liquid instability associated with micron porous membranes, nanofiltration membranes were used which greatly reduce the instability problem, but potentially at the expense of increased gas transport resistance provided by the NF membranes [38]. Experimental stability tests demonstrated that the impregnated ionic liquids did not discharge from the NF membrane structure even under a high transmembrane pressure of 10 bars. The test results suggest that SILMs supported on nanofiltration membranes offer a good stability able to sustain high gas phase pressure relevant to practical applications.

5. Process Intensification Using SILM

The combination of biocatalysed reactions with supported ionic liquid membranes has been shown to be a promising integrated reaction/separation process for the synthesis of fine-chemical and pharmaceutical products. One of the most interesting applications of these systems is the kinetic resolution of racemic alcohols, since these systems integrate the enantioselective catalytic action of the enzyme and the selective permeability of the compounds through the SILM, which allows the conversion and separation of the enantiomer in a single step. The first example of process integration using SILMs were reported by Miyako *et al.* (2003) [70], who carried out the lipase-facilitated transport of (S)-ibuprofen through an ionic liquid-based SLM. In this system, the feed phase contains (rac)-ibuprofen and lipase from Candida rugosa and the enantioselective esterification takes places in this phase. Afterwards, esters selectively pass through the SLM to the receiving phase where they are further hydrolyzed by lipase from porcine pancreas. More recently, others kinetic resolutions have been carried out using this reaction/separation process, for example, the kinetic resolutions of *rac*-2-pentanol [71] and *rac*-1-phenyletanol [72] by transesterification with a vinyl ester catalysed by a commercial immobilized *Candida antarctica* lipase B (CaLB). In this system, the enantioselective esterification takes place in the feed phase, and the non-reacted isomer of the alcohol diffuses through the receiving phase. Figure 6 depicts the time course of the reaction/separation process in the feed and receiving phases in the case of the kinetic resolution of rac-phenylethanol using vinyl laurate as acylating agent.

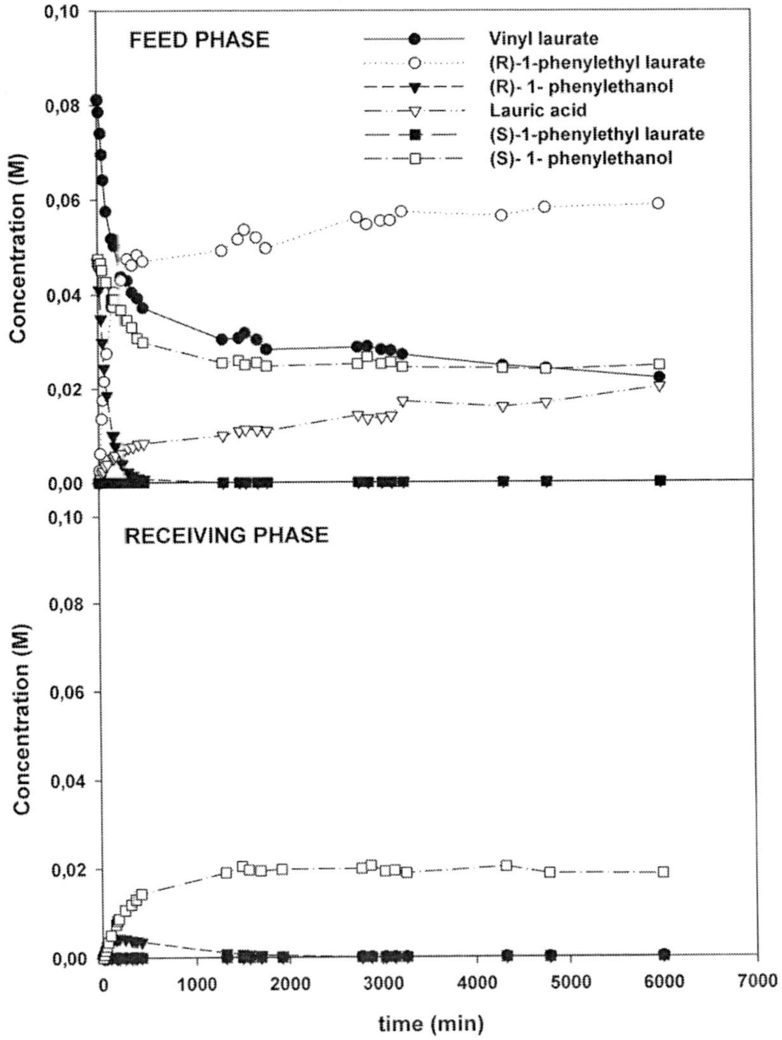

Figure 6. Time courses for (R)-1-phenylethanol, (S)-1-phenylethanol, vinyl laurate, (R)-1-phenylethyl laurate and lauric acid in the feed and receiving phase using the Nylon membrane with [$bmim^+$][BF_4^-] when vinyl laurate were used as acyl donor (conditions: initial solute concentrations, 100 mM; amount of enzyme, 80 mg; temperature, 30 °C; time, 72 h.)

As can be seen from this figure, the (S)-1-phenylethanol isomer in the feed compartment did not react with vinyl laurate since (S)-1-phenylethyl laurate was not detected, so the enantioselectivity exhibited by the enzyme was very high (e.e.$_{Product}$> 99.9%). Therefore, the (S)-1-phenylethanol diffused to the receiving phase until their equilibrium concentration was reached. In contrast, the (R)-1-phenylethanol reacted with the vinyl ester in the feed compartment to form (R)-1-phenylethyl laurate. In the early stage of the experiment the unreacted (R)-1-phenylethanol diffused to the receiving phase but, when the concentration of this substrate in the feed phase was lower than that in the receiving phase, the (R)-1-phenylethanol presented in the receiving phase returned to the feed phase, where it reacted with the vinyl laurate until no (R)-1- phenylethanol was detected. After 24 h of operation, the (S)-phenylethanol is nearly completely separated in the receiving phase. These results demonstrate that the coupling of lipases with an SILM provides a promising basis for the development of environmentally friendly methodologies for practical production of enantiomerically pure or enriched compounds.

6. Conclusions

Supported ionic liquid membranes (SILMs) have been applied to a wide range of separation processes in different fields of application due to the possibility of design suitable *ionic liquids* for *specific* separation problems. The emergence of a considerable number of new ILs, more and more currently task-specific ionic liquids, will open up new field of application of these SILMs. Stability studies on SILMs have corroborated their high operational and structural stability in liquid/liquid and gas/gas separations, even at relatively *high-temperature/high-pressure conditions, which makes possible their industrial application.* The use of this new technology in industrial processes could also lead to minimize waste generation and savings in energy consumption, while providing high product quality. As has been evidenced in this review, the supported ionic liquid membrane technologies have been incessantly progressing during the past ten years and no limit in the future progress is currently in sight.

References

[1] Crespo, JG; Coelhoso, IM; Viegas, RMC. Encyclopedia of Separation Processes, Academic Press, *San Diego*, 2000, 3303-3311.

[2] Muthuramam, G; Palanivelu, K. Transport of textile dye in vegetable oils based supported liquid membrane, *Dyes Pigm*, 2006, *70*, 99-104.

[3] Happel, S; Streng, R; Vater, P; Ensinger, W. Sr/Y separation by supported liquid membranes based on nuclear track micro filters, Radiat. *Meas*, 2003, 36, 761-766.

[4] Lozano, LJ; Godínez, C; Alguacil, FJ. Facilitated transport of vanadium (V) by supported liquid membranes, *Hydrometallurgy*, 2005, 80, 196-202.

[5] Yang, XJ; Fane, AG; Soldenhoff, K. Comparison of liquid membrane processes for metal separations: permeability, stability and selectivity, Ind. Eng. *Chem. Res.*, 2003, 42, 392-403.

[6] Teramoto, M; Sakaida, Fu, SS; Ohnishi, N; Matsuyama, H; Fukui, T; Arai, K. Attempt for the stabilization of supported liquid membrane, *Sep. Purif. Technol.*, 2000, 21, 137-142.

[7] Kemperman, AJB; Bargeman, D; Boomgaard, T; Strathmann, H. Stability of supported liquid membranes: state of the art, Sep. *Sci. Technol.*, 1996, 31, 2733-2762.

[8] Takeuchi, H; Takahashi, K; Goto, W. Some observations on the stability of supported liquid membranes, *J. Membr. Sci.*, 1987, 34, 19-31.

[9] de los Rios, AP; Hernandez-Fernandez, FJ; Tomas-Alonso, F; Palacios, JM; Gomez, D; Rubio, M; Villora, G. A SEM–EDX study of highly stable supported liquid membranes based on ionic liquids, *J. Membr. Sci.*, 2007, 300, 88-94.

[10] Fortunato, R; Afonso, CAM; Benavente, J; Rodriguez-Castellon, E; Crespo, JG. Stability of supported ionic liquid membranes as studied by X-ray photoelectron spectroscopy, *J. Membr. Sci.*, 2005, 256, 216-223.

[11] Fortunato, R; Afonso, CAM; Reis, MAM; Crespo, JG. Supported liquid membranes using ionic liquids: study of stability and transport mechanism, *J. Membr. Sci.*, 2004, 242, 197-209.

[12] Brennecke, JF; Maginn, EJ. Ionic liquids: Innovative fluids for chemical processing, *AIChE J*, 2001, 47, 2384-2389.

[13] Branco, LC; Crespo, JG; Afonso, CAM. High selective transport of organic compounds by using supported liquid membranes based on ionic liquids, Angew. *Chem. Int. Ed.*, 2002, 41, 2771-2773.

[14] Matsumoto, M; Inomoto, Y; Kondo, K. Selective separation of aromatic hydrocarbons through supported liquid membranes based on ionic liquids, *J. Membr. Sci.*, 2005, 246, 77-81.

[15] Marták, J; Schlosser, Š; Vlčková, S. Pertraction of lactic acid through supported liquid membranes containing phosphonium ionic liquid, *J. Membr. Sci.*, 2008, 318, 298-310.

[16] Hernández-Fernández, FJ; de los Ríos, AP; Rubio, M; Tomas-Alonso, F; Gomez, D; Víllora, G. A novel application of supported liquid membranes based on ionic liquids to the selective simultaneous separation of the substrates and products of a transesterification reaction, *J. Membr. Sci.*, 2007, 293, 73-80.

[17] Scovazzo, P; Kieft, J; Finan, DA; Koval, C; DuBois, D; Noble, R. Gas separations using non-hexafluorophosphate anion supported ionic liquid membranes, *J. Membr. Sci.*, 2004, 238, 57-63.

[18] de los Ríos, AP; Hernández-Fernández, FJ; Rubio, M; Tomas-Alonso, F; Gomez, D; Víllora, G. Prediction of the selectivity in the recovery of transesterification reaction products using supported liquid membranes based on ionic liquids, *J. Membr. Sci.*, 2008, 307, 225-232.

[19] de los Ríos, AP; Hernández-Fernández, FJ; Lozano, LJ; Sánchez, S; Moreno, JI; Godínez, C. Removal of Metal ions from aqueous solutions by extraction with ionic liquids, *J. Chem. Eng. Data*, 2009, *In press.*

[20] Ruiz, A; de los Rios, AP; Hernández-Fernández, FJ; Janssen, MHA; Schoevaart, R; van Rantwijk, F; Sheldon, RA. A cross-linked enzyme aggregate of *Candida antarctica* lipase B is active in denaturing ionic liquids, Enzyme Microb. *Technol.*, 2007, 40, 1095-1099.

[21] de los Rios, AP; Hernandez-Fernandez, FJ; Rubio, M; Gomez, D; Villora, G. Stabilization of native penicillin G acylase by ionic liquids, *J. Chem. Technol. Biotechnol.*, 2007, 82, 190-195.

[22] Feher, E; Illeova, V; Kelemen-Horvath, I; Belafi-Bako, K; Polakovic, M; Gubicza, L. Enzymatic production of isoamyl acetate in an ionic liquid–alcohol biphasic system, *J. Mol. Catal. B: Enzym.*, 2008, 50, 28-32.

[23] Frater, T; Gubicza, L; Szollosy, A; Bakos, J. Enantioselective hydrogenation in ionic liquids: recyclability of the [Rh(COD)(DIPAMP)]BF4 catalyst in [bmim][BF_4], *Inorg. Chim. Acta*, 2006, 359, 2756-2759.

[24] Sheldon, R. Catalytic reactions in ionic liquids, *Chem. Commun.*, 2001, 23, 2399-2407.

[25] Earle, MJ; Seddon, KR. Ionic liquids. Green solvents for the future, Pure Appl. *Chem.*, 2000, 72, 1391-1398.

[26] Scott, K; Basov, N; Jachuck, RJJ; Winterton, N; Cooper, A; Davies, C. Reactor studies of supported ionic liquids rhodium-catalysed hydrogenation of propene, *Chem. Eng. Res. Des.*, 2005, 83, 1179-1185.

[27] Branco, LC; Grespo, JG; Afonso, CAM. Studies on the selective transport of organic compounds by using ionic liquids as a novel supported liquid membranes, *Chem. Eur. J*, 2002, 8, 3865-3871.

[28] Bonhote, P; Dias, AP; Papageourgio, N; Kalayanasundaram, K; Gratzel, M. Hydrophobic, highly conductive ambient-temperature molten salts, Inorg. *Chem.*, 1996, 35, 1168-1178.

[29] Welton, T. Room-temperature ionic liquids. Solvents for synthesis and catalysis, *Chem. Rev.*, 1999, 99, 2071-2083.

[30] Seddon, K. Ionic liquids: designer solvents for green synthesis, *Chem. Eng.*, 2002, 730, 33-35.

[31] Scovazzo, P; Visser, AE; Davis Jr., JH; Rogers, RD; Koval, CA; DuBois, DL; Noble, RD. Supported ionic liquid membranes and facilited ionic liquid membranes, in: R.D. Rogers, K. R. Seddon (Eds.*), Ionic Liquids: Industrial applications to Green Chemistry*, ACS Symposium Series 818, *Am. Chem. Soc.*, Washington, DC, 2002, 68.

[32] Hernández-Fernández, FJ; de los Ríos, AP; Tomás-Alonso, F; Gómez, D; Víllora, G. Kinetic resolution of 1-phenylethanol integrated with separation of substrates and products by a supported ionic liquid membrane. *J. Chem. Technol. Biotechnol*, 2007, *82*, 190-195.

[33] Neves, LA; Nemestothy, N; Alves, VD; Cserjesi, P; Belafi-Bako, K; Coelhoso, IM. Separation of biohydrogen by supported ionic liquid membranes, *Desalination*, 2009, 240, 311-315.

[34] Luis, P; Neves, LA; Afonso, CAM; Coelhoso, IM; Crespo, JG; Garea, A; Irabien, A. Facilitated transport of CO_2 and SO_2 through Supported Ionic Liquid Membranes (SILMs), *Desalination*, 2009, 245, 485-493.

[35] Fortunato, R; González-Muñoz, MJ; Kubasiewicz, M; Luque, S; Alvarez, JR; Afonso, CAM; Coelhoso, IM; Crespo, JG. Liquid membranes using ionic liquids: the influence of water on solute transport, *J. Membr. Sci.*, 2005, 249, 153-162.

[36] Hernández, FJ; de los Ríos, AP; Tomás-Alonso, F; Gómez, D; Víllora, G. Preparation of supported ionic liquid membranes: Influence of the ionic liquid immobilization method on their operational stability, *J. Membr. Sci.*, 2009, 341, 172-177.

[37] Ropel, L; Belvèze, LS; Aki, SNVK; Stadtherr, MA; Brennecke, JF. Octanol–water partition coefficients of imidazolium-based ionic liquids, *Green Chem.*, 2005, **7,** 83-90.

[38] Gan, Q; Rooney, D; Xue, M; Thompson, G; Zou, Y. An experimental study of gas transport and separation properties of ionic liquids supported on nanofiltration membranes, *J. Membr. Sci.*, 2006, 280, 948-956.

[39] Hernández, FJ; de los Ríos, AP; Tomás-Alonso, F; Gómez, D; Víllora, G. Kinetic resolution of 1-phenylethanol integrated with separation of substrates and products by a supported ionic liquid membrane, *J. Chem. Technol. Biotechnol*, 2007, 82, 190-195.

[40] de los Ríos, AP; Hernández, FJ; Tomás-Alonso, F; Gómez, D; Víllora, G. Synthesis of flavour esters using free *Candida antarctica* lipase B in ionic liquids, *Flavour Fragrance Journal*, 2008, 23, 319-322.

[41] de los Ríos, AP; Hernández-Fernández, FJ; Rubio, M; Tomás-Alonso, F; Gómez, D; Víllora, G. Prediction of the selectivity in the recovery of transesterification reaction products using supported liquid membranes based on ionic liquids, *J. Memb. Sci.*, 2007, 293, 73-80.

[42] Hernández, FJ; de los Ríos, AP; Tomás-Alonso, F; Gómez, D; Víllora, G. Improvement in the separation efficiency of transesterification reaction compounds by the use of supported ionic liquid membranes based on the dicyanamide anion, *Desalination*, 2009, 244, 122-129.

[43] de los Ríos, AP; Hernández-Fernández, FJ; Rubio, M; Gómez, D; Víllora, G. Highly selective transport of transesterification reaction compounds through supported liquid membranes containing ionic liquids based on the tetrafluoroborate anion, *Desalination*, 2010, 250, 101-104.

[44] de los Rios, AP, Hernandez-Fernandez, FJ; Presa, H; Gomez, D; Villora, G. Tailoring supported ionic liquid membranes for the selective separation of transesterification reaction compounds, *J. Membr. Sci.*, 2009, 328, 81-85.

[45] Matsumoto, M; Inomoto, Y; Kondo, K. Comparison of solvent extraction and supported liquid membrane permeation using an ionic liquid for concentrating penicillin G, J. *Membr. Sci.*, 2007, 289, 92-96.

[46] Scovazzo, P; Havard, D; McShea, M; Mixon, S; Morgan, D. Long-term, continuous mixed-gas dry fed CO2/CH4 and CO2/N2 separation performance and selectivities for room temperature ionic liquid membranes, *J. Membr. Sci.*, 2009, 327, 41-48.

[47] Bara, JE; Lessmann, S; Gabriel, CJ; Hatakeyama, ES; Noble, RD; Gin, DL. Synthesis and Performance of Polymerizable Room-Temperature Ionic Liquids as Gas Separation Membranes, Ind. Eng. *Chem. Res.*, 2007, 46, 5397-5404.

[48] Hanioka, S; Maruyama, T; Sotani, T; Teramoto, M; Matsuyama, H; Nakashima, K; Hanaki, M; Kubota, F; Goto, M. CO_2 separation facilitated by task-specific ionic liquids using a supported liquid membrane, *J. Membr. Sci.*, 2008, 314, 1-4.

[49] Cserjesi, P; Nemestothy, N; Vass, A. Zs. Csanadi, K. Belafi-Bako, Study on gas separation by supported liquid membranes applying novel ionic liquids, *Desalination*, 2009, 245, 743-747.

[50] Myers, C; Pennline, H; Luebke, D; Ilconich, J; Dixon, JK; Maginn, EJ; Brennecke, JF. High temperature separation of carbon dioxide/hydrogen mixtures using facilitated supported ionic liquid membranes, *J. Membr. Sci.*, 2008, 322, 28-31.

[51] Gan, Q; Rooney, D; Xue, M; Thompson, G; Zou, Y. An experimental study of gas transport and separation properties of ionic liquids supported on nanofiltration membranes, *J. Membr. Sci.*, 2006, 280, 948-956.

[52] Ilconich, J; Myers, C; Pennline, H; Luebke, D. Experimental investigation of the permeability and selectivity of supported ionic liquid membranes for CO_2/He separation at temperatures up to 125 °C, *J. Membr. Sci.*, 2007, 298, 41-47.

[53] Park, YI; Kim, BS; Byun, YH; Lee, SH; Lee, EW; Lee, JM. Preparation of supported ionic liquid membranes (SILMs) for the removal of acidic gases from crude natural gas, *Desalination*, 2009, 236, 342-348.

[54] Izák, P; Köckerling, M; Kragl, U. Solute transport from aqueous mixture throught supported ionic liquid membrane by pervaporation, *Desalination*, 2006, 199, 96-98.

[55] Izák, P; Ruth, W; Fei, Z; Dyson, PJ; Kragl, U. Selective removal of acetone and butan-1-ol from water with supported ionic liquid–polydimethylsiloxane membrane by pervaporation, *Chem. Engin. Journal*, 2008, 139, 318-321.

[56] Matsumoto, M; Ueba, K; Kondo, K. Vapor permeation of hydrocarbons through supported liquid membranes based on ionic liquids, *Desalination*, 2009, 241, 365-371.

[57] Sun, P; Armstrong, DW. Ionic liquids in analytical chemistry, Analytica Chimica Acta 2010, In press.

[58] Peng, J; Liu, J; Hu, X; Jiang, G. Direct determination of chlorophenols in environmental water samples by hollow fiber supported ionic liquid membrane extraction coupled with high-performance liquid chromatography, *J. Chromatogr. A*, 2007, 1139, 165-170.

[59] Tao, Y; Liu, J; Hu, X; Li, H; Wang, T; Jiang, G. Hollow fiber supported ionic liquid membrane microextraction for determination of sulfonamides in environmental water samples by high-performance liquid chromatography, *J. Chromatogr. A*, 2009, 1216, 6259-6266.

[60] Hsieh, YN; Huang, PC; Sun, IW; Whang, TJ; Hsu, CY; Huang, HH; Kuei, CH. Nafion membrane-supported ionic liquid–solid phase microextraction for analyzing ultra trace PAHs in water samples, *Analytical Chim. Acta*, 2006, 557, 321-328.

[61] Zhao, F; Meng, Y; Anderson, JL. Polymeric ionic liquids as selective coatings for the extraction of esters using solid-phase microextraction, *J. Chromatogr. A*, 2008, 1208, 1-9.

[62] López-Darias, J; Pino, V; Anderson, JL; Graham, CM; Afonso, AM. Determination of water pollutants by direct-immersion solid-phase microextraction using polymeric ionic liquid coatings, *Journal of Chromatography A*, 2010, In press.

[63] Hernández-Fernández, FJ; de los Ríos, AP; Tomás-Alonso, F; Palacios, JM; Víllora, G. Preparation of supported ionic liquid membranes: Influence of the ionic liquid immobilization method on their operational stability, *J. Membr. Sci.*, 2009, 341, 172-177.

[64] de los Ríos, AP; Hernández-Fernández, FJ; Tomás-Alonso, F; Palacios, JM; Víllora, G. Stability studies of supported liquid membranes based on ionic liquids: Effect of surrounding phase nature, *Desalination*, 2009, 245, 776-782.

[65] Bijani, S; Fortunato, R; Martínez de Yuso, MV; Heredia-Guerrero, FA; Rodríguez-Castellón, E; Coehloso, I; Crespo, J; Benavente, J. Physical–chemical and electrical characterizations of membranes modified with room temperature ionic liquids: Age effect, *Vacuum*, 2009, 83, 1283-1286.

[66] Cserjési, P; Nemestóthy, N; Vass, A. Csanádi, Zs; Bélafi-Bakó, K. Study on gas separation by supported liquid membranes applying novel ionic liquids, *Desalination*, 2009, 245, 743-747.

[67] Hanioka, S; Maruyama, T; Sotani, T; Teramoto, M; Matsuyama, H; Nakashima, K; Hanaki, M; Kubota, F; Goto, M. CO2 separation facilitated by task-specific ionic liquids using a supported liquid membrane, *J. Membr. Sci.*, 2008, 314, 1-4.

[68] Ilconich, J; Myers, C; Pennline, H;. Luebke, D Experimental investigation of the permeability and selectivity of supported ionic liquid membranes for CO_2/He separation at temperatures up to 125 °C, *J. Membr. Sci.*, 2007, 298, 41-47.

[69] Myers, C; Pennline, H; Luebke, D; Ilconichb, J; Dixon, JK; Maginn, EJ; Brennecke, JF. High temperature separation of carbon dioxide/hydrogen mixtures using facilitated supported ionic liquid membranes, *J. Membr. Sci.*, 2008, 322, 28-31.

[70] Miyako, E; Maruyama, T; Kamiya, N; Goto, M. Use of ionic liquids in a lipase-facilitated supported liquid membrane, *Biotechnol. Letters*, 2003, 25, 805-808.

[71] Hernández-Fernández, FJ; de los Ríos, AP; Tomás-Alonso, F; Gómez, D; Víllora, G. On the development of an integrated membrane process with ionic liquids for the kinetic resolution of *rac*-2-pentanol, *J. Membr. Sci.*, 2008, 314, 238-246.

[72] Hernández-Fernández, FJ; de los Ríos, AP; Tomás-Alonso, F; Gómez, D; Rubio, M; Víllora, G. Integrated reaction/separation processes for the kinetic resolution of rac-1-phenylethanol using supported liquid membranes based on ionic liquids, *Chem. Engin. Process*, 2007, 46, 818-824.

In: Desalination: Methods, Cost and Technology
Editor: Irena A. Urbonienė pp. 229-245
ISBN: 978-1-61668-909-4

Chapter 7

A NONCONVENTIONAL REVERSE OSMOSIS PLANT POWERED BY STORAGE OF WIND ENERGY AS GRAVITATIONAL POTENTIAL ENERGY

E.A.F.A. Fadigas [*] ***and J.R. Dias***
Department of Electrical Automation and Energy Engineering, São Paulo University,
Av. Luciano Gualberto, Travessa 3, No. 158, Cidade Universitária,
São Paulo, Brazil, 05508-900

ABSTRACT

This chapter proposes two alternative systems for conventional reverse osmosis (RO) desalination plants by incorporating the use of gravitational potential energy. In the proposals, this kind of energy is obtained by storing water to be desalinated (sea or brackish -water).

The primary source of energy of both systems is wind. According to the chosen device (windmill or wind generator) a percentage of the available wind energy is used to pump water to a reservoir located at a certain height (less than 20 m). The stored water has the potential energy that will make a special plunger pump work. The piston of this special plunger pump is designed so that high pressure (over 55 bar) can be achieved in a different way from the conventional one. Conventionally, RO plants use a high-pressure pump, powered by electricity or fossil fuel. The function of the pump is to send a flux of saline water to a group of semi-permeable membrane modules, capable of "filtering" the dissolved salts.

In the alternative systems, here proposed, to lift water to the reservoir, the first one uses multi-bladed windmills and the second one uses wind generators that power an electric pump. To pump water to the RO membranes, the special pistons go downwards due to their own weight, and are lifted, through a system of pulleys, with a counterweight filled with water obtained from the reservoir. The technical viability of the alternatives was theoretically proven by deductions based on physics and mathematics and with a prototype.

[*] Corresponding author: Tel. +55 (011) 3091-5349; Fax: +55 (011) 3091-5349; email: eliane.fadigas@poli.usp.br

Case studies are presented and, along them, the two systems are compared based on some chosen performance parameters such as specific consumption of energy, occupied area, availability etc.

Keywords: Wind energy; Gravitational potential energy; Reverse osmosis; Water desalination.

1. Introduction

Desalination is a process through which pure water is obtained from water with a high concentration of dissolved salt (seawater or brackish water). In general, in a desalination plant there is an inlet flux of salted water and two outlet fluxes, Figure 1.

One of the outlet flux (the brine) has a concentration higher than the inlet flux and the other has a much lower concentration than the inlet flux (drinking water) To achieve that high reduction of solute concentration, it is necessary to deliver a certain amount of energy. The consumption of energy of a desalination plant is generally evaluated in terms of the quantity of energy used for each cubic meter of drinking water obtained. This parameter is called the specific energy consumption expressed in kWh/m^3. Among all desalination processes we will briefly describe the reverse osmosis process for it is the one used with the system proposed in this chapter.

2. Reverse Osmosis-Ro

Reverse osmosis is the reversal of a process that occurs naturally named osmosis. Osmosis occurs when two solutions with different concentration are separated by a semi-permeable membrane, Figure 2.

In such circumstances, there is a natural tendency to achieve the chemical equilibrium. Thus, the solvent crosses the membrane from the side of less concentration to the other side until the concentrations on both sides are the same. To reverse this natural tendency, a pressure (higher than the osmotic pressure) must be exerted on the side originally with higher concentration. This is what is done to achieve desalination by reverse osmosis and why, generally, there is a high specific energy consumption. Figure 3 shows a schematic of a RO desalination plant.

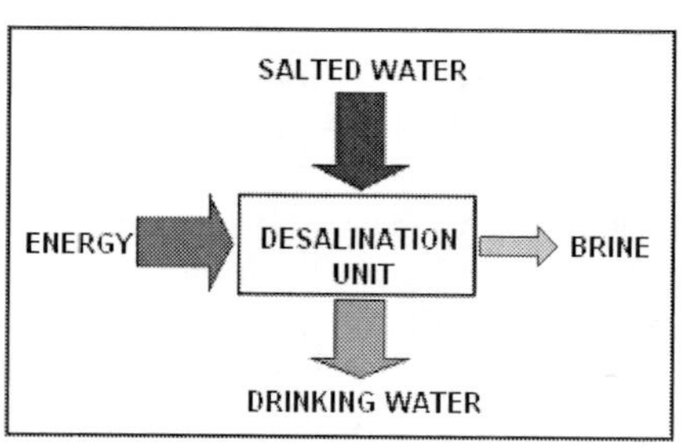

Figure 1. Simplified scheme of a desalination unit

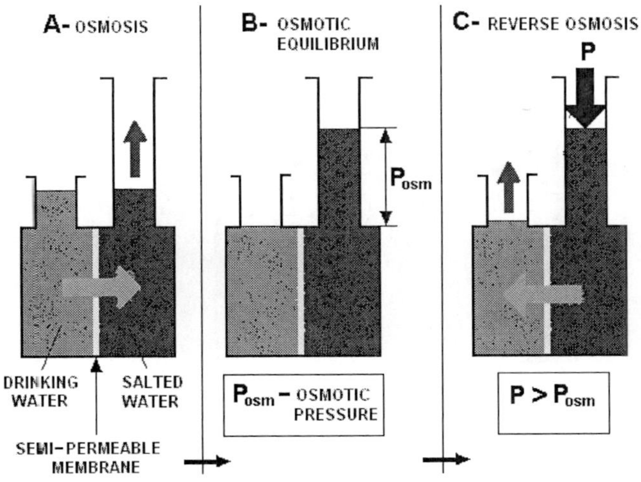

Figure 2. Fluxes by Osmosis, and by Reverse Osmosis

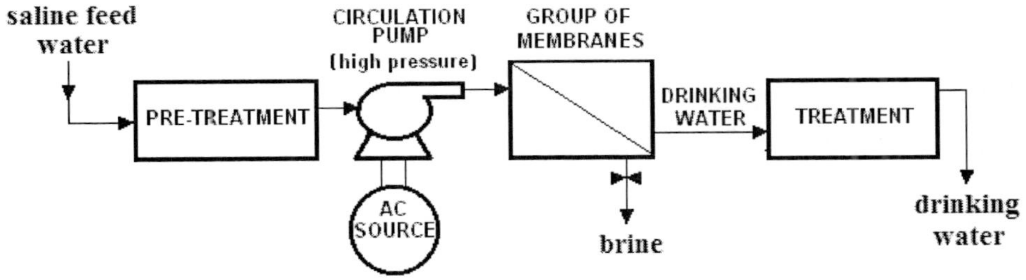

Figure 3. Scheme of a reverse osmosis desalination plant

In this chapter it is proposed an alternative RO system powered by wind and gravitational potential energy- GPE. The proposed system would basically replace the high pressure circulation pump shown in the configuration of Figure 3.

3. The Proposed System

Figure 4-A shows the general configuration of the proposed system which desalinates water by RO using wind energy, which is stored as potential gravitational energy in a reservoir at a certain height (less than 20 m). The water is pumped to the reservoir either by a traditional windmill (system 1-S1) or with an electric pump powered by a wind turbine (system 2- S2). A subsystem (the hydraulic power column-HPC) of the model uses the gravitational potential energy of the stored water to achieve the necessary pressure (about 55 bar for seawater) to feed the reverse osmosis membranes so that fresh water is obtained. Notice that S1 does not use any electricity throughout the whole process.

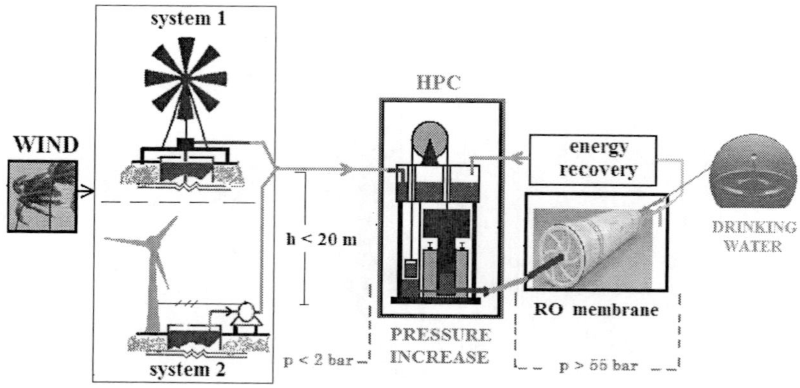

Figure 4 A. The Hydraulic Power Column renewable system

Figure 4-B shows the block diagram of the model outlining the Hydraulic Power Column-HPC where the increase in pressure occurs (from less than 2 bar to over 55 bar).

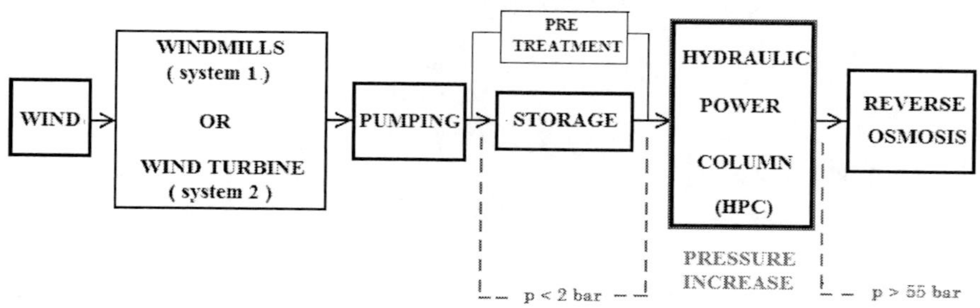

Figure 4-B. block diagram of the renewable system

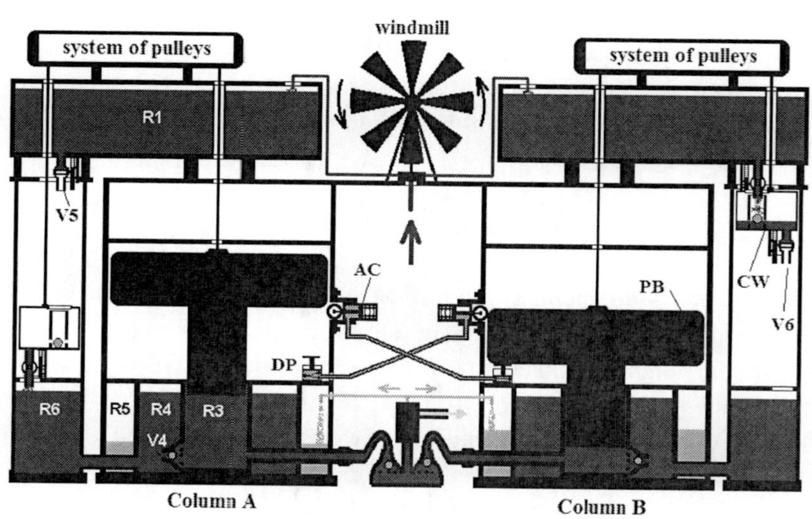

Figure 5. parallel operation of two hydraulic pressure columns. R1 fed with a windmill:S1

For seawater desalination, one needs to have an operation pressure of around 55 bar while for brackish water it is around 20 bar. With the proposed system, this magnitude of pressure is given by a special piston working inside the HPC.

Figure 5 shows the proposed configuration for the HPC. The model is composed of two HPC's which, separately, can be seen as individual special piston pump. The two individual pumps are connected through a hydraulic automation system in order to send a continuous flux of salted water to the modules of membrane where the desalination occurs.

The pre-treatment of seawater namely filtering is designed to occur by gravity in reservoir R1. After the filtering process chemicals may be added if necessary. Note that if a reverse osmosis desalination system works with a low recovery rate (e.g. 20%) chemicals may not be needed. Recovery rate is the percentage of drinking water obtained from a volume of sea water that passes the membrane module (the rest is brine at high pressure).

With the system here proposed, to recover the brine energy a Clark pump is used. It is a kind a flow-work device. Brine at high pressure enters the Clark pump leaves it at atmospheric pressure and delivers work through a double piston pump. The system uses this work to pump water to the top reservoir R1.

Calculations have shown [3;10] that it is possible to reach high pressures with a special piston of moderate dimensions. The counterweight gets the necessary volume of water (GPE) from the reservoir R1 to make the special piston go upwards. The system of pulleys is designed so that it yields a reasonable force reduction to lift the piston. When the CW empties, the special piston goes downwards exerting pressure on the water in chamber R3. Consequently water is pumped into the reverse osmosis module under the designed and desired pressure making reverse osmosis happen.

Shortly, during the parallel operation of columns A and B, (see Figure 5) there are the following events:

1) When the piston PB of column B reaches the its bottom course, two events occur:
   The counterweight CW is filled up because valve V5 opens.
   The bottom of the piston rests on the dampers DP, pressing the fluid inside them. Simultaneously the actuators AC of column A retract due to the signal (fluid under pressure) received from DP, and the piston of that column is released.
   Reciprocally, when the piston of column A reaches its bottom course, two identical events occur:
2) When the piston of column B, raised by the counterweight, reaches the upper course two other events occur:
   It is braked by the actuators and will remain in that position until, after filled up with water, it receives the signal sent by the actuators of column A
   Its CW empties because valve V6 opens.

An identical process takes place when the piston of column A comes to its upper course.

The hydraulic automatism guarantees the continuous function of the system as long as there is enough water in the reservoirs R1. The continuous flux of salted water to the modules of membrane is guaranteed by the two-ways valve V7 that connect the two columns.

So far, a small prototype has been constructed at the polytechnic school of the *Universidade de São Paulo-USP in Brazil.* The prototype has proved that the mechanism of the HPC works. Since osmosis reverse is already a solid technique, the focus of the prototype was not to obtain drinking water because it would not be reasonable to reach 55 bar because of the size of the special piston.

Therefore the objective of the prototype was to prove that the HPC pumps water using the gravitational potential energy obtained from the water in R1 and that the necessary increase of pressure is achieved due to the shape of the special piston. The flux under the pressure exerted by the piston was used to drive a small Pelton wheel supported with magnetic levitation. This was done so, also to prove that the system (specially S1) may be used for electricity generation which is another goal of the HPC under study.

4. THE SPECIAL PISTONS AND THEIR HIGH PRESSURE

Figure 6 shows the shape of the special piston that works inside the HPC. The HPC may work either with a massive special piston (piston B) or with one (piston A) where the increase of pressure is achieved mainly with the weight of water obtained from R1.

The pressure P_A exerted by the special piston A on its circular bottom of diameter d_e is a function of the geometric parameters illustrated in Figure 6. According to results already obtained [3], it is possible to achieve pressures high enough (over 55 bar) to desalinate seawater with a special piston of moderate dimensions (fitting inside a 5 m height HPC).

The expression for pressure P_A exerted by the special piston A on its bottom of diameter ***de*** is given by the total weight of the piston ***Pt***, divided by the circular area $\mathbf{A_e}$.

$$P_A = Pt / A_e \qquad (1)$$

Starting with Eq 1 and knowing that ***Pt*** is the sum of the weight of steel, P_{steel} plus that of the water P_{water} in R2 and the filling up material P_{fill}, and that $A_e = \pi/4\ de^2$, after some algebraic manipulation, one gets the following relation [3; 10]

$$P_A = \frac{g}{de^2}\,[\rho_{stl}\,(De^2 b + de^2 a)] - Di^2(b-2t) - di^2 a] + \rho_{wat} Di^2 n + \rho_{fill}\, di^2 a] \qquad (2)$$

Where:

- **g** is the acceleration of gravity [m/s^2]
- ρ_{stl} is the density of steel [kg/m^3]
- ρ_{wat} is the density of seawater (or brackish water) [kg/m^3]
- ρ_{fill} is the density of the of the filling up material [kg/m^3]
- and **Di=De -2t** [m], being **t** the thickness of the piston′s wall.

A similar approach with special piston B leads to the following expression for P_B:

$$P_B = \frac{g}{de^2}\,\rho_{stl}\,(De^2 b + de^2 a) \qquad (3)$$

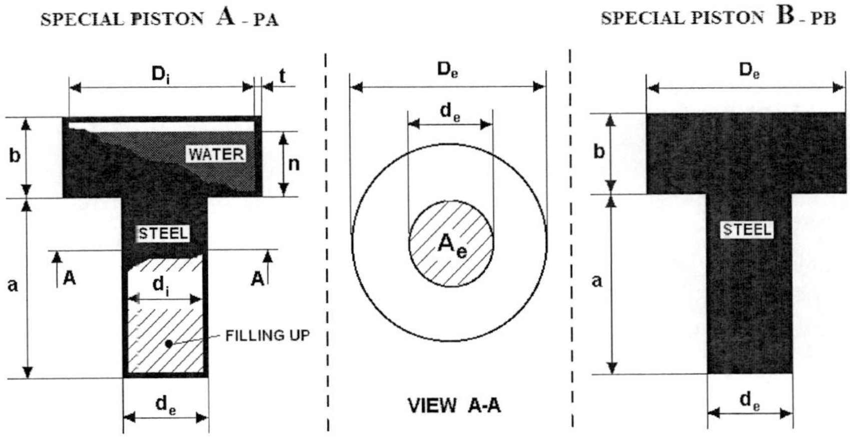

Figure 6. Scheme of the special piston

A very interesting aspect of using special piston A is the filling up material may be anything available where the system will operate such as stones, sand, or even recycled metals. This low cost characteristic is useful for remote areas because it would not be necessary to transport a very heavy structure.

5. The System's Specific Energy Consumption

The analysis of a configuration for an underground HPC, Figure 7, led us to some conclusions about the specific consumption of energy [3;10]. Notice that there is no need to pump water to R1 and the energy consumption is used only to lift the piston.

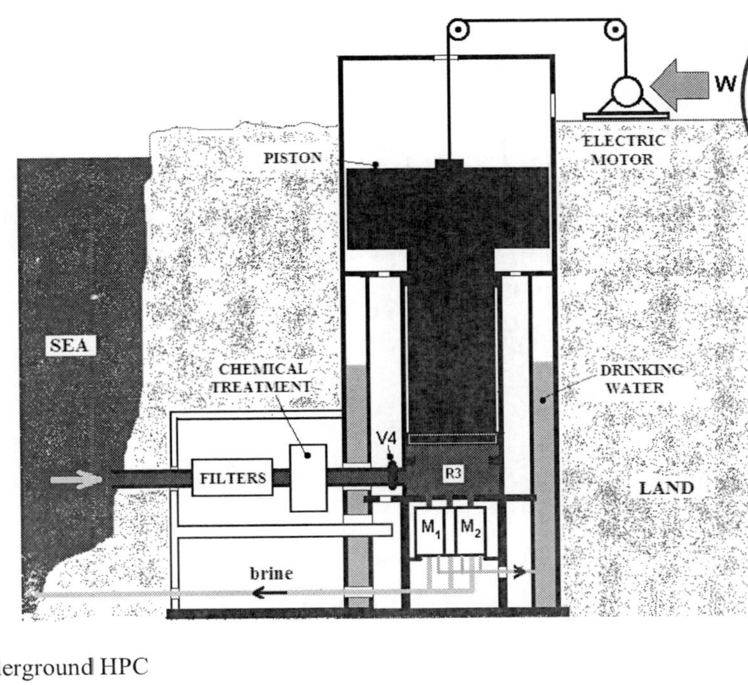

Figure 7. Underground HPC

considering a frictionless piston with an operating pressure of 20, 3 bar the system would consume around 0,563 kWh for each cubic meter of drinking water which is the minimum allowed by Nature [5].

considering a frictionless piston with an operating pressure of 55 bar the HPC would consume around 1,528 kWh for each cubic meter of drinking water .

For an operating pressure of 55 bar and taking into account acceptable efficiencies ("actual" values) for the components of the HPC, it has been shown that the predicted "actual" specific energy consumption is about 2,811 kWh for each cubic meter of drinking water. This is a very good result compared to the conventional RO systems for which this parameter is generally above 4 kWh/m^3.

6. FEEDING R1 WITH TWO DIFFERENT SYSTEMS

The HPC is designed so that it works as long as there is enough water stored in the reservoir R1 regardless of how water was put in it. This characteristic yields a certain versatility to the whole system. This means that one could feed R1 with other sources of energy such as wind, sea-waves, sun, etc. However, in this chapter we cope with two ways of feeding R1 with wind energy: one with windmills (system 1) and the other with a wind turbine (system 2) Figure 8.

Figures 9 and 10 show the block diagrams of system1 and system2. As mentioned before the energy recovery subsystem is a Clark pump which is a device that converts the flow of brine under high pressure into work which is used to pump water to R1. Thanks to the energy recovery subsystem, the system has a predicted low specific energy consumption.

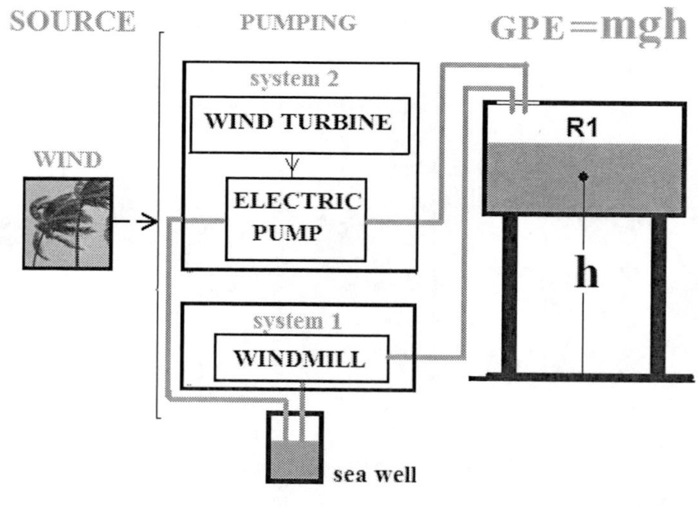

Figure 8. Feeding of Reservoir R1 with system 1 or system 2

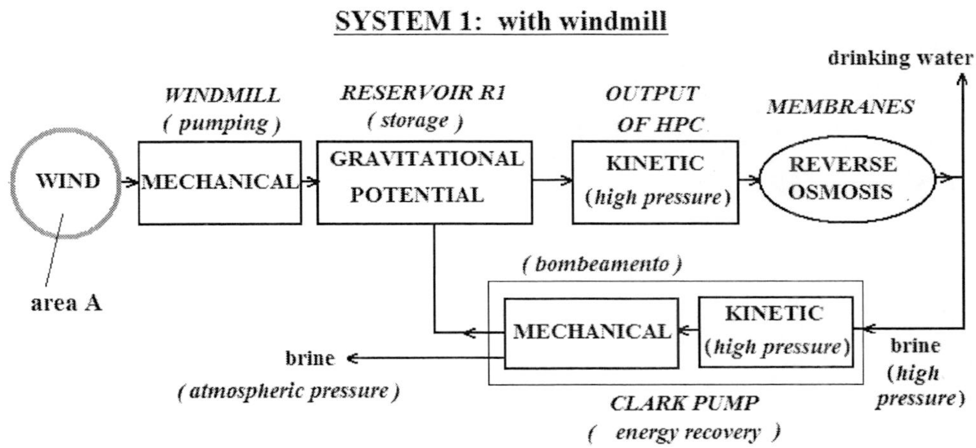

Figure 9. Block diagram of system 1

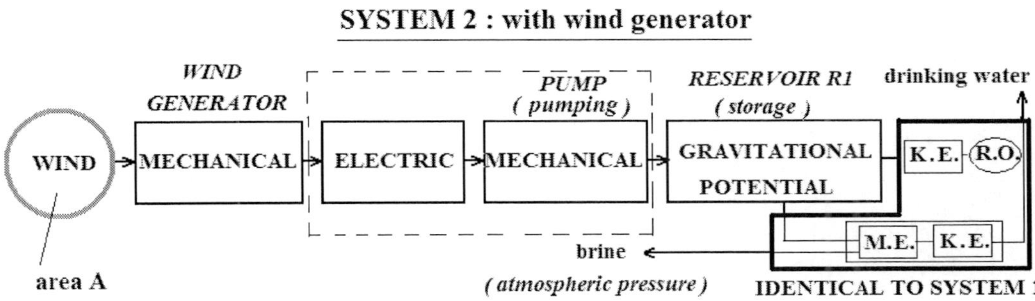

Figure 10. Block diagram of system 2

7. Calculation of Pumping Power for Both Systems

Figures 11 and 12 shows the chain of energy transformations that occur in S1 and S2, starting from wind energy until the storage of gravitational potential energy (GPE) in the reservoir R1 on top of the hydraulic power column.

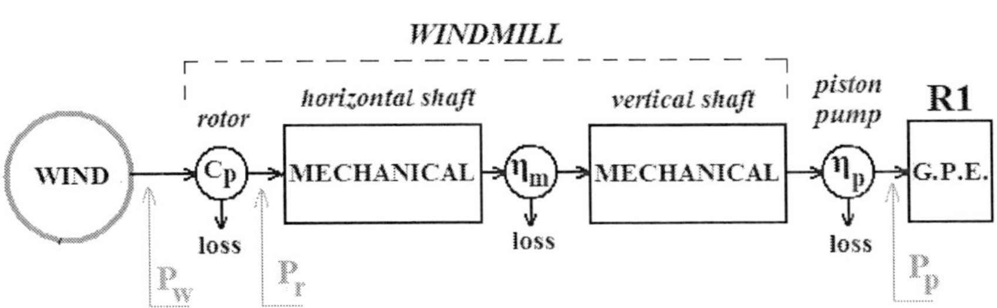

Figure 11. Energy conversions of system 1

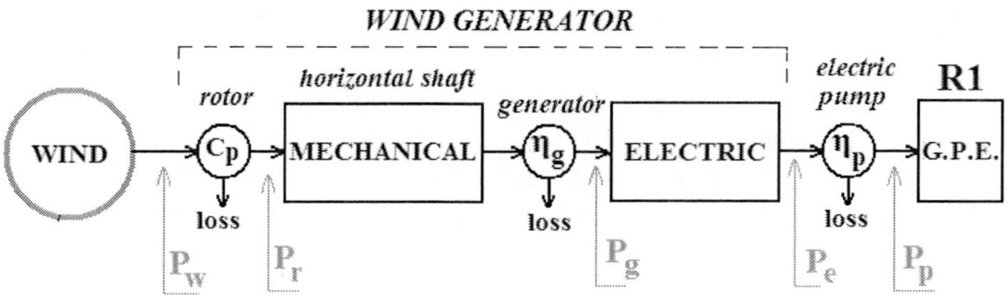

Figure 12. Energy conversions of system 2

In both block diagrams of energy transformation, it is illustrated the conversion of wind power P_w into pumping power P_p that will be used to feed R1 with water to be pumped from a well. Along those conversions, there are power losses due to the inefficiencies of the devices (pumps and generator) and mechanical connections (shafts, gears etc). Further in this chapter, it will be shown how the pumping power P_p is converted into a volume of water V_w, stored in R1.

The power of wind flow P_w is given by [11]:

$$P_w = 1/2\ \rho\ A\ V^3 \qquad (4)$$

Where:

ρ is the air density.
A is the frontal area through which the air flows.
V is the wind speed.

A wind machine (windmill or wind turbine) can be used to harvest this wind power and convert it into its rotor power P_r. Betz has proved that 59% of P_w is the maximum P_r that an ideal rotor would be able to harvest from the wind [11]. Observing Figures 11 and 12 one concludes that for system 1 (windmill), P_r is used to drive a piston pump which will feed R1, while in system 2 (wind turbine) it will be converted into an electric power P_e that will power an electric pump which feeds R1.

Back to the rotor power, its magnitude is:

$$P_r = C_p\ 1/2\ \rho\ A\ V^3 \qquad (5)$$

Where C_p is the power coefficient of the machine. C_p can generally be expressed as a function of the tip speed ratio, λ define by [11]:

$$\lambda = \text{bladetip speed/ wind speed} = \Omega R / V \qquad (6)$$

Where Ω is the angular velocity of the wind rotor and R is the radius of the wind rotor.

Figure 13 illustrates the aerodynamic efficiency of a windmill and a wind turbine which is characterized by its power coefficient C_p. It shows that a wind turbine has a better C_p than a windmill.

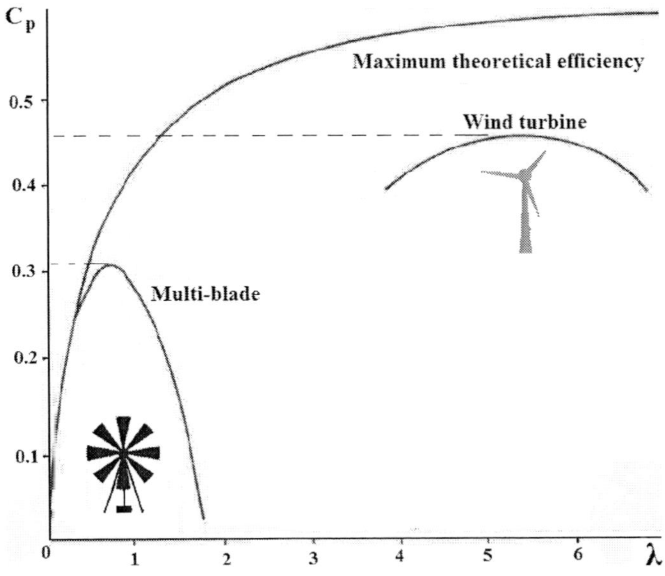

Figure 13. aerodynamic efficiency of wind machines: Source [11] with our drawings

For S1, according to Figure 11, taking into account the efficiencies of the gear η_m and that of the piston pump η_p the pumping power P_p is given by:

$$P_p=P_r.\ \eta_m.\eta_p \tag{7}$$

So, the pumping power of the system with windmill is:

$$P_p= C_p\ \ 1/2\ \rho\ \ A\ V^3\ \eta_m.\eta_p \tag{8}$$

For S2, according to Figure 12, taking into account the efficiencies of the generator η_g and that of the electric pump η_p the pumping power P_p is given by:

$$P_p=P_r.\ \eta_g.\eta_p \tag{9}$$

So the pumping power of the system with wind turbine is:

$$P_p= C_p\ \ 1/2\ \rho\ \ A\ V^3\ \eta_g.\eta_p \tag{10}$$

Eqs 8 and 10 represent the theoretical pumping power P_p at the end of the chain of conversions shown in Figures 11 and 12 for system 1 (S1) and system 2 (S2). However, in practice one can resort to other equations that are based on the characteristics of each wind machine used by the system. Since windmills are generally used to pump water, for S1, it is sufficient to determine its pumping power P_p. For S2, first it is necessary to determine the electric power P_e delivered by the wind turbine, and then estimate the efficiency of its connection with the electric pump in order to find the pumping power.

Following this approach, and starting with S2, to estimate the electric power $\mathbf{P_e}$ of a wind turbine one has to resort to its power curve together with statistical methods of wind distribution such as Weibull's [11]. The power curve yields the electric power delivered by the wind turbine as a function of wind speed at the height of its hub. For a rough estimate one can resort to other simple methods. Direct use of collected data is a method that can be used for that estimation. It tells us that if one has a series of N wind speed observations and V_i is an average of each time interval Δt, the average wind speed V_{av} for the entire period of data collection is [11]:

$$V_{av} = \frac{1}{N}\sum_{i=1}^{N} V_i \tag{11}$$

And the average electric power of the wind turbine P_{av} is given by:

$$P_{av} = \frac{1}{N}\sum_{i=1}^{N} P(V_i) \tag{12}$$

Where P (V_i) is the power given by the power curve of the wind turbine.

Knowing P_{av} which in Figure 12 corresponds to P_e, one can find the pumping power with the following equation:

$$P_p = P_{av} \cdot \eta_p \tag{13}$$

For a windmill (S2) a rough estimate of its pumping power based on practice may be found using the following equation [9]:

$$P_p = 0{,}1 * A * V^3 \tag{14}$$

8. From Pumping Power to Potential Gravitational Energy Stored in R1

After finding the pumping power P_p, one has to find the flux of pumped water to R1 as a function of the manometric height Hm (head) as well as the volume of water V_w stored in R1, Figure 14. During the pumping process, the work done by the pump is equivalent to energy transferred (E_{tra}) to the water which will be stored in R1 as gravitational potential energy (GPE).

The energy transferred E_{tra} to the water during the pumping process is:

$$E_{tra} = P_p.\Delta t. \tag{15}$$

Where Δt is time (in seconds) corresponding to the period of operation of the pump [hour(s), day(s), month(s), year(s)…].

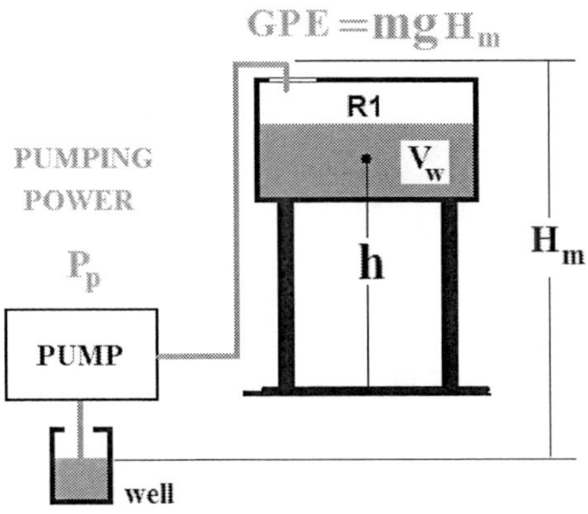

Figure 14. Storage of gravitational potential energy in reservoir R1

As mentioned, the transferred energy is stored as potential gravitational energy (GPE) in R1. Therefore one has:

$$E_{tra} = GPE, \tag{16}$$

or

$$P_p.\Delta t = m.g.H_m. \tag{17}$$

Where H_m (the head) is the height that the pump must overcome in order to force water from the well to the top reservoir R1, Figure 14.

Since mass of pumped water is the product the specific mass for its volume ($m = \rho_w V$), the Eq 17 becomes:

$$P_p.\Delta t = \rho_w V_w \; g \; H_m. \tag{18}$$

Solving Eq 18 to obtain the volume of a pumped water one has:

$$V_w = P_p.\Delta t /(\rho_w.g.H_m) \tag{19}$$

Where:

V_w is the volume of pumped water (stored in R1).
P_p is the pumping power.
ρ_w is the water density.
g is the acceleration of gravity
H_m is the manometric height.

If one considers the time of operation of one hour ($\Delta t = 3600$ s), Eq 19 becomes:

$$V_w = 3600 \, P_p. / (\rho_a.g.H_m) \tag{20}$$

With $\rho_w = 1025\ kg/m^3$ corresponding to the density of seawater to be pumped to R1, and , $g = 9.81$, Eq 20 reduces to:

$$V_w = P_p / (2{,}790\ H_m \tag{21}$$

Eq. 21 yields the volume of pumped water per hour. Therefore the result of this equation should be multiplied by 24 to obtain the volume of water stored in R1 during a day. Once there is enough water in R1, to make the hydraulic power column work, as already proved with the prototype, the reverse osmosis process will occur as long as the special piston exerts the necessary pressure. For that, the pressure is previously calculated according to the design of the piston and using Eqs 2 and 3.

9. Some Qualitative Comparisons between System 1 and System 2

Considering Eq 21, $\mathbf{V_w = P_p / (2{,}790\ H_m)}$ which gives the volume of water stored in R1, one deduces that if a windmill (system 1-S1) and a wind turbine (system 1-S2) with the same swept area A are used to feed R1, it will be more efficient the system whose chain of conversion of energy (sea Figures 11 and 12) delivers more pumping power $\mathbf{P_p}$.

S1 has the advantage of being simpler and having less specialized maintenance than S2. One can deduce that the installation and maintenance costs are lower with S1 because a wind turbine is more expensive than a windmill of the same capacity and its maintenance demands more specialized personnel.

Windmills can operate with wind speed higher than the cut-out speed, for which wind turbines are stopped for mechanical reasons and protection of the generator. This an interesting advantage of S1 since it may harvest more power from strong winds and gusts.

With S2 it is possible to locate the wind turbine where the wind resource is higher (up in a hill for instance) and so it is not necessary to have it close to the source of water or the well.

Besides that, with S2 there is not the need for digging many wells (as it is the case with windmills), since the wind turbines are easily interconnected to power a unique pump close to a unique well. In a word, S2 is more versatile than S1.

As to the occupied area, let us suppose that a combination of a wind turbine and an electric pump (S2) is able to pump more water than a windmill and its piston pump (S1). Let us imagine that with S2 we will have four times more water in reservoir R1. To overcome this difference with S1, one could use a rotor of larger diameter in order to harvest more wind power with the windmill. However, since the swept area A varies with the square of the diameter, it would be necessary only a slight increase. This slight increase would not be enough to make an appreciable change in the base of the tower (probably the same tower would have enough strength to hold the new rotor with higher diameter). So we can deduce that the difference in the occupied area between S1 and S2 is not significant. A wind turbine has a higher C_p than that of a windmill so, this may be an advantage for S2.

10. CASE STUDIES

10.1. Case 1: System 1 for Big Capacity

Along the academic work [3] done in 2004 it was studied the case of using the HPC to produce 5000 m^3 per day of drinking water. The study was based on a conventional reverse osmosis desalination plant installed in Santiago which is one of Cape Verde islands. This RO plant works with an centrifugal electric pump (800 kW-6300 V) pumping seawater at 67 bar to a set of 560 spiral wound membranes. About 45% of the volume of seawater (with a salinity of 36 000 ppm) that is fed to the membranes is converted to drinking water with less than 500 ppm. The brine energy recovery is done by a Pelton wheel connected to the shaft of the electric motor. Thanks to this energy recovery, the plant has a specific energy consumption of 4.12 kWh/m^3.

The objective of the study was to design a hypothetical RO plant with the same production of drinking water working with the Hydraulic Power Column using windmills (system 1) to pump seawater to reservoir R1. Therefore, the special piston was designed to exert 67 bar. The following results were achieved:

It would be necessary to have 97 windmills with 16 m rotor diameter to feed 11 111 m^3 of seawater to R1 located at 7 m height.

The hub of the rotors would be at a height of 25 m where the average wind speed would be 8,08 m/s according to wind data of Santiago.

The specific energy consumption would be 2,811 kWh/m^3 thanks to 93% energy recovery, achievable with the Clark pump [6].

A rough economic estimate of the pay-back time was made. The procedure was to use all the capital cost of all electric part of the considered conventional RO plant to build the HPC and buy the windmills. The estimate showed that the pay-back time of the HPC plant would be around 3 years.

10.2. Case 2: System 1 for Low Capacity

Since to produce 5 000 m^3/day of drinking water would cause difficulties because of the size and number of the windmills, another situation was took into consideration. Supposing that the HPC reverse osmosis plant (system 1) was to be used to produce drinking water for a small village. The results are:

Two windmills of 3 m rotor diameter working under the same conditions as in case 2, would be enough to produce 1 m^3 of drinking water.

If, in this case, the HPC was installed in a semi-arid region, where (according to the World Health Organization and the United Nations) it is recommended at least 20 liters of water per capita per day for all needs, two windmills would feed 50 persons.

10.3. Case 3: System 2 versus System 1 for Low Capacity

In this case we intend to predict the performance of system 2 with wind turbine. Since the HPC works regardless of how its top reservoir R1 is fed of water, it is sufficient to compare how much water would be pumped per day with the combination of the wind turbine and the electric pump. This will be done taking into account case 2 and a new technology recently announced by the company *Southwest Windpower*. We learned of that technology through their booklet entitled *"WIND-ELECTRIC WATER PUMPING : a revolution in water pumping"*. According to that company (with our outlining):

> "New technology finally provides a more powerful source of water pumping than the traditional mechanical windmill... *Just imagine, four times the water in the same wind"*
>
> The so called Wind-Electrical Water-Pumper uses any conventional off-the-shelf ¾ hp three-phase 230 VAC centrifugal submersible or surface pump.

Therefore, one infers that using Wind-Electrical Water-Pumper-WEWP to feed R1 the HPC would have the configuration of system 1. Furthermore, since, as claimed, it pumps four times the water pumped by a windmill in the same conditions, it would deliver 4 times more gravitational potential energy stored in R1 than a windmill. So, as in case 2, one WEWP working with the HPC would produce enough water for 100 persons in a semi-arid region. This deduction is plausible since the HPC,s specific energy consumption does not depend on the amount of water in R1. Beside the hydraulic and mechanical losses, it only depends on the pressure exerted by the special piston. In this case, depending on the rate the water is used by the counter weight (CW) to lift the massive special piston B or by the CW and the reservoir within piston A, the top reservoir R1 would have to be bigger in order to hold four times the amount of water.

11. Conclusion

From what has been done so far, we conclude that the model is technically and economically practicable. The economic viability is supported, not only by the fact that the model uses a renewable "free" source of energy, but also because of a predicted low specific energy consumption (around 2.811 kWh/m^3) regardless of which system is used to feed the top reservoir R1. Inherent to the use of wind gravitational potential energy is the reduction of emissions of CO_2 which is an environmental advantage compared to conventional RO plants. Another advantageous fact is a low cost of maintenance due to the simplicity of the hydraulic power column. The major drawback of the model (system 1) is the need for a big area to install the windmills if one needs to desalinate a large quantity of seawater. This disadvantage can be overcome with the use of wind turbines (system 2) or using a hybrid source of energy feed R1 such as wind-sea-sun combination. The recently constructed prototype has proved that both systems are practicable. Qualitative comparison between system 1 (S1) and system 2 (S2) infers that S2 is more versatile than S1 although it is more expensive and has higher maintenance an operation costs.

References

[1] Sourirajan, S. *Reverse Osmosis*: Academic Press, New York, 1970.

[2] Parekh, Bipin S. Reverse osmosis technology: Applications for high-purity-water production, *Marcel Dekker Inc.*, New York, 1988.

[3] Dias, JR. Dessalinização de água por osmose reversa usando energia potencial gravitacional e energia eólica. *Dissertação de Mestrado*. Defendida na Universidade de São Paulo, Brasil, 2004, 212.

[4] *Spectra Watermachines*. Available on: http://www.spectrawatermakers.com, accessed in 10/4/2004.

[5] *Isentropic Desalination* (at minimal energy cost) of seawater. Available in: http://mshades.free.fr/isentropiques/desalination.html, acessed on 15/03/2004.

[6] *Batteryless Photovoltaic Reverse Osmosis Desalination System*. Available in: http://test.netgates.co.uk/nre/pdf/SP200305.pdf, accessed on 15/03/2004.

[7] Medina San Juan, José António; Desalación de águas salobres y de mar. Osmosis Inversa. Mundi-Prensa, 2000.

[8] Mierzwa, José Carlos and Hespanhol Ivanildo. *Água na Indústria. Uso racional e reuso.* Oficina de Textos, 2005.

[9] Silva, CD; Seraphim, OJ; Teixeira, NM. Potencial eólico para bombeamento de água na fazenda lageado.

[10] Fadigas, EAFA; Dias. JR. Desalination of water by reverse osmosis using gravitational potential energy and wind energy. *Desalination*, 2009, 237, 140-146. Available on line at www.sciencedirect.com.

[11] Manwell, JF; McGowan, JG; Rogers, AL. wind energy explained. *University of Massachusetts, Amherst, USA* JOHN WLEY & SONS, LTD

In: Desalination: Methods, Cost and Technology
Editor: Irena A. Urbonienė pp. 247-261
ISBN: 978-1-61668-909-4

Chapter 8

VAPOUR SIDE CORROSION (VSC) OF COPPER-NICKEL CONDENSER TUBES IN MSF PLANT: COPPER-NICKEL DUST FORMATION

A. Sabah Abdul-Wahab*
Mechanical & Industrial Engineering Department, Sultan Qaboos University

ABSTRACT

There are few reports in the open literature describing the vapor side corrosion (VSC) of condenser tubes in multistage flash desalination (MSF) plants. This chapter describes findings of a laboratory investigation on the vapor side corrosion of copper-nickel base alloys, which are extensively used in desalination plant. The materials under investigation were Cu-10%Ni-base alloy and Cu-30%Ni-base alloy. The experiments comprised study of the behavior of these two copper-nickel alloys in condensing-vapour conditions under real operating conditions similar to those in MSF plants. Some of the experiments were conducted to test pure water vapour whilst others were done to examine the effects of injection of carbon dioxide and air into the water vapour. The findings were discussed in terms of the weight-loss measurements and microscopically examination of specimens. The experimental tests in pure water vapour indicated that the Cu-30%Ni-base alloy was more corrosion resistance than Cu-10%Ni-base alloy. When carbon dioxide was injected into the impinging water vapour, the severity of the corrosion of both Cu-10%Ni and Cu-30%Ni alloys was observed to be increased. Moreover, the corrosion of Cu-30%Ni specimen appeared to be of a similar magnitude to that of the Cu-30%Ni specimen. Hence, the superiority of Cu-30%Ni-base alloy over Cu-10%Ni-base alloy, which was seen with pure water vapour, appeared to be diminished. The microscopically examinations of specimens demonstrated a good agreement with the weight loss measurements of the two alloys. Also, the results indicated similar features on the corroded Cu-10%Ni tube removed from the MSF plant and the laboratory Cu-10%Ni specimens. The usefulness of this work comes from the importance of the MSF

* Corresponding author: Mechanical & Industrial Engineering Department, Sultan Qaboos University, P.O. Box 33, Al Khoud, P.C. 123, Muscat, Sultanate of Oman, Tel. +968 24141360; Fax +968 24141316; email: sabah1@squ.edu.om

desalination plants since they are considered as one of the main sources of drink water in the Gulf countries.

Keywords: Carbon dioxide; condenser tubes; copper-nickel alloys; multistage flash desalination (MSF) plant; non-condensable gases; vapour-side corrosion.

1. INTRODUCTION AND PROBLEM FORMULATION

All the Arabian Gulf States suffer from a shortage of fresh water resources. This deficiency is overcome by means of desalination of the seawater, mainly by the multi-stage flash (MSF) technique [1]. According to Al-kawari and Al-Bastaki [2], more than 50% of the world installations of MSF were situated in the Arabian Gulf region. Multi stage flash desalination plants consist of many complex parts and equipment. In general, the process of producing water passes through many steps until the final product, which is distillate water, is obtained [3]. The durability of existing and future MSF plants depends on the persistence of the materials of construction employed in designing the plants [2].

Copper–nickel alloys have a good resistance to corrosion in seawater and they are extensively used in modern desalination plants. They have been specified for seawater use for over 50 years [4]. The excellent corrosion protection behaviour of the copper–nickel alloys is due to the formation of a thin, compact and adherent corrosion protective layer mainly consisting of cupric and cuprous oxides which separates the alloy's surface from the sea water and decrease the electric conductivity [5]. In addition, copper–nickel alloys have a remarkable high resistance to biofouling in seawater, mainly due to the release of copper ions [6]. Therefore, in many cases an additional chlorination is dispensable. Another important feature of copper–nickel alloys is its ability to be readily welded and fabricated. Therefore, they are an obvious first choice for seawater service as piping material for heat condensers, desalination equipment, boat hulls, water supply lines, tube applications and in other structures engineered for marine use [4-5].

In the past decade or so, a new problem found to appear in some MSF desalination plants around the world. There were a number of reported instances of serious vapour-side corrosion (VSC) problems of copper/nickel alloy tubes in the high-temperature stages of some MSF desalination plants [7]. In these cases, the corrosion comprising the attack of copper–nickel alloys on the tube vapour side. Consequently, the tube failures were occurring after quite a number of years of operation but after operational periods much shorter than projected plant lifetimes [8]. As the term implies, VSC starts from the vapour side and leads to the thinning of tube walls until they yield to the pressure of the inside water [1,9]. According to Al-Subaie and Hodgkiess [8], the VSC problems were occurred largely in MSF plant but also involving in multi-effect distillers (MED).

The VSC problem was associated with a unique uniform layer of fine brown dust covering a number of copper–nickel alloy condenser tubes used in MSF desalination plants. The corrosion attack was first recognized by the higher-normal Cu^{+2} content of the distillate and, at advanced stages, by the increase in electrical conductivity of the product water. Attack took place at the inlets of condenser tubes in the first high temperature cells and was located mainly at the lower part of the tube bundle [9].

The attack of copper–nickel alloys on the tube vapour side vapour side was found to depend upon the actual chemical and physical environment to which these tubes were exposed in the condensing vapour. The vapor side corrosion of condenser tubes was found extremely complex and indeed difficult to measure and hence to characterize. According to Al-Subaie and Hodgkiess [8] and Al-Thubaiti et al. [10], it was found very difficult to identify the actual chemical and physical environment to which the corroded tubes were exposed in the condensing vapour.

The failures of the tubes of the condenser were located in the high-temperature stages but not necessarily in the top-temperature ones [10]. However, information is still lacking in ascertaining the actual cause of VSC of the condenser tubes. In principle, a number of possible agents present in the vapour space (e.g., bromine, chlorine, hydrogen sulphide, ammonia) were spring to mind to cause the failure of condenser tubes that were exposed in the condensing vapour. Also, other studies indicated that the locations of the corroded tubes was associated with areas of inadequate venting which leads to a strong implication of non-condensable gases (NCG) in the corrosion process [7]. As a result, effective venting systems to remove NCG were a major effort towards preventing VSC [2]. Therefore, a particular area of concern in the published literature was the venting system and the main effort in counteracting VSC in affected MSF plants was towards modification of the venting system [10]. Nevertheless, the outcome of such venting arrangements might not become apparent for some time because tube failures might continue as a result of wall-thickness loss in earlier phases of operation [8].

Hodgkiess and Mantzavinos [7] mentioned that one obvious common factor on attacking the tube vapour side was the presence of the carbon dioxide (together with oxygen) due to the usage of scale-control inhibitors for alkaline-scale control. The affected MSF desalination plants are mainly additive-dosed units. Therefore, they possessed the common characteristic of the presence of CO_2 in the vapour spaces as a result of thermal decomposition of the bicarbonate constituent of the feed in the flashing/boiling process [7-8]. There was also the likelihood of the presence of oxygen either as a result of incomplete deaeration of the feed or from in-leakage that was also involved in failure of condenser tubes [10].

Water-side corrosion of condenser tubes in MSF distillers was extensively studied and was well documented [9]. There was a wide range of mechanisms and explanations that were generally related to both the type of materials and the surrounding environment [1]. In spite of the extensive studies of the water-side corrosion of condenser tubes in MSF units, unfortunately there were very few reports appeared in the open literature describing the vapour side corrosion of condenser tubes. As a result, very limited number of investigations was focused on the VSC issue. Hence, more attention should be devoted to vapor side corrosion in order understand the mechanisms of the VSC processes.

The present chapter was undertaken to shed more light on the importance issue of the influence of carbon dioxide and oxygen in affecting the vapor side corrosion of copper-nickel alloys. For this purpose, experiments in controlled conditions was undertaken in order to permit reasonably unambiguous interpretation without the complexities associated with findings linked to long-term operation. The principle objectives were to compare the behavior of copper-nickel alloys containing 10% and 30% nickel and to assess the influences of CO_2 and O_2.

2. Previous Published Related Studies

Shams El Din and Mohammed [9] described the vapour side corrosion of 70 Cu- 30 Ni condenser tubes in some MSF distillers. A simple experimental set-up to simulate VSC in the laboratory was described. The study aimed at obtaining basic understanding of the mechanisms of the VSC processes.

Failure of condenser tubes due to non-condensable gases, in the desalination plant (number 1) of the Al-Ghubrah power at Sultanate of Oman, was reported by Abu Dayyeh et al [11]. The work described the study of corrosion in the plant that was commissioned in the year 1976. The plant had experienced severe corrosion in the vapour space, and more in the high temperature stages. The evidence in the tube sample and distribution of failure indicated that vapour side corrosion as the cause of the problem. It was analyzed that corrosion was due to build-up of non-condensable gases in these high temperature stages. The main component of these gases is carbon monoxide, which is created by thermal decomposition of bicarbonate in sea water. The study also described the modifications of the venting system and preventive measures being carried out to prevent the accumulation of non-condensable gases.

Asrar et al. [12] reported the early failure of 90/10 Cu-Ni condenser tubes in unit #16 of stage 4 Jeddah-4 MSF desalination plant. They discussed the causes of 90/10 Cu-Ni alloy condenser tubes. The failure of condenser tubes was attributed mainly to corrosion caused by the dissolution of CO_2 along with some sulphur containing compounds in condensate water vapour. The authors recommended some operational modifications in the vent line to avoid the dissolution of non-condensable gases in condensate water.

Hodgkiess and Mantzavinos [7] focused on the behaviour of copper-nickel alloys in distilled water with or without the presence of carbon dioxide. The objective of the study was not to address the operational problem directly but to provide some basic information on the corrosion behaviour of two relevant commercial alloys in distilled water. The main component of the work involved electrochemical measurement. Microscopic examination of specimens revealed susceptibility to pitting attack and, to a lesser extent, crevice corrosion after extended exposures at 20-80 °C.

Mantzavinos et al. [13] studied the corrosion of condenser tube materials that are related to problems of vapour-side corrosion of copper/nickel alloys in multistage flash desalination plant. The main objective of their work was to obtain basic information and understanding of the corrosion behaviour of copper/nickel alloys in distilled water over a range of relevant environmental conditions. Some tests were also devoted on alternative materials such as titanium and an austenitic stainless steel. The experiments were conducted in controlled laboratory conditions with assessment of corrosion behaviour mainly involving electrochemical monitoring. The results demonstrated a clear superiority of the Cu/30Ni alloy over Cu/10Ni material at ambient temperatures and also non-protective behaviour of both alloys upon initial exposure to distilled water at 80 °C.

The vapour side corrosion in a multi effect (ME) desalination plant at Jebel Dhana, Abu Dhabi, was described by El-Dahshan et al. [14]. The analysis of the failure of the condenser tubes was discussed. The type of failure, correlating it to material, and working conditions were also given. The corrosion products were analysed by X-ray, EDAX, and EPMA techniques. The 90/10 copper nickel alloy tubes were found to have suffered from vapour side corrosion. However, there was no sign of any copper oxide deposition, as has often been seen in failed tubes from MSF distillers. The failure took two forms: corrosion fatigue cracking, and pitting corrosion.

Al-Subaie and Hodgkiess [8] presented the results of a laboratory investigation relating to vapour-side corrosion problems. The experiments comprised study of the behaviour of two copper-nickel alloys (90/10 Cu/Ni and 70/30 Cu/Ni) in condensing-vapour conditions at 90-95 °C. Some of the tests involved pure water vapour whilst others examined the effects of injection of carbon dioxide and air into the condensing vapour. The corrosion process was studied by means of weight-loss measurements. The results showed that samples of 90/10 Cu/Ni alloys suffered from greater weight loss compared to 70/30 Cu/Ni when exposed to droplets of pure condensing vapor. However, when CO_2 was injected, the weight loss of both alloys increased. When CO_2 was present, the higher superiority of the higher-nickel alloy was not observed. The corrosion of the Cu-30Ni-base material appeared to be of a generally similar magnitude to that of the Cu-10Ni-base alloys. The corrosion process was also studied by means of microscopical examination of specimens which revealed interesting features of the morphology of the corrosion attack that was discussed in relation to the basic mechanisms of corrosion.

In the Kingdom of Bahrain, brown copper dust (powder) formation was observed in the distillate channels of all of the heat recovery stages constructed with 90/10 Cu/Ni material in Al-Hidd MSF units that was constructed in 1997 [2]. The quantity of the powder varied from stage to stage. On the other hand, no copper dust was observed in the stages constructed with the 30/70 Cu/Ni material. Samples of the copper powder were taken for chemical analysis. The results confirmed that copper was the main constituent (85% of the material was copper). A possible source of the dust was due to the disturbance of the equilibrium in the 90/10 Cu/Ni alloy as a result of a combination of the operating conditions and chemicals present in the seawater. In addition to CO_2, which may be produced by thermal decomposition of bicarbonates, pollutants which may be present in the intake water could be the sources of the detrimental chemical.

Al-Thubaiti et al. [10] described an experimental study of the detailed influences of three major environmental parameters (temperature, carbon dioxide and oxygen) on the vapour side corrosion of Cu/Ni. The Cu/Ni materials in the investigation contained between 10 and 64% nickel but with emphasis on Cu/10 Ni-base and Cu/30 Ni-base alloys which are widely employed for heat transfer tubes. Electrochemical techniques were used to investigate the corrosion behaviour of these alloys in distilled water at temperatures of 20-90 °C with, and without, the injection of carbon dioxide and oxygen. The investigation was undertaken to shed more light on the important issue of the influence of carbon dioxide and oxygen. The study also included a higher nickel alloy, Monel 400, in order to provide more understanding of the role of nickel in the practically-used, lower-nickel materials.

Abouswa et al. [4] reported that copper-based alloy tubes were found to suffer from severe localized corrosion at 6 o'clock position (corrosion throughout the pipe wall thickness). In order to establish the main cause of corrosion and failure of the copper-based alloy tubes, several samples from the failed tubes were subjected to metallographic examination and electrochemical test in the simulated working environment (i.e. chloride content, temp. non evacuated system).

3. Experimental Setup and Procedures

The chapter focused on the corrosion behaviour of two copper-nickel alloys (Cu-10%Ni-base alloy and Cu-30%Ni-base alloy). For each alloy type, flat specimens measuring 1.6 x 1.6 cm were used in the experiments. These specimens were sheared from unused condenser tubes samples (25 mm diameter with 1 mm thickness) which were obtained from a MSF desalination plant in Oman (Figure 1). The specimens were cut from the same lot to insure uniform composition. Then the specimens flattened ground and polished to a 1 micron finish.

The experiments comprised testing the performance of Cu-10%Ni and Cu-30%Ni under real operating conditions similar to those high temperature stages of heat recovery section in MSF plant. Therefore, the experimental tests involved exposing specimens of Cu-10%Ni and Cu-30%Ni to droplets of condensing vapor impinging on the surface of specimens.

Figure 2 shows a photograph of the experimental setup used to simulate vapor side corrosion in the laboratory. This set-up was based on that used by Al-Subaie and Hodgkiess [8]. Pure water was used as a feed to a heated flak which functioned as an evaporator. As a result water vapour was produced. A glass condenser was used to condense this water vapour. Both the flows of make-up water and the cooling water of the condenser were under well controlled conditions in order to maintain a constant temperature of the condensate droplets striking the surface of the specimens. The specimens were positioned in a manner to allow condensate maintaining a thin distillate film on the surface of the specimens. The PH of the condensate film was measured regularly by a PH measuring device.

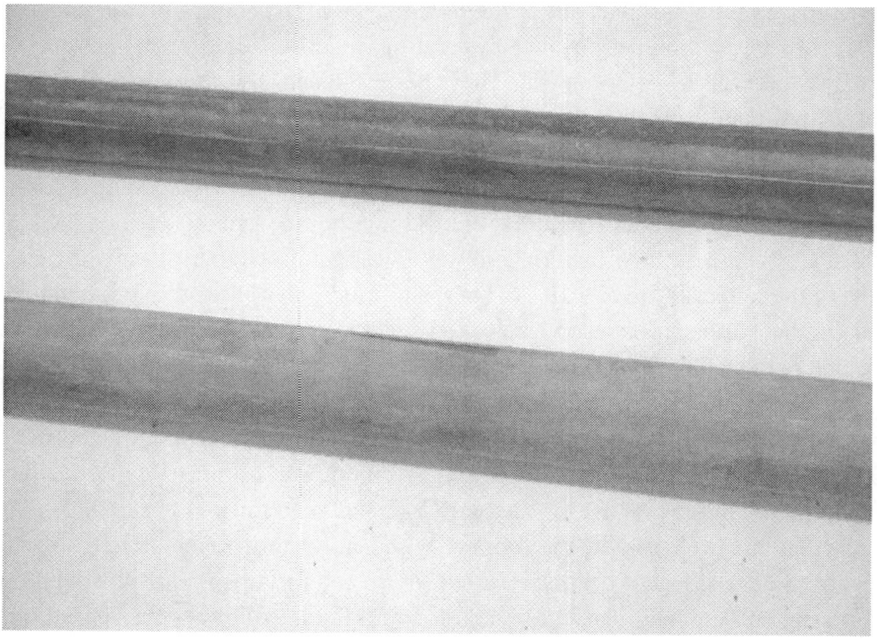

Figure 1. Unused condenser tubes samples (Cu-10%N, top and Cu-30%Ni, bottom) that were obtained from MSF desalination plant in Oman, from which the specimens were sheareda

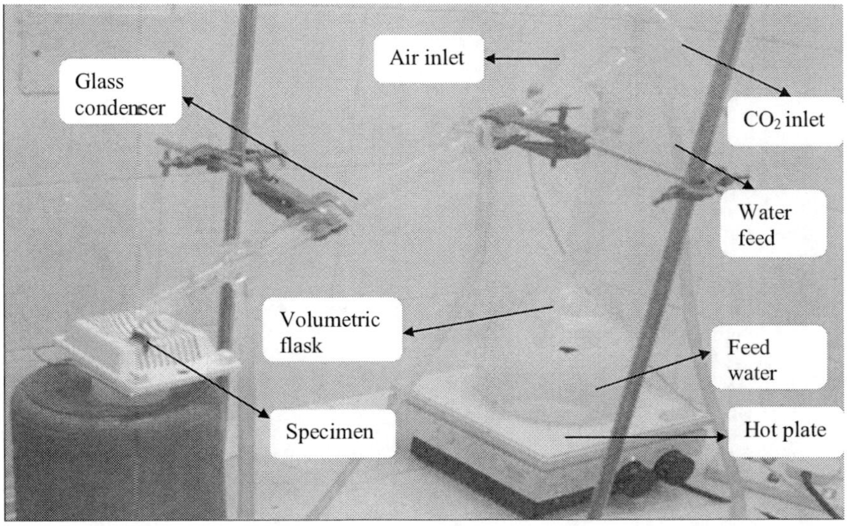

Figure 2. Experimental setup

Experiments were carried out at three specific operating conditions that are similar to those described by Al-Subaie and Hodgkiess [8]. In the first operating condition, the experimental tests were conducted so as the Cu-10%Ni and Cu-30%Ni specimens were exposed only to pure water vapour. In the second operating condition, the influence of CO_2 on the vapour side corrosion process was examined by bubbling CO_2 gas continuously into the space of the evaporator flask. As a result, the water vapour mixed with CO_2 (i.e., non-condensable gas) before its condensation in the condenser. In the third operating condition, the influence of CO_2 with air was tested by injecting air in addition to CO_2 into the space of the evaporator flask so as the produced water vapour mixed with CO_2 and air before its condensation. The concentration of CO_2 was controlled by measuring frequently the PH of the droplets of the condensate. Table 1 summarises the three conditions of the experimental tests.

At the ending of each experiment, the specimens of Cu-10%Ni and Cu-30%Ni were washed in methanol, dried in air and examined as corroded state in the microscope (JEOL 840 Scanning Electron Microscope). The specimens were then immersed in an inhibited acid for 1-2 seconds. The inhibited acid was mainly used in order to remove corrosion products from the specimens without attack on their underlying metals (Figure 3). The specimens were then washed in water and methanol, dried, reweighed immediately to determine the weight loss in the experimental test and examined again under the microscope.

Table 1. Various conditions of the experiments

No.	Alloy	Condition	Duration time (hour)
1	Cu-10%Ni	Pure water	24
2	Cu-30%Ni	Pure water	24
3	Cu-10%Ni	CO_2	72
4	Cu-30%Ni	CO_2	72
5	Cu-10%Ni	CO_2+ Air	72
6	Cu-30%Ni	CO_2+ Air	72

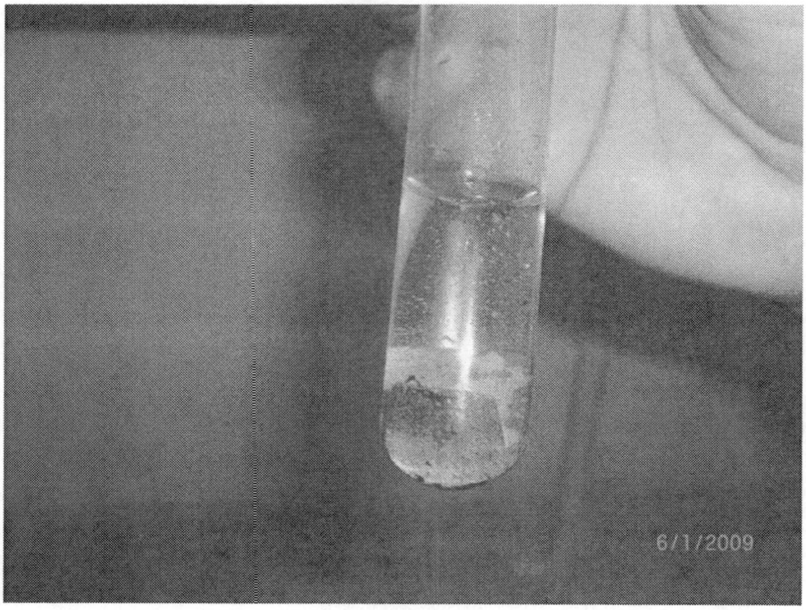

Figure 3. Specimen immersed in corrosion inhibitor

Table 2. Weight loss measurements

No.	Alloy	Condition	Time (hour)	Weight before (mg/cm^2)	Weight after (mg/cm^2)	Weight loss (mg/cm^2)
1	Cu-10%Ni	Pure water	24	2920.1	2920.0	0.1
2	Cu-30%Ni	Pure water	24	3209.0	3209.0	0.0
3	Cu-10%Ni	CO_2	72	3400.7	3400.0	0.7
4	Cu-30%Ni	CO_2	72	2670.6	2670.1	0.5
5	Cu-10%Ni	CO_2+ Air	72	3488.3	3487.9	0.4
6	Cu-30%Ni	CO_2+ Air	72	3085.1	3084.8	0.3

4. RESULTS AND DISCUSSION

4.1. Weight Loss Analysis

The results of the weight losses recorded on Cu-10%Ni and Cu-30%Ni specimens at the three operating conditions are presented in Table 2. With only impinging pure water, it can be seen that the weight loss for the Cu-10%Ni is greater than Cu-30%Ni. In fact, there was a very small weight loss in the Cu/10Ni specimen while there was no record of any weight loss for the Cu/30Ni specimen during the non-stop 24 hours of the experiment. Measured PH values in pure water were in the range 5.7-5.9.

With the injection of CO_2, the weight losses of both Cu-10%Ni and Cu-30%Ni specimens increased compared to that in impinging pure water. Also, the apparent superior performance of Cu-30%Ni specimen over Cu-10%Ni in impinging pure water was not repeated in the injection of CO_2. Hence, the superiority of Cu-30%Ni alloy appeared to be largely lost. This explains exactly what happened in real desalination plant where VSC failures of both Cu-10%Ni and Cu-30%Ni were experienced. Al-Subaie and Hodgkiess [8] pointed out the choice of Cu-30%Ni in preference to Cu-10%Ni for the higher temperature stages of MSF plant is beneficial from a VSC resistance standpoint only if the presence of CO_2 is avoided (i.e., by effective venting of non-condensable gases).

With the simultaneous addition of air plus CO_2, the corrosion rates exhibited a decrease compared to the situation with CO_2 alone. However, Cu-10%Ni specimen still suffered a greater material loss than the Cu-30%Ni. This result was in good agreement with the study of Al-Subaie and Hodgkiess [8] who also found lower weight losses with the case of injected CO_2 plus air than compared with the case of injection of CO_2 alone. The PH values during the injection experiments (CO_2 alone or CO_2 plus air) were in the range 4.5-5.3, indicating the expected increase in water acidity in the presence of injected CO_2.

Thus the results indicated an increase in aggressiveness of both Cu-10%Ni and Cu-30%Ni specimens associated with the case of injection of CO_2. According to Al-Subaie and Hodgkiess [8], this finding was expected as the injection of CO_2 made the impinging water more acidic and so more corrosive. Also, no attempt was made in the experiments to exclude air from the condensing water on the specimen surface. Thus, oxygen would have been present in the water film even in tests involving injection of CO_2 alone into the vapour.

4.2. Microscopically Examination

Figures 4.1 and 4.2 show the surfaces of Cu-10%Ni and Cu-30%Ni specimens after 24 h in impinging condensate pure water, respectively. It can be seen that there was no damage occurred on Cu-30%Ni whereas there was a little damage occurred on Cu-10%Ni alloy. Moreover, some dark points were seen to appear in Cu-10%Ni alloy.

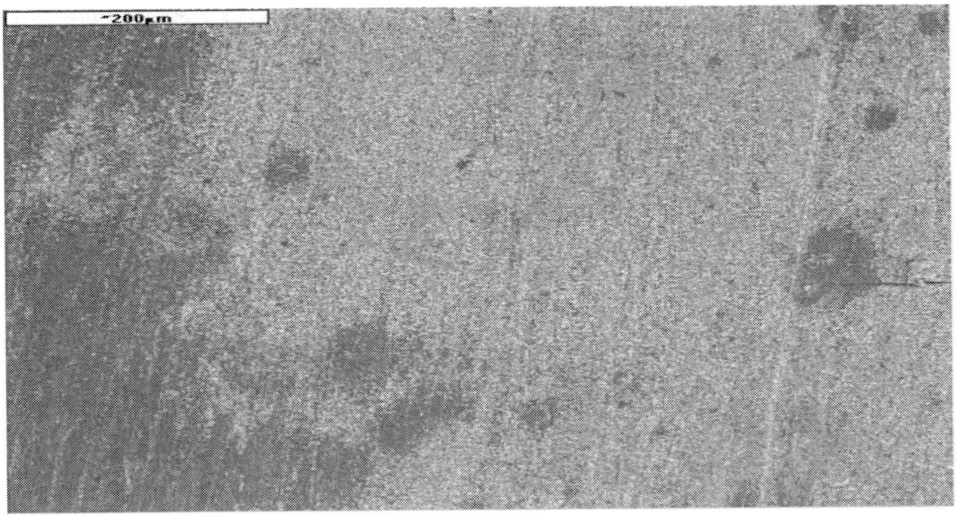

Figure 4.1. Cu/10 Ni after 24 h in impinging condensate pure water

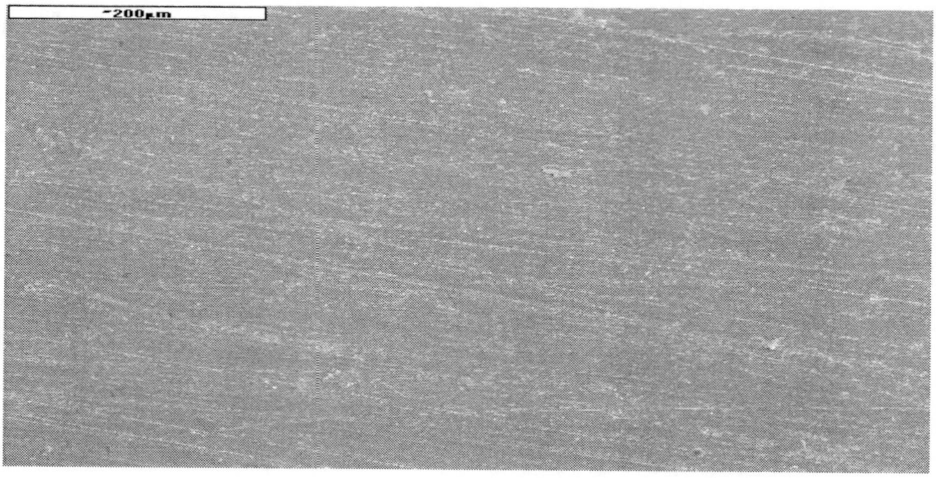

Figure 4.2. Cu/30 Ni after 24 h in impinging condensate pure water

Figures 5.1 and 5.2 show the surfaces of Cu-10%Ni and Cu-30%Ni specimens after 72 h in condensate pure water with injection of CO_2, respectively. These figures revealed that there was damage on both specimens with the case of injection of CO_2. In both specimens, there were corrosion pits on the surfaces but these pits were differing in shape and deep. In Figure 5.1 (Cu-10%Ni), the pits were narrow and deeper whereas in Figure 5.2 (Cu-30%Ni), the pits were shallow. In general, the damage on the surface of Cu/10 Ni was more than the damage on Cu/30Ni specimen. Thus, the microscopically examinations showed an increase in aggressiveness associated with the presence of injected CO_2. This result was consistent with the weight loss measurements. Hence, the microscopically examinations yielded important confirmatory evidence to the weight loss measurements.

Figure 5.1. Cu/10 Ni after 72 h in impinging distilled water with injection of CO_2

Figure 5.2. Cu/30 Ni after 72 h in impinging distilled water with injection of CO_2

Figures 6.1 and 6.2 show the surfaces of Cu-10%Ni and Cu-30%Ni specimens after 72 h in condensate pure water with injection of CO_2 plus air, respectively. It can be seen that there were damage happened to both Cu-10%Ni and Cu-30%Ni specimens. However, with the simultaneous addition of air plus CO_2, the damages on both alloys were decreased compared to the situation with CO_2 alone. This was well consistent with the weight loss measurements. Hence, both the result of weight loss measurements and microscopical examinations demonstrated an increase in aggressiveness associated with the presence of injected CO_2. It can also be seen that the damage on Cu/10Ni was more than that of Cu/30Ni specimen. The surface on Cu-30%Ni specimens (Figure 6.2) was appeared to be husky and there were some corrosion pits also appeared with different size and deep. In the other hand, the surface of Cu-10%Ni (Figure 6.1) had pits with narrower size and much deeper than seen in Cu-30%Ni specimens.

Figure 6.1. Cu/10 Ni after 72 h in impinging water with injection mixture of CO_2 and air

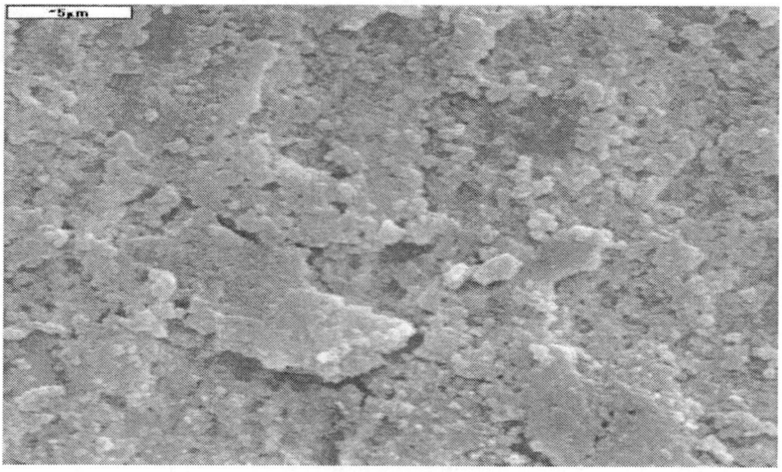

Figure 6.2. Cu/30 Ni after 72 h in impinging water with injection mixture of CO_2 and Air

Figure 7.1. picture of the condenser tubes of an operating MSF plant in Oman

Figure 7.1 demonstrates a picture of the condenser tubes of an operating MSF plant in Oman. Figures 7.2 shows severely corroded condenser tube removed from an operating MSF plant in Oman which had been suffering from VSC tube failures. The picture shows the effect of VSC on Cu-10%Ni alloy that resulted in thinning of the condenser tube from the outer side. Figure 7.3 provides the microscopy of the same condenser tube (Cu/10Ni) shown in Figure 7.2. It can be seen different sizes of corrosion pits on the external tube surface. Some of these pits were large and shallow, others were large and deep. Whereas, some of these pits were small and deep. Many of these pits were penetrated throughout the tube wall thickness and they were visible. Moreover, the surface included some crakes. Furthermore, Figure 7.3 provides an evidence of the relevance of the experimental conditions (i.e., laboratory Cu-10%Ni specimen) to the practical VSC problems (i.e., corroded Cu-10%Ni tube removed from the MSF plant).

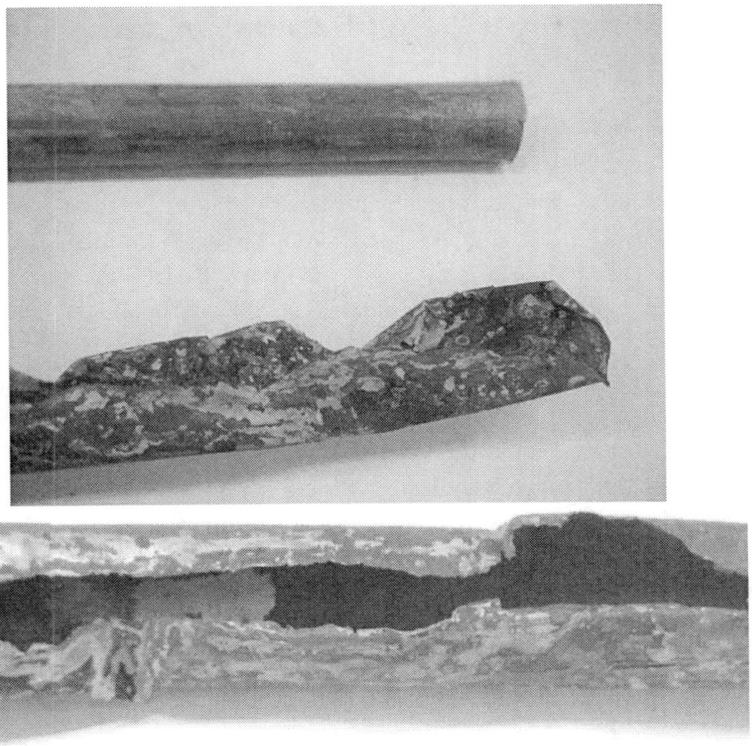

Figure 7.2. Severely corroded condenser tube removed from an operating MSF plant in Oman. The picture shows VSC of Cu-10%Ni that are leading to tube thinning of the condenser tube

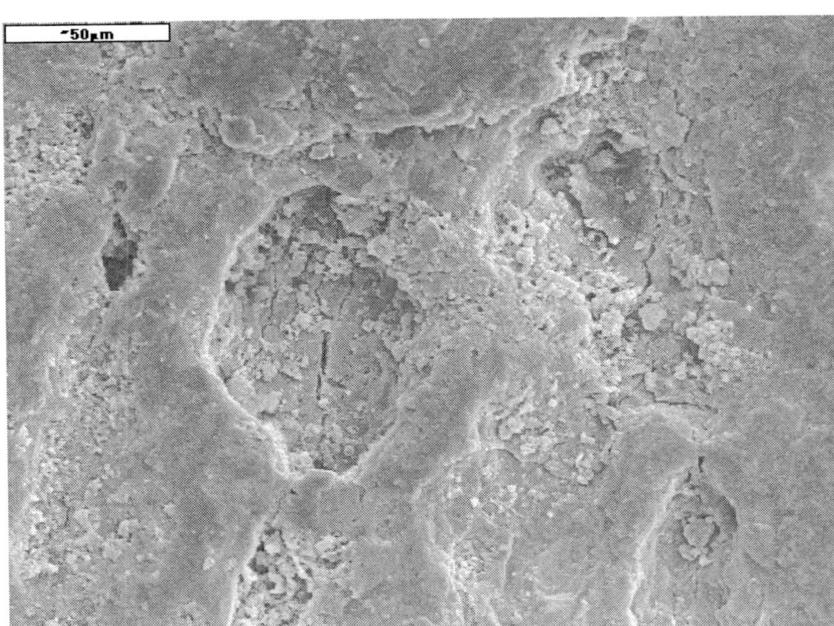

Figure 7.3 Real sample from AL-Ghubrah desalination plant of condensing pipes of Cu/10Ni

CONCLUSION

Vapour side corrosion of the condenser tubes in MSF desalination plants can cause a variety of undesirable consequences. This includes loss of the condenser tubes, unplanned shutdowns, expensive repairs, contamination of distillate products, serious health problems as well as the discontinuity of these plants and the early failure of their condenser tubes. This chapter was directed to investigate the corrosion behavior of Cu-10%Ni-base alloy and Cu-30%Ni-base alloy. For this purpose, the behavior of these two copper-nickel alloys in condensing-vapour conditions was investigated under real operating conditions similar to those in MSF plants. Some of the experiments were carried out with only pure water vapour whilst others were conducted to examine the effects of injection of carbon dioxide and air into the water vapour. The corrosion process was assessed by means of weight-loss measurements and microscopically examination of specimens.

The results of the chapter indicated that the Cu-10%Ni-base alloy suffered more corrosion (greater weight loss) compared to Cu-30%Ni-base alloy when the two alloys were exposed to droplets of condensing water vapour impinging on their surfaces in the absence of CO_2. When CO_2 was injected, the weight loss and so the corrosion rates of both alloys were seen to be increased. Also, the corrosion of Cu-30%Ni specimen appeared to be of a similar magnitude to that of the Cu-30%Ni specimen. Furthermore, the test of the two alloys using microscopical examination confirmed the observation obtained from the weight loss measurements.

Similar features was demonstrated between the corroded Cu-10%Ni tube removed from the MSF plant and the laboratory Cu-10%Ni specimen. This yielded an important confirmatory evidence of the relevance of the experimental findings to practical VSC problems in MSF desalination plant. Hence the early failure of the condenser tubes of the MSF desalination plant was attributed to severe corrosion caused by non-condensable gases that were loaded with the vapours and hit the condensing tubes on their condensation side, causing their corrosion. These effects are differing from a desalination plant to another depending on the design and the efficiency of the venting system in the plants. Poor venting system and its bad design will allow the non-condensable gases such as CO_2 to stay and interact with materials of the condenser tubes for longer time. Hence, operational modifications in the vent line should be done to avoid the dissolution of non-condensable gases in condensate water.

ACKNOWLEDGMENTS

The author is grateful to the Al Ghubrah Power & Desalination Company SAOC at the Sultanate of Oman. Special thanks go to Engineer Ali Mohammed Al-Balushi, Operation Manager, Al Ghubrah Power & Desalination Company SAOC.

REFERENCES

[1] El-Dahshan, M. E. (2001). Corrosion and scaling problems present in some desalination plants in Abu Dhabi. *Desalination*, *138*, 371-377.

[2] Al-Kawari, M. & Al-Bastaki, N. (2004). Copper dust formation in a large scale MSF desalination plant. *Desalination*, *166*, 153-156.

[3] Abdul-Wahab, S. A., Elkamel, A., Al-Weshahi, M. A. & Al Yahmadi, A. S. (2007). Troubleshooting the brine heater of the MSF plant using a fuzzy logic-based expert system. *Desalination*, *217*, 100-117.

[4] Abouswa, K., Elshawesh, F., Elragei, O. & Elhood, A. (2007). Corrosion investigation of Cu-Ni tube desalination plant. *Desalination*, *205*, 140-146.

[5] Todd, B. (2001). Copper-nickel alloys in desalination systems, *Marine Applications of copper-nickel alloys*, http://marine.copper.org.

[6] Powell, C. A. (2001). *Copper-nickel alloys: Resistance to corrosion and biofouling, Marine Applications of copper-nickel alloys*, http://marine.copper.org.

[7] Hodgkiess, T. & Mantzavinos, D. (1999). Corrosion of copper-nickel alloys in pure water. *Desalination*, *126*, 129-137.

[8] Al-Subaie, K. Z. & Hodgkiess, T. (2003). Corrosion of copper-nickel alloys in simulated vapourside environments. *Desalination*, *158*, 43-50.

[9] Shams El Din, A. M. & Mohammed, R. A. (1998). Contribution to the problem of vapour-side corrosion of copper-nickel tubes in MSF distillers. *Desalination*, *115*, 135-144.

[10] Al-Thubaiti, M. A., Hodgkiess, T. & Ho, S. Y. K. (2005). Environmental influences on the vapourside corrosion of copper-nickel alloys. *Desalination*, *183*, 195-202.

[11] Abu Dayyeh, A. R., Hamdan, R., Mukerjee, P. K. & Vijay Kumar, P. A. (1998). Rehabilitation of a long-tube parallel flow evaporator Al-Ghubrah power plant and desalination station. *Desalination*, *120*, 61-72.

[12] Asrar, N., Malik, A. U., Ahmed, S., Al-Sheikh, A. H., Al-Ghamdi, F. & Al-Thobiety, M. A. (1998). Early failure of cupro-nickel condenser tubes in thermal desalination plant. *Desalination*, *116*, 135-144.

[13] Mantzavinos, D., Hodgkiess, T. & Lai, S. L. C. (2001). Corrosion of condenser-tube materials in distilled water. *Desalination*, *138*, 365-370.

[14] El-Dahshan, M. E., Al Malek, S. & Al katheeri, E. (2003). Vapour side corrosion in a ME desalination plant at Jebel Dhana. *Desalination*, *155*, 67-78.

In: Desalination: Methods, Cost and Technology
Editor: Irena A. Urbonienė 263-324
ISBN: 978-1-61668-909-4

Chapter 9

FRESHWATER SCARCITIES AND DESALINATION: EVIDENCE FROM MOROCCO, SPAIN, THE OCCUPIED PALESTINIAN TERRITORIES AND SOUTH INDIA

Gregor Meerganz von Medeazza*
Institute for Environmental Sciences and Technology (ICTA)
Autonomous University of Barcelona, Spain

ABSTRACT

Direct per capita availability of freshwater resources decreases as world population continues its growth and human activities further intensify. This fact threatens the well-being and ultimately the survival of humanity as a whole. Among the scientific community and dominant political establishment, the belief that knowledge and human ingenuity will ultimately solve water scarcity problems through the efficiency improvement of freshwater production is widely held. In this vision, seawater desalination seems very promising and especially reverse osmosis technologies are increasingly portrayed as a panacea for much of the world's water woes. However, the cost of such technology in terms of energy and its environmental impacts must be taken into account, together with the different purposes served by desalinated water.

The primary aim of this chapter is to question the long-term sustainability of the desalination technology as a solution in tackling the global freshwater crisis. A North-South approach is used to raise certain questions on the significance of *scarcity*. The issue of water services for tourism is far removed from water scarcities for the Poor: water satisfying luxury or maintaining livelihood? An approach based on the "basic needs" scenario is relevant to address these issues. The city of Laâyoune (Moroccan Sahara), the Canary Island of Lanzarote (Spain), the Gaza Strip (Occupied Palestinian Territories) and the urban slums of Chennai (South India) are taken as case studies to illustrate the different purposes served by the desalination technology: alleviation of scarcity situations where the dominant component are either physical, socially-

* A version of this chapter was also published in Desalination Research Progress, edited by Daniel J. Delgado and Pablo Moreno, Nova Science Publishers. It was submitted for appropriate modifications in an effort to encourage wider dissemination of research.

constructed, politically-constructed or structural. Beyond technological *change*, this chapter focuses on technological *equity* as a cornerstone for debate.

INTRODUCTION

Over the past decades, society has become increasingly aware of the role water plays as a critical resource as well as of its vulnerability in terms of quality. This raised key questions as far as water supply, access, and management are concerned, especially in the context of demographic growth and increasing living standards, triggering ever more unquenchable water demands. Despite its vital importance, poor resource management has lead to a dramatic deterioration of aquatic environments, resulting in decreasing quality and quantity of readily available freshwater. Among the scientific community and dominant political establishment, the belief that knowledge and human ingenuity will ultimately solve water scarcity problems through the improvement of efficiency in freshwater production, is widely held. In this view, sea- and brackish water desalination seems very promising and reverse osmosis technologies are increasingly portrayed as a panacea for much of the world's water woes. However, with ever likely decreasing cheap energy resources, desalination technologies, like other energy-intensive water supply systems, may well fail to fulfill their long-term expectations. In order to fully embrace the concept of Integrated Water Resource Management (IWRM), one should look beyond the limits of technology. As Kranzberg's first law states: "technology is neither good, nor bad; nor is it neutral" [1], the question "to which scarcity situations (physical, social, political, etc.) can desalination ethically be applied" needs to be explored. This chapter distinguishes between the various components of freshwater scarcity and goes, therefore, beyond the scope of the 'desalination technology' as such: it questions the sustainability of the desalination alternative and opens up to broader debates on consumption theory and economic growth.

This chapter primarily discusses how the (un)sustainability of the desalination technology is (un)able to alleviate freshwater scarcities at global level. The question then begs to be asked: what is the meaning of "scarcity"? The issue of "water services for tourism" may indeed seem far removed from water scarcities for the Poor in the South. The underpinning intention is to reflect upon initial drivers of consumption and how they correlate with various water management trends. This leads to discuss to what extent do improvements in energy-intensive technologies enable the mitigation of scarcity situations embedded in our oil-based socio-economic system.

The Contribution of the Present Chapter is Fivefold:

Firstly, it sheds light upon various societal water perceptions and how those trigger distinct consumption patterns. In Lanzarote for instance, a sudden change in the insular water culture occurred just 40 years ago when the desalination technology was introduced, making vast amounts of water available to a community formerly living in an acute physical scarcity environment.

Secondly, this chapter addresses the issue of the energy-intensiveness and the often non-renewable energy dependency of the desalination option. Can a process translating a non-renewable resource (i.e. tending towards physical scarcity) into a socially scarce one, subject to growing demands at high environmental costs, ever be sustainable?

Thirdly, beyond the energy issue lies the debate of the purpose served by the desalination technology. The objective is to emphasize the various water uses: satisfying luxury, securing livelihood or maintaining well-established power structures. In the case of Lanzarote, water production soon became the decisive agent to serve tourist services, enabling accelerated economic development. This is far removed from water uses as observed in the Saharan city of Laâyoune. A desalination scheme under Palestinian control may enhance political autonomy. In the case of Chennai, under current circumstances, the new desalination project is not likely to improve the inequitable water governance slums suffer from.

Fourthly, if we assume a technological trajectory of decreased monetary costs, decreased energy costs per cubic meter, and moreover increased share of renewable energies in desalination (a kind of win-win-win scenario), does this mean that water for urban use of poor people in the world will cease to be a problem? Will not the energy costs remain too high? An approach based on the "basic needs" scenario is relevant to address these questions. Indeed, human water needs lie at the interface between what will be defined as *physical* and *anthropogenic* water scarcities. The conjunction of these issues ultimately leads to a debate on the interpretation of those needs and the associated quality of life within the economic growth paradigm.

Finally, the *problématique* of assessing the sustainability of the desalination technology, leading to a differentiation and categorization of scarcity situations, inevitably results in concerns about equity in water management. The present chapter establishes links between sustainability, scarcity and equity, arguing that "putting equity first" enhances sustainability (in its *strong* sense) and leads to more efficient ways of alleviating scarcity.

This chapter is structured in six sections. In section 1, a discussion on the significance of "scarcity" sets the scene to raise the question of "technological equity". Section 1 ends by introducing the desalination technology, its applications and concerns. The second part presents the case studies (sections 2, 3 and 4). Section 5 analyses the intertwined nature of the various "scarcities" putting forward the concept of "technological equity" to by-pass what will be argued to be a normative deficit of applied ethics in addressing the link between technology and equity. Finally, section 6 analyses the drift from physical to social scarcity.

1. Human Appropriation of Water Resources, Scarcity and the Desalination Response

From an ecological economic point of view [2] and according to the second law of thermodynamics, the economic system is analyzed as a system that requires the entry of energy and matter and at the same time produces wastes in the form of dissipated heat or degraded energy and waste matter, partly recycled by the market or by nature. The major material difference between humans and other species is that, in addition to our biological metabolism, the human enterprise is characterized by an industrial metabolism. In ecological terms, all our toys and tools (the "capital" of economists) are "the exosomatic equivalent of organs" –as Alfred Lotka [3] would say- and, like bodily organs, require continuous flows of energy and matter to and from "the environment" for their production and operation [4]. For instance, 400,000 liters of water are required to manufacture a car, 3800 liters to produce 500 grams of aluminum, 3000 liters for 500 grams of grain-fed beef, 100 liters per kilogram of paper, 18 liters to produce 1 liter of fuel and 9 liters for 1 liter of *Cola* drinks; also around 10% of the total withdrawn water is spent on power plant cooling.

The hydrological cycle (a mechanism operating from 15 km above terrestrial surface to 5 km below ground level), powered by solar and gravitational energy, continuously collects, purifies and distributes the fixed supply of naturally available freshwater through evaporation, transpiration, condensation, precipitation, infiltration, percolation and runoff. This purification process is the greatest natural desalination process occurring on Earth, constantly running a distillation and pumping system. It provides plenty of freshwater as long as we do not either overload it with persistent pollutants and non-degradable wastes or withdraw it from underground supplies faster than it is replenished. Unfortunately, we are currently doing both.

Water covers about 71% of the earth's surface. However, in the planet's water budget, out of the 1,38 billion km^3 of water found on earth only a tiny fraction of the world's total water supply is directly available freshwater. Indeed, out of the entire water mass present on earth 97.41% is salty; from the remaining 2.59%, 1.984% is trapped in ice caps and glaciers, another 0.592% is stored as groundwater, resulting that only 0.014% can be accounted as readily accessible freshwater [5].

This relatively tiny fraction of freshwater that constitutes our survival support, satisfied during millennia, mankind's development. World population literally exploded for the past couple of centuries. This demographic boom, combined with current consumption patterns and technological choices triggered the major environmental and social conflicts we are experiencing worldwide today. Since living standards also increased, general consumption patterns and in particular, demand for freshwater grew rapidly, yet faster than population. Figure 1 shows the comparative growth tendencies of both world population and total water consumption; it emphasizes that in one century, water consumption almost increased ten-fold, while the world population quadrupled.

While both population and average per capita water withdrawal increase, total natural water stocks are fixed or in decline. As a consequence, the annual reduction of water availability is 0.6% in Europe and an alarming 1.8% in North Africa for instance [6]. Thus, the biosphere is being required to accommodate not just more, but in Catton's terms, effectively 'larger' people [7].

The demographic factor combined with the rising per capita demand, the increasing frequency of extreme events such as droughts and floods and the decreasing readily available resources result in the era of global freshwater scarcity we are facing today. Indeed, quantity of the world water fundamentally stays unchanged; what is changing is the available unpolluted quantity, due to altered quality through withdrawal, consumption and contamination. We are not any longer using the water flow made naturally available through the hydrological cycle, but the non-regenerative water capital, rendering life supporting ecosystems highly vulnerable. Also, 30,000 people die every day, including one child dying every 8 seconds because it drank contaminated water [8]; 80% of all known diseases are related to water misuse [9]. Despite internationally acknowledgement of freshwater's vital importance, over 1.2 billion people do not have access to safe drinking water and the 25 countries currently suffering from water stress are mostly located in the Middle East and North Africa. This number is expected to rise to 90 by 2025, and with about three quarters of the world's major river systems being transnationally shared, water is likely to become one of the burning foreign policy and environmental security issues of this century. This global freshwater scarcity trend will be further exacerbated by global warming, altering rainfall patterns which might lead to extreme flood and drought events, as well as water supply disruptions [10]. One of the most important challenges humanity will face during this century is certainly to provide the world's population with sufficient freshwater.

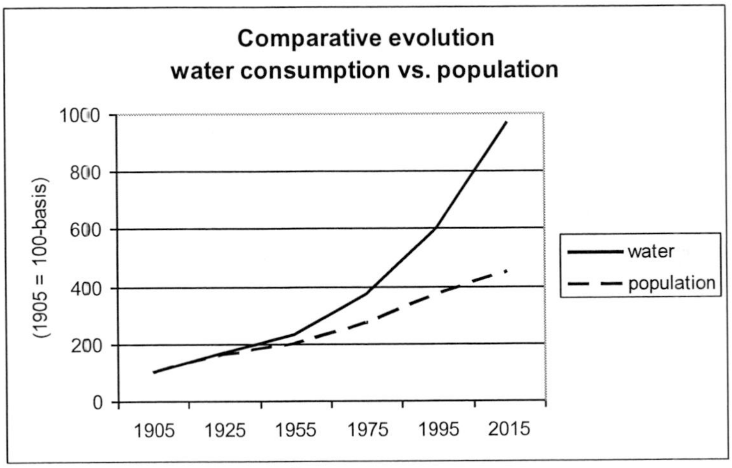

Source: Own elaboration.
Figure 1. Comparative evolution of global water consumption and population growth

1.1. What *Scarcity* Are We Talking about?

The UN differentiates between "water stress" when annual water supplies are below 1,700 m^3/person, and "water scarcity" when below 1,000 m^3/person. The latter is considered as the benchmark level below which water lack is said to impede development. This chapter analyses such situations from two different perspectives: its naturally occurring climatic-geophysical origin and its anthropogenic-based factor. On the one hand, spatial and temporal distribution of freshwater stock and flows is hugely uneven around the globe. Indeed, for climatic and geophysical reasons, differences in average annual precipitation divide the world into water scarce and water sufficient regions. On the other hand, besides the climatic factor, there seems to be a direct correlation between poverty and water scarcity, as communities suffering from lack of access to water resources largely overlap with the 1.3 billion people living under the "extreme poverty line" of US$1/day. Brazil, for instance, is the country accounting the greatest surface water volume yet 55 million people of this nation do not have access to it. Also, while a Belgian household spends on average 0.6% of its revenues on water bills, a Zambian family could spend as much as half of its monthly earnings. This situation has less to do with physical scarcity than with socio-economic scarcity, i.e. a man-made inequality. The average US citizen consumes over 500 liters per person per day (lpd). This value rises to 4,100 lpd in the garden-watered California where 560,000 swimming pools are to be found. California now faces a serious (social) water scarcity and plans to make an extended use of the desalination technology.

Notwithstanding, as explained by Naredo [11], it is the human demand for available water in a given region and its associated consumption patterns that eventually turns a *physical* scarcity (of climatologic and territorial origin) into a *social* one, perceived by the local population. In the middle of the Sahara desert, there is an evident physical water scarcity; yet not a social one if nobody inhabits that particular spot. Oppositely, the Metropolitan area of Barcelona in Spain does not, relatively speaking, present a fierce physical scarcity. However, important demography, combined with high water demands and heavy contaminating uses led to the construction of a desalination plant in order to provide sufficient supplies and to overcome the socially-constructed scarcity. In this case, scarcity can be seen as the lack of enough resources to satisfy insatiable wants as opposed to "basic needs" that Max-Neef [12] define as finite and few.

Viewed from a different perspective, the notion of "water scarcity" can also be distinguished between its two senses, in a similar way to how Alfred Lotka [3] separated endo- and exosomatic uses of energy. The minimum amount of endosomatic energy required to keep a human body alive cannot be substituted by other goods. The necessary daily endosomatic energy consumption is around 2000 kcal. Besides this necessary endosomatic energy, exosomatic use exceeds, in rich countries, on average 100,000 or 200,000 kcal per person/day in order to support our modern life styles. This amount does not constitute a basic need as such, but rather a socially-constructed one, which could better be called *want*. Exosomatic energy uses have significantly increased through history. Similarly, scarcity below 20 or 30 lpd is a physical scarcity where health indicators, such as cholera, may appear. Tourists in Lanzarote or in Laâyoune require 460 and over 310 lpd respectively, compared to the average daily consumption of the local population of Lanzarote (120 lpd) and Laâyoune (61 lpd, 58% of which consume 29 lpd).

The principal imbalance between water availability and its uses originate when human activities are imported to zones without any consideration for their inherent capability to host such social habits. Playing golf in the lush vegetation of the humid European North might be a pleasant activity. Yet, sprinkling hectares of arid soils in Lanzarote and Morocco with desalinated water in order to satisfy a desire for entertainment is highly questionable both environmentally and ethically. In Lanzarote's golf club for instance, each play (i.e. 18 holes per person) requires over 10,000 liters of water. The same argument applies when growing tomatoes in the desert of Almeria or importing the *Hamham* culture into the Sahara. These communal hot bathing facilities, which represent a key custom of the Moroccan culture, have been practically inexistent in the southern regions before 1975 due to their important water requirements. Indeed, one single *Hamham* in Laâyoune consumes on average around 9,500 liters per day.

The first "anthropogenic-type of scarcity" will, therefore, be labeled "socially-constructed scarcity", where a situation of freshwater scarcity is triggered essentially by excessively increased per capita consumption rates. This arises primarily from economic and socio-cultural factors. The case of the Canary island of Lanzarote will contextualize this notion, comparing it to the case of Laâyoune where the "physical scarcity" notion will be exemplified.

This chapter distinguishes between four types of freshwater scarcities: one of a physical order, three others of an "anthropogenic" order. While the first "anthropogenic-type of scarcity", just described above, can be seen as *demand*-induced -essentially triggered by increasing per capita activity- the other two anthropogenic scarcity types can be understood as being essentially *supply*-induced. In this context, supply-induced refers to scarcity occurring due to a) the decrease in renewable resource because of resource degradation or depleted faster than replenishment; but above all b) due to a deficit in the supply system itself.

Although the latter can have many forms and origins, this chapter presents two major cases with relevant illustrations: the Gaza Strip and the urban slums of the Indian metropolis of Chennai. In the first instance, the deficit in the supply system is mainly caused (and maintained) by the geopolitical situation in the Middle East and the military Occupation. In the second one, the deficit is essentially a result of the ill-suited and, above all, inequitable water governance schemes. The first has been labeled "politically-constructed" scarcity; the second one, "structural scarcity". These notions and the terminology chosen to describe them will be further explained during the presentation of the case studies. Although these two scarcity types have been portrayed as being primarily supply-induced, it must nonetheless be added, that both cases also have of course a significant demand-induced factor; this time, evidently not so much triggered by increasing per capita activity than by important population growth rates. Nonetheless, they remain primarily supply-induced in essence as, in both cases, the supply issue is at the origin of the water crisis as well as the departing point to design viable solutions. Figure 2 helps to visualize the way the different types of freshwater scarcities will be presented in this chapter.

When differentiating between these four scarcity "labels", it is important to emphasize that they do not pretend to provide an absolute or definite representation of a given situation. In this chapter, when we speak of "physical scarcity" for instance, what is meant is the description of a scarcity situation where the *dominant* component is of physical (i.e. climatic and geophysical) origin. Similarly, the label "politically-constructed scarcity" has to be read as "a situation of scarcity where the *dominant* component is political", but is certainly not to dismiss the extremely low precipitation rates -a climatic and geophysical factor- of the Gaza Strip, for instance.

1.2. Introducing the Desalination Technology

Desalination: A Growing Sector

Considering the immense volume of available seawater, from a purely technical point of view, the supply capacity of desalination is virtually infinite. In 1961, U.S. President John F. Kennedy noted that if humanity could find an inexpensive way to get fresh water from the oceans, that achievement would really dwarf any other scientific accomplishments. The desalination technology embodies such hope and for the past 30 years desalination has been increasingly perceived as a feasible solution to meet ever growing freshwater demands. Desalination offers indeed a great potential to the 2.4 billion people living in coastal areas. What is more, 42 out of the 70 cities with more than 1 million inhabitants suffering from water scarcity, are coastal towns. As a result, over the past 15 years, the daily desalinated water production went from approximately 13 million m^3 to 32 million m^3 in over 15,000 desalination plants operating worldwide and supplying about 160 million people [13].

Desalination plants are mostly found in water-poor but energy-rich countries boarding the Persian-Arabian Gulf where desalination accounts for 40% of the municipal and industrial water used. Bahrain for instance totally relies on desalinated water and Saudi Arabia, with over 4 million m^3 of desalted water produced daily, accounts on its own for around 1/3 of the global total production. Nevertheless, desalination plants start mushrooming around the globe in not necessarily arid countries: Japan, Kazakhstan, India, Taiwan or South Africa. The technology is rapidly taking off also in the European Union, notably in Spain, where Europe's biggest reverse osmosis desalination plant is located.

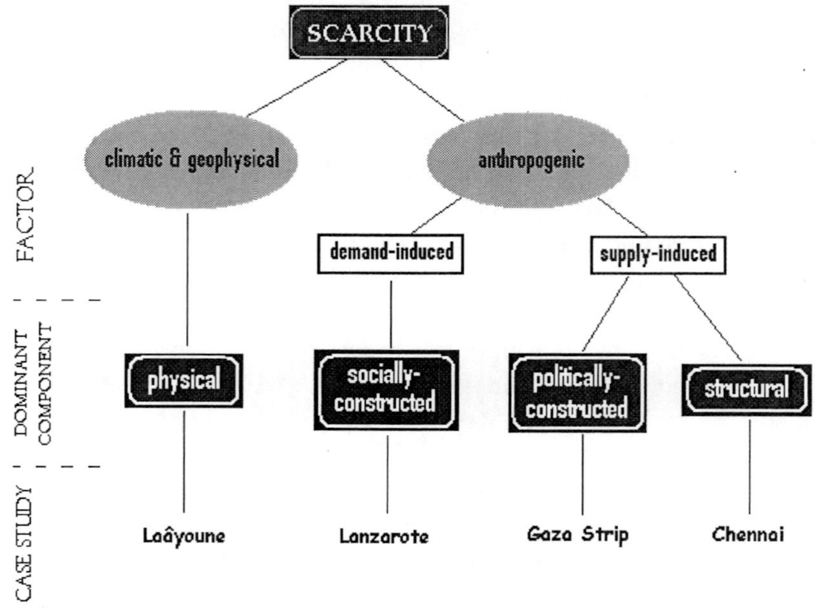

Source: Own elaboration.

Figure 2. The dominant components of scarcity situations as presented in this chapter

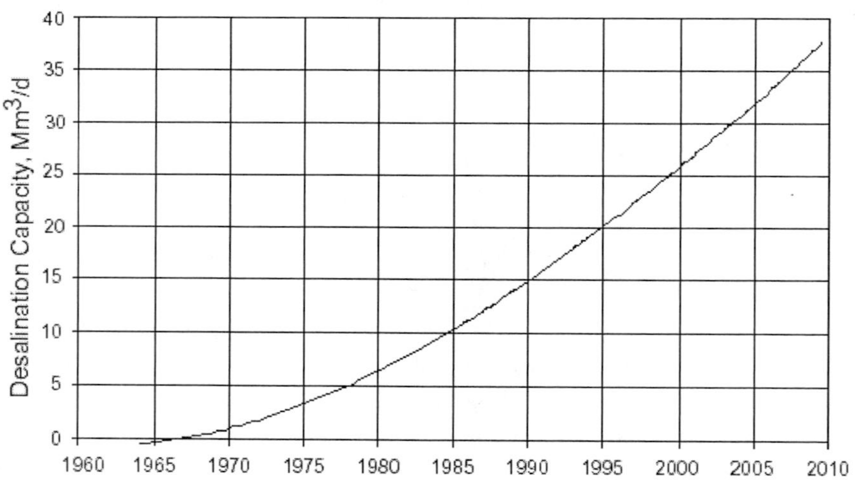

Source: [13].

Figure 3. Projected growth of desalination plant capacity

This growth seems indeed promising for the desalination industry that estimates 10 years as a doubling time for desalinated water demand (Figure 3); especially around the Mediterranean rim, with demand growing to 10 million m^3/d by 2010. The desalination industry represents a global market of over US$ 30 billion that is expected to reach US$ 120 billion by 2015[2].

[2] For updated projections refer to www.globalwaterintel.com.

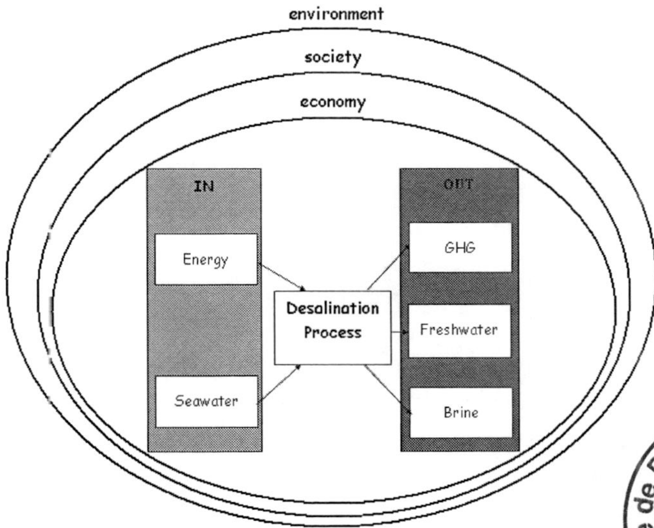

Source: own elaboration.

Figure 4. Desalination metabolism

Desalination processes: A material and energy flow approach

Although various desalination methods exist -most commonly, distillation (thermal separation) and membrane technologies (reverse osmosis, RO)- the overall process metabolism is generally the same. Figure 4 shows that, when considering the desalination process as a black-box, the overarching principle is to turn a saline solution into freshwater. This requires an energy input and triggers the output of a concentrated saline brine as well as greenhouse gas emissions. The surrounding circles are to emphasize that this technological process forms part of a given economic process, influencing its social context and ultimately embedded in, and having an impact on, the local and global environment (resource extraction, waste disposal).

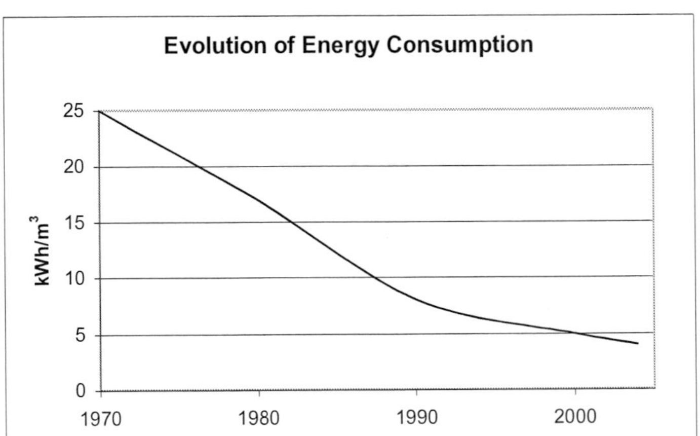

Source: Own elaboration.

Figure 5. Decreasing trend of energy consumption to desalinate water

Energy considerations

The operational pressure to force seawater through the membrane remains around 75 bars. The energy requirements to overcome the osmotic pressure of a saline solution are, therefore, quite considerable and represent still the major cost of the desalination process. From thermodynamics, the theoretical minimum to obtain freshwater is around 0.7 kWh/m^3 [14]. Back in the 1970's, when the technology started to be commercialized, over 20 kWh were required to desalinate one cubic meter of water. As shown in Figure 5, this value has constantly decreased over the past decades and currently ranges between 3 and 20 kWh/m^3, with the older distillation plants at the top end.

As a comparison, 6 kWh are required to lift one cubic meter of water by 1,800 meters, i.e. higher than any worldwide currently undertaken bulk water transfer. So, despite concentrated efforts and real efficiency improvements, the desalination process still remains expensive for most poor nations in need. At US$ 0.5 to US$ 2 per m^3, turning brackish- or seawater into a freshwater is still around two to eight times more expensive than the average cost of urban water supplies in most poor countries and at least 5-20 times what farmers –the world greatest water users- are used to pay. Consequently, desalination accounts approximately for only 0.3% of the total water volume used around the world. Notwithstanding, state-of-the-art desalination progressively lowers its cost down to around US$ $0.6/m^3$ and thus becomes largely competitive with, say, shipping surface water over long distances or inter-basin transfer schemes. It also becomes competitive for many wealthy countries where average urban water production costs have gradually increased with rising pollution and are now in the range of US$ $0.45/m^3$. Potential for further growth is immense.

Environmental impacts

Although environmental impacts of desalination are intrinsically related to the energy efficiency of the system (which improved), some significant flaws remain. Moreover, when considered in absolute terms, overall impacts have dramatically increased due to the steady growth of the desalination industry around the world.

One major drawback is the production of greenhouse gases as well as other contaminating airborne emissions associated with the required power generation. Environmental loads and airborne emissions associated to RO are one order of magnitude lower than those corresponding to thermal MSF and MED processes. Environmental loads can be considerably reduced through water-energy cogeneration and the use of renewable energies. Arid regions lacking of drinkable water often correspond with areas of high availability of renewable energy such as solar irradiation and wind speed. According to Raluy et al. [15], the best alternative is the integration of RO desalination with wind energy: environmental impacts reduction up to 75%.

Another environmental impact is related to the discharge of hypersaline brine, the unavoidable desalination sub-product, which heavily affects marine flora and fauna. With massive developments along the Gulf coast, brine-induced environmental damages rapidly became evident [16]. Furthermore, during pre- and post-treatment processes a variety of chemical agents are added to enhance flocculation or prevent foaming, others to avoid membrane deterioration. Eutrophication, pH value variations, accumulation of heavy metals as well as sterilizing properties of disinfectants have the most pronounced effects on receiving waters and their ecosystems.

2. Physical and Socially-Constructed Scarcity: Laâyoune and Lanzarote

2.1. Brief Description of Laâyoune's Case

Laâyoune is the main town of the Moroccan Sahara. With an annual precipitation average of around 60 mm/m^2 and its increasing water demand, desalination is, according to the Moroccan National Office for Water, the only solution for cities like Laâyoune [17]. Morocco started its desalination experience in 1976 with the brackish water distillation plant of Tarfaya and developed, since then, a fair expertise in this field. Laâyoune now features Morocco's greatest facility, with 7,000 m^3 of desalinated water produced daily. Mixed with an additional 5,600 m^3/d obtained from the local *Foumoued* aquifer, it satisfies the water demand of its 180,000 inhabitants. The aquifer has, however, practically nil recharge and under current over-extraction regimes is likely to dry out in less than 50 years. This is a case of non-renewable groundwater "mining".

The desalinated volume is produced with an average energy cost of approximately 5.3 kWh/m^3 and the average per capita water consumption of Laâyoune is 61 liters, with 58% of the inhabitants actually consuming less than 30 lpd. In comparison with Lanzarote, where as described later, water mainly sustains services for tourism, Laâyoune could probably be depicted as a case where desalination serves to maintain livelihood. Before starting the analysis of the second case study, Table 1 provides an overview of Laâyoune's main characteristics.

Table 1. Laâyoune's main characteristics

<table>
<tr><td rowspan="10">Water</td><td>Precipitation</td><td></td><td colspan="2">~ 60 mm/y</td></tr>
<tr><td rowspan="9">Desalination</td><td rowspan="7">Production</td><td colspan="2">Total 13,600 m^3/day</td></tr>
<tr><td>Salinity</td><td>1988 µS/cm (~ 1,510 ppm)</td></tr>
<tr><td>Process</td><td>RO</td></tr>
<tr><td>Energy cost</td><td>5.3 kWh/m^3</td></tr>
<tr><td>Energy source</td><td>95% fossil fuel
5% hydro & wind</td></tr>
<tr><td>Monetary cost</td><td>US$ 4.2/m^3</td></tr>
<tr><td>Price paid by consumers</td><td>US$ 0.55/m^3</td></tr>
<tr><td rowspan="2">Consumption</td><td>Residents</td><td>61 lpd
(58% of residents only 29 lpd)</td></tr>
<tr><td>Tourists</td><td>310 lpd</td></tr>
<tr><td rowspan="3">Popu-ation</td><td>Residents</td><td colspan="3">180,000 inhabitants</td></tr>
<tr><td rowspan="2">Tourism</td><td colspan="3">110 tourists/day
"tourism" is represented up to 90% by the United Nation representatives</td></tr>
<tr><td colspan="2">Accommodation capacity (beds)</td><td>1000 (in 2003)
50 (in 1975)</td></tr>
</table>

Source: Own elaboration.

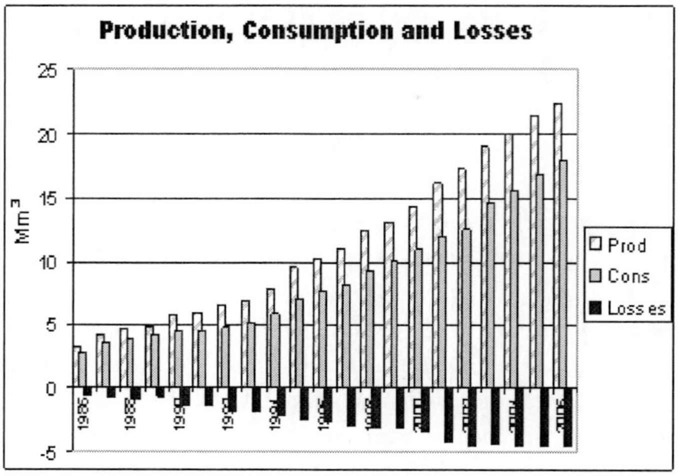

Source: Own Elaboration Based on [18].

Figure 6. Evolution of Lanzarote's water production

2.2. The Case of Lanzarote

The island of Lanzarote is located in the Canary archipelago (Spain), 120 km from the African coast and 200 km from Laâyoune. Its precipitation rates are less than 200 mm/y, i.e. among the lowest in the EU. The island is of volcanic nature and since the first human settlements, Lanzarote's inhabitants had to adapt to the aridity of the insular environment by developing various innovative agricultural techniques. The population was around 20,000 during the 1920s and it was only in the 1970s that migration trend reversed from predominantly emigration to immigration. Lanzarote's total population is now around 170,000, with almost one third tourists. The main factor, which triggered this sudden demographic take off was the implementation of the desalination technology; in Lanzarote there truly has been a *pre-* and a *post*-desalination era. The technology now covers almost 100% of the insular freshwater demand.

After desalination: Water abundance within absolute scarcity

The trend of Lanzarote's water production can be seen in Figure 6.

As a consequence of the increase of both network losses and unitary consumption, in 15 year the water production increased by over 425% while the resident population grew by approximately 108% and the tourists by 250%. In terms of incremental trends, demographic growth slows down, while water production speeds up. At current rates, the water production will double within the next ten years. The unquestionable (or rather unquestioned) success of desalination in Lanzarote produced the sensation that water problems have disappeared from Lanzarote for ever. As this conception spread amongst the island's population and tourist industry, the traditional culture of water use prudence progressively eroded, generating rising consumption patterns. Due to its absolute dependency on the desalination process, Lanzarote runs high risks in case of technical breakdown, heavy seawater contamination or interruption of the energy flow. This situation creates a serious contradiction in which a "water squander" culture emerges in a natural context of absolute geophysical scarcity. This situation should be criticized for at least three reasons: firstly, the current desalination trajectory -ever increasing and primarily based on fossil fuels- supposedly accepts an external strategic dependency, which may trigger some serious problems, especially due to the irreversible technological and ideological "lock-in". Secondly, this energy-intensive technology is responsible for greenhouse gas emissions, the environmental costs of which remain uncovered. Availability of water contributes also indirectly to increasing tourist travel, implying high emissions. Finally, the desalination technology makes the former water saving culture redundant, nourishes the false impression of abundance, responsible for the ever increasing demands, subsequently alimenting the two previously mentioned issues. Reviewing historical data does indeed show how increased supplies raise demands.

Water services for tourism

The island's population is around 120,000 with an additional 50,000 tourists [19]. While "sustainable tourism strategies" recommend a 10/1 resident to tourist ratio [20], in Lanzarote an approximately 2 to 1 proportion can be observed: a situation, which suffers many organizational drawbacks as far as insular resource and service management is concerned. In 1998, the net consumption per resident was 138 lpd, while tourists consumed 230 lpd. These figures evolved to 120 lpd and 460 lpd respectively by 2001 [21]. However, the latter figure represents only official monitored data. Another 3-5 Mm^3/y of unmonitored water production has to be accounted for; this is the estimated total yield of the private desalination facilities that can be found in the islands hotels [22]. All in all, luxurious hotels require daily well over 1000 liters per tourist to function. It is striking to see how tourism and the water desalination industry simultaneously developed on the island. Desalination production rose from 0.45 Mm^3 in 1968 to 17.21 Mm^3 in 2002, while the number of yearly tourists passed from 10,205 to over 1.87 million. From Figure 7 it can be clearly seen how tourism took off in the 1970s, after the introduction of the desalination technology and parallel to water production increase.

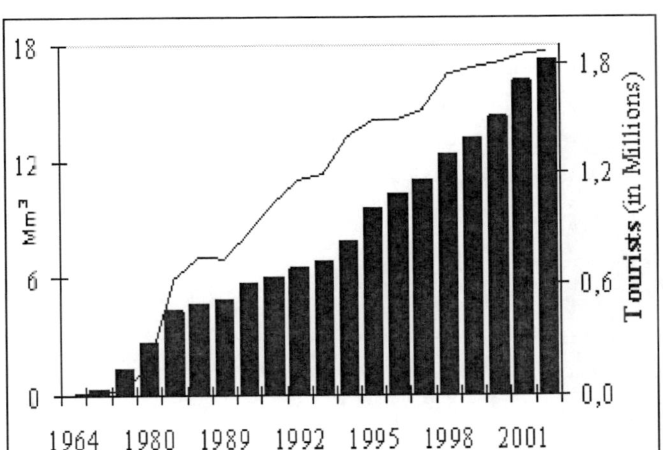

Source: Own elaboration based [18, 19, 23].

Figure 7. Evolution of yearly water production (bars) and number of tourists (line)

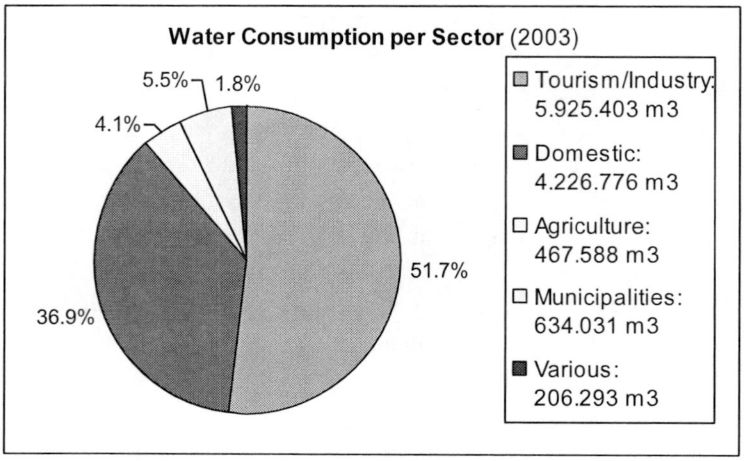

Source: Own elaboration based on [18].

Figure 8. Tourist appropriation of water production

Furthermore, Figure 8 shows how the water production is massively dedicated to the tourist industry and its directly related service sector.

The tourist industry is, thus, highly dependent on desalination and ultimately the availability of oil, the main energy provider. Furthermore, Lanzarote's water market is somewhat distorted in the sense that domestic water consumption is highly subsidized, whereas the industry sector (and tourism in the first place) is subject to certain eco-taxes. As a consequence, illegal or unregulated small-scale desalination plants are built by the private sector to meet the high water demands of the tourist industry. This results in an unmonitored water consumption, uncontrolled brine disposal and high induced social costs [24].

Oil for water

The desalination industry is the island's largest energy consumer, accounting for over 15% of the total consumption [19]. The specific average energy consumption for the year 2003 was 5.26 kWh/m^3 [18]; however, to this figure another 10,717 MWh should be added for distribution and pumping services, resulting in 5.88 kWh/m^3 delivered to consumers. Table 2 summarizes the most relevant data about water desalination and energy consumption in the island of Lanzarote.

Table 2. Summary of Lanzarote's desalination industry (2002)

	Water produced (Mm^3)	Water consumed (Mm^3)	Water Loss (%)	Total Electricity consumed in desalination (GWh)	Proportion of Lanzarote's electricity consumption (%)	Electricity generated by oil, consumed by desalination (GWh)	Oil / total electricity consumed in desalination (%)	Electricity generated by wind, consumed by desalination (GWh)	Wind/total electricity consumed in desalination (%)
2002	17,210	11,461	33.4	101.3	15.4	90.1	88.9	11.2	11.1

Source: Own elaboration based on [19, 25].

The 18.3 Mm^3 of desalinated water produced in 2003, triggered 97,017 tons of CO_2 emissions[3]. What is more, approximately 22 Mm^3 of hypersaline brine are discharged every year; this represents 1,540,000 tones of salt material and vast amounts of chemicals.

Lanzarote's water metabolism

After the main input and output data of the insular "water cycle" have been commented, a summarizing metabolism sketch can be presented (Figure 9). Note that over 1/5 of the total insular CO_2 emitted for electricity production are to be accounted to the desalination sector.

What is more, the same technology serves different purposes in Lanzarote (*water services for tourism*) than in Laâyoune (*water for basic needs*). Table 3 emphasizes the differences in consumption patterns for both case studies, pointing out that "luxury" consumption still remains relatively cheap compared to the important environmental impacts.

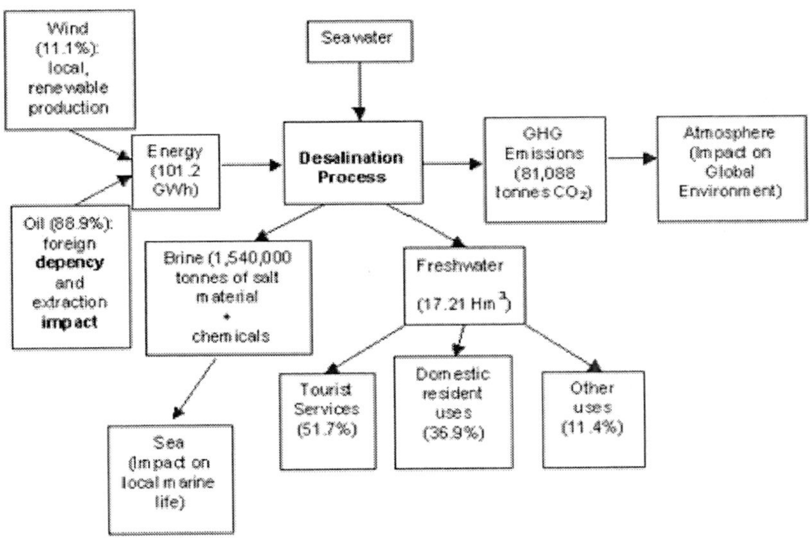

Source: Own elaboration based on [19, 25].

Figure 9. Material and energy flow diagram of Lanzarote's desalination process (year 2002).

Table 3. Energy and monetary costs of consumed water and the associated CO_2 emissions

			lpd	kWh/p/day	US$/p/day	kg CO_2/day
Lanzarote	Resident		120	0.71	0.19	0.64
	Tourist	Standard	460	2.7	1.25	2.4
		Luxury	1000	5.88	2.7	5.3
Laâyoune	Resident	Average	61	0.32	0.035	0.23
		Poorest 60%	29	0.15	0.018	0.11
	Tourist		310	1.64	0.173	1.18

Source: Own elaboration based on [17, 18, 19, 25].

[3] assuming a equivalent rate of 5.3 kg CO_2/m^3 and a wind share of 11.1%.

To conclude this section, it can be said that desalination remains an energy-intensive solution to provide freshwater, heavily loaded with environmental costs. Moreover, the massive introduction of the technology subsequently triggered a fundamental cultural change both in Laâyoune and Lanzarote, transforming a rare good into an apparently endless resource. In the case of Lanzarote, one example is the agricultural sector, where watering techniques had adapted to the scarce conditions. This lay knowledge has been replaced by unconsidered water over-consumption and squander. The first "anthropogenic-scarcity type" that has been described in this chapter was labeled "social (or socially-constructed) scarcity", precisely because it relates to freshwater scarcity situations triggered essentially by excessively increased per capita consumption rates, arises primarily from economic and socio-cultural factors.

3. Politically-Constructed Scarcity: The Gaza Strip

Water related conflicts proliferate around the world and the Middle East offers one of the most emblematic illustrations. Along with the ever-lasting dispute over borders, settlements and returnees, water is a major issue of the Israeli-Palestinian conflict. As Yitzhak Rabin once said, "without agreement on water, there will be no agreement". The present case study analyses the water crisis in the Occupied Palestinian Territories (OPT), and in the Gaza Strip in particular.

From a purely technical point, desalination holds the potential to bring the long water dispute to an end. However, water in the Middle East is a highly political matter and its scarce nature is not only *physical* but predominantly *political*. This is what the UNDP refers to as the "limitations imposed on the development of new water resources and facilities resulting in artificial suppression of the demand on water."[4]

Around half of the water in the West Bank is purchased from Mekorot, Israel's national water utility. However, Mekorot routinely cuts services and 75% of communities in the OPT have reported a reduction in or, in some cases, a complete halt of their water supply [26]. Moreover, Mekorot installed a dual tariff structure, resulting in un-payable water bills for Palestinians while Israeli settlers, occupying land in the OPT, enjoy lower rates [27]. With their lower incomes, water is consequently 20 times more expensive for Palestinians than for Israelis [28]. According to the UNEP [29], Israeli settlers use per capita, four times more water than Palestinians and, before the disengagement in August 2005, settlers in the Gaza Strip used nearly ten times more. These settlements further exacerbated the hydro-stress on the Coastal Aquifer while, the Oslo II Accord required that additional water demand due to settlement growth is to be met by Israeli sources, i.e. outside the Strip.

The Middle East features an evident (growing) discrepancy between the water volumes theoretically demanded for domestic, industrial and agricultural needs and the water quantities readily available to be supplied. In the OPT, this water gap is widening to even greater proportions as water supplies are artificially constricted by the occupational status. This situation of chronically unmet demands has serious adverse effects on socio-economic development and will further exacerbate the political situation if left unresolved.

---

[4] www.papp.undp.org

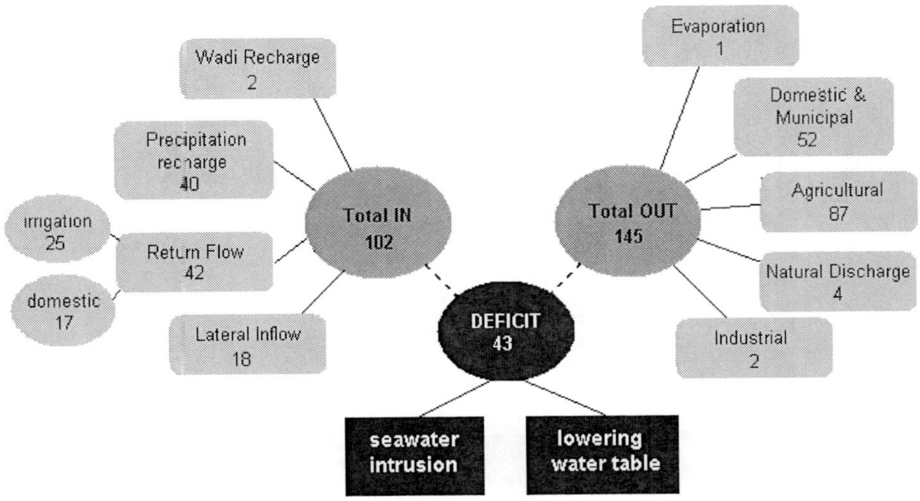

Source: Own elaboration based on [30, 31].

Figure 10. In- and output flow (in Mm3/y) of the Coastal Aquifer system for the Gaza Strip

3.1. Water Crisis in the Gaza Strip

The Gaza Strip is 45 km long with an average width of about 8 km. With a population of around 1.43 million, it is one of the most densely populated places on earth. The main –and almost only- water source of the Gaza Strip is the Coastal Aquifer. Figure 10 presents the input-output balance of the Coastal Aquifer showing the induced hydro-deficit. The over-extraction rates are responsible for the lowering of the groundwater table and seawater intrusion.

The Gaza Strip is the second poorest region in the world as far as annual per capita water availability is concerned (52 m^3/p/y). Moreover, the appalling quality of the water available to the population poses some serious health issues. Water with Total Dissolved Solids (TDS) concentrations over 1000 mg/l is considered saline. In the Gaza Strip, TDS concentrations sometimes exceed 3000 mg/l. The dramatic deterioration of the Coastal Aquifer is caused by untreated wastewater discharge, infiltration of chemical fertilizers, and above all, by over-extraction rates exceeding about three times sustainable yields [29]. Seawater intrusion and saline up-coning imperils the hydrogeological resilience. Due to nitrates and chloride levels substantially exceeding WHO standards, 90% of the water delivered from municipal wells is considered unsuitable for human consumption. Consequently, although the average availability is 71 lpd, the real consumption rates in the Gaza Strip is to be considered as only 15 lpd when quality and supply intermittences are taken into account. In certain communities of the OPT, consumption drops below 10 lpd. Figure 11 summarizes the projected demands for the municipal, industrial and agricultural sectors, and the resulting gap between demand and supply.

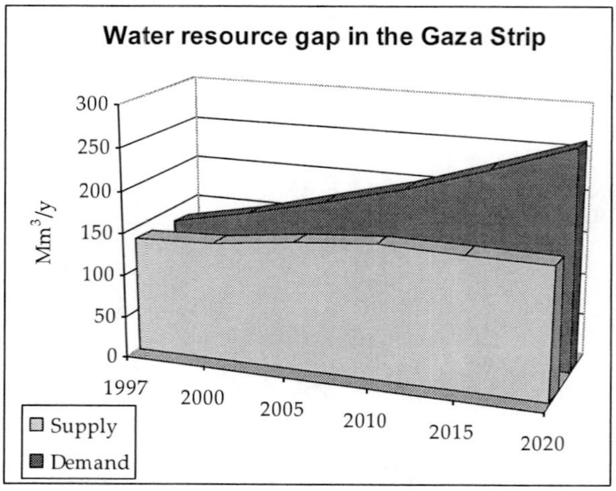

Source: Own elaboration based on [32, 33].

Figure 11. Evolution of the total water demand and the available resources in the Gaza Strip

3.2. Politically-Constructed Water Scarcity

Before suggesting possible options to alleviate Gaza's water crisis, the political situation and its implications for the water sector must be understood. As Swyngedouw explains [34], the Middle East offers the most notorious illustration of a precarious socio-spatial water balance, which keeps extending its geopolitical reach while intensifying its complex and conflict-ridden condition. To its battle for keeping control over the Mountain Aquifer –the region's largest groundwater source- Israel is getting bogged down in tedious negotiations with Syria about the occupied Golan Heights, Israel's most important surface water source, and monopoly over the waters from the Jordan river has been a conflictive ground with Jordan and Palestinians even before the creation of its State. As Roy shows [35], Israel's occupation imposed a process of "de-development"; conducting, in other words, a "deliberate, systematic deconstruction of an indigenous economy by a dominant power". "De-development", has to be distinguished from "underdevelopment", precisely due to the intentional consequences. Israel's de-development strategy methodically dispossessed the OPT by dismantling the existing economic infrastructure and simultaneously expropriating and assimilating its man power, land, water and other vital resources. Structural discrimination has been an undeclared but very apparent means of occupation.

Just as much as the West Bank Mountain Aquifer, surface water not only of the -for Palestinians inaccessible- Jordan river or of the highly contaminated Wadi Gaza but also of the "rivers of sewage» spilled from Israeli settlements towards Palestinian cities are battlegrounds of the conflict. And, so is the unbearable water crisis originating from a highly over-stressed Coastal Aquifer in the Gaza Strip. The initial *physical* scarcity (i.e. related to climatic and geophysical factors) is strongly exacerbated by the geopolitical situation and by military occupation (i.e. related to anthropogenic factors), turning it into what has been coined in this chapter as "*politically-constructed* scarcity"; i.e. a situation of scarcity where the dominant component is political.

This situation very much exacerbates the conflict and is used as a political instrument and warfare weapon. The most obvious example is the newly erected Separation Wall that, once completed, will stretch for some 730 km and will *de facto* annex around 50% of the West Bank most fertile land; the appropriation of the coveted resources of the Mountain Aquifer is very much part of this strategy. This can best be understood by observing how the wall's path swirls around wells and other Palestinian water sources to seize them. A total of 1096 km^2 of the Western Aquifer Basin will be annexed; this represents 65% of the recharge area of the aquifer located inside the West Bank. In fact, 76% of the total unconfined area of the aquifer will be controlled by Israel; although the majority of the water comes from within the West Bank, it will primarily be consumed outside of it [26].

Water, power and control: Inequitable allocation of water resources

Effectively, before the disengagement, Israeli settlers in Gaza consumed about 8.5 times more (of fresh, clean water) per capita than the (brackish and polluted) water consumed by the Palestinian population. Roberts and Emel [36] question the traditional view, which depicts groundwater depletion as the *Tragedy of the Commons*, entailed by the open access to a resource. According to them, groundwater depletion should be rather seen as the result of uneven development opportunities between 'users'. Equitable allocation of resources is a fundamental issue the Palestinian community is struggling for. As Gleick posits:

> "Although many Israelis argue that efforts should focus on enlarging the "pie" so that existing uses can be maintained, Palestinians insist that a discussion of reallocation of water rights and control over the existing supply must precede any major efforts to enhance total availability." [37: p.13]

No community on earth would be willing to accept seeing their neighbors within the same borders monopolizing almost ten times more than themselves from the same limited resource; especially, when considered one's own – and the "sharing" therefore becomes "theft". Frustration arises when different rights apply to one group or another. Palestinians want water resources under their control. This argument is political and based on security reasons, but refers also to the emotional freedom-laden dimension associated with the values of this vital resource. Likewise loosing land, giving up the "water battle" would be considered as a betrayal of their struggle for finally regaining dignity, identity - in brief, the right to exist as a community. As long as this is not acknowledged, not only the water dispute, but the entire conflict will remain in a deadlock sclerosis.

In his essay on "The Politics of Verticality", Weizman [38], explains the three-dimensional character of the territorial planning at the core of military occupation. Just as (illegal) settlements and military watchpoints were purposefully built on hilltops, gaining control over the surroundings through their Foucaultian panoptical attributes, the sub-terrain and the aquifers also become conflict zones. Those underground battlefields are both affecting and affected by the political situation enfolding above. As Halper argues [39], the water pipes and the sewage running through the West Bank and Gaza are also part of a meticulous matrix of control. From the Oslo and Camp David negotiations it becomes clear that Israel would never be willing to lose sovereignty upon groundwater in any permanent resolution. Since 1967, Palestinians are restricted from digging new wells, while settlements were granted permission to bore unrestrictedly. On the one hand, the number of Palestinian wells operational in the OPT went down from 774 to 328 [40], while the population dramatically increased and the water quality worsened. On the other hand, the hundreds of Israeli wells built within the limits of the 1949 Armistice Line account for around 85% of the water pumped annually from the aquifers of the OPT [40]. As a result, Israelis settlers consume nearly 40% of the West Bank's water, while representing less than 15% of its population; the remaining water has to suffice for the 2.3 million Palestinians living on their plundered land.

3.3. Options to Alleviate Gaza's Physical and Politically-Constructed Scarcity

To narrow the supply-demand gap, both the technical and political aspects have to be considered. Although technically speaking, options are readily available, the bottleneck is indeed often the political allowances to enable the Gaza Strip to have a secured and permanent water source. Politically, any future solution should include "reciprocal supply"; that is, water provided by Israel to the OPT and by the Palestinian Authority (PA) from the Mountain Aquifer to Israel. This would, however, require solving the long-lasting dispute between Israel and the PA on the Water Rights of the Mountain Aquifer. This "political" solution would be the most economic option to meet, at least partly, the Gazian water gap. Under the Oslo II Accord, the current supply regime provided from Israel to the Gaza Strip should be approximately doubled.

Figure 12 summarizes the main options available to tackle Gaza's freshwater crisis, differentiating between the two management approaches: demand and supply. Each option is marked with a comment tag. As seen, only water conservation, treated wastewater reuse and desalination are readily implementable and politically desirable, whilst contributing to reduce the hydro-stress on the damaged Coastal Aquifer. Other options, such as allowing the shifting of agricultural water towards the domestic and industrial sector through increasing virtual water trade as well as undertaking an equitable inter-regional reallocation scheme (including water transfers from Israel, the Mountain Aquifer and neighboring countries) are desirable strategies to be implemented in the future. The latter, however, involve important political negotiations.

Put simply, in the case of the Gaza Strip, adding more water to the system seems indeed far from superfluous; hence the need for a scarcity management strategy that should adopt predominantly the "Increased Supply" approach. On the supply side, options include an inter-regional reallocation of water resources, to further (over-)exploit of the Coastal Aquifer, to import bulk water, to harvest rainwater, to recycle treated wastewater or use desalination. Out of these, only the last two are feasible options and must go hand in hand.

Recycle treated wastewater

Although over half of Gaza's population is reported to be connected to a sewerage network, only 20% to 30% of the collected wastewater is currently treated. As a general estimate, half the initially supplied freshwater volume can be re-used after undergoing appropriate treatment. This means that the potential for water savings are great and waste water reuse programs are feasible, especially for agricultural uses [31]. Existing treatment plants should be repaired and extended. Figure 13 shows the projected evolution of Gaza's wastewater treatment capacity.

Desalination

The Gaza Strip has a long experience in desalination with its first unit set up in the 1980s. 29 small-scale brackish desalination plants exist in the Gaza Strip, having a total capacity of 6580 m^3/d [41]. However, the actual yield is estimated to be much lower as most of these plants operate under reduced productivity, many of them being in poor condition or out of order. Overall, the seawater and brackish water desalination only contributes to around 2% of daily water supplies in the Gaza Strip.

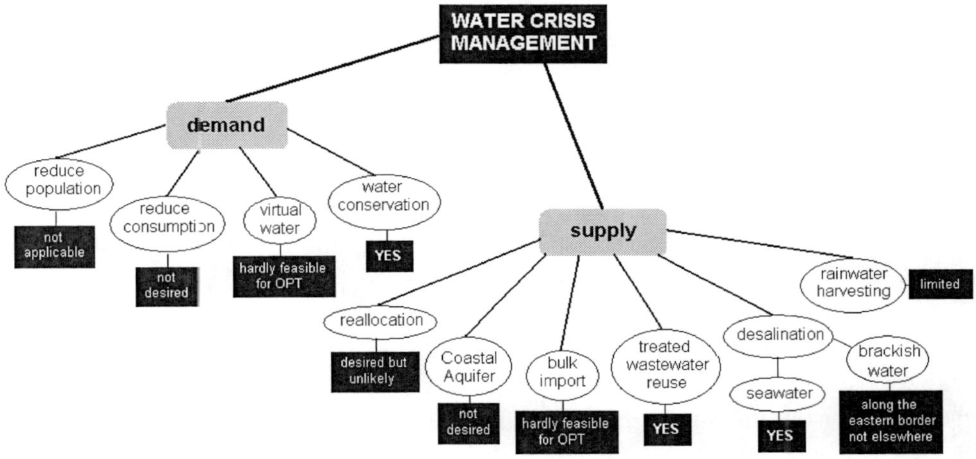

Source: Own elaboration.

Figure 12. Main options available to tackle Gaza's freshwater crisis and summary of their appreciation

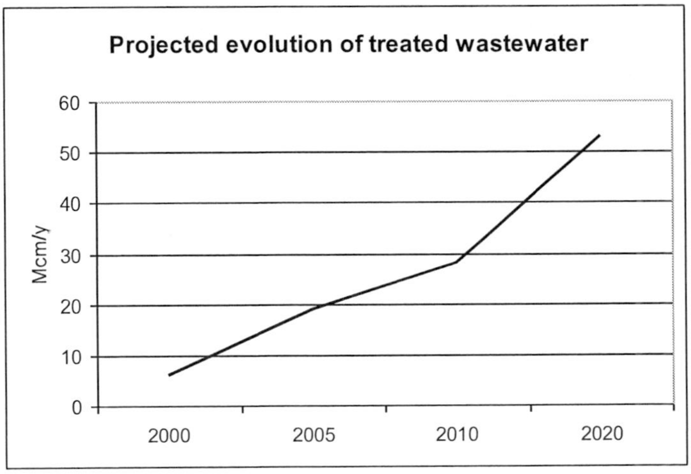

Source: Own elaboration based on [33].

Figure 13. Projected evolution of Gaza's wastewater treatment capacity

Leading scientists have emphasized the growing necessity for a large-scale desalination project to reduce the hydro-stress put on Gaza's groundwater resources while producing additional and much needed water quanta [42]. Brackish and seawater desalination are officially considered as an essential component of the strategy elaborated by the Palestinian Water Authority (PWA) to provide its people not only with a safe but also a politically secured drinking water source. The UNEP also agrees with this view:

> "Considering the improving economic viability of desalination and the severe difficulties of meeting the demand for water using available natural resources in Gaza, it seems likely that desalination will become [an] increasingly important element of future strategies [...] drawing on renewable sources of energy where possible." [29: p.29, p.131]

The much awaited "Regional" plant

A few years ago, USAID launched a program to build a US$ 70 million "regional" desalination plant with a water network from North to South of the Gaza Strip. Its final production capacity was planned to be about 150,000 m^3/d. Unfortunately, the program was put on hold during the second *intifada*, with the 2003 shooting targeting an American diplomatic convoy. Gaza truly needs such type of plant capacity and the population is still hoping for the frozen program to be revived.

The other advantage of such large plant, would be the extended autonomy and security offered by desalination *vis-à-vis* Israel. As former Palestinian President Al-Hamedi emphasizes, a desalination plant in the Gaza Strip would be a major step toward the independence of the Palestinian people. In order to meet its water demands until the year 2020, the PWA estimates that a total desalinated water capacity of 155 Mm^3/y will have to be developed in the Gaza Strip. Of this volume, 100 Mm^3/y should originate from brackish water desalination; the other 55 Mm^3/y from seawater desalination. Such facilities would require a capital investment of over US$ 220 million [43].

With an estimated energy consumption of around 1000 MWh/d, an additional thermoelectric plant of approximately 40 MWe power capacity will have to be built. A sound desalination scheme would thus require a tremendous effort also from the Palestinian energy sector. Almost all the energy required by the OPT is currently imported. This situation of absolute external dependency causes great uncertainties in securing the present and future livelihood of the Palestinian people. Much hope lies in Gaza's recently discovered offshore gas fields. They, however, remain a politically sensitive issue and are still under Israeli control. Notwithstanding, Gaza's energy opportunity lies in its high solar radiation, with an average of 222W/m^2.

Cost of the desalination option in the Gaza strip

The operation but above all, the investment costs necessary from the construction of desalination plants are very high indeed. The calculated cost for a medium size brackish water desalination unit is around US$ 5000 for each m^3/h. Such cost is exorbitant for the bankrupt PA, especially when compared to the construction cost of an average well of similar yield capacity, five to ten times cheaper. Users are currently charged between US$ 0.25 and 0.5 per m^3. Once the Strip shifts to desalination, bills are likely to increase to US$ 1/m^3. Nevertheless, a recent study shows that the Palestinian population is ready to pay twice more for good quality water [41]. It should be emphasized that the costly, yet essential desalination option, inevitably requires substantial technical and financial aid from donor countries interested in promoting peace in the region.

Desalination and wastewater treatment must go hand in hand

To recapitulate: in order to narrow Gaza's water gap, the additional resources required has in fact to be "doubly additional": some extra water to increase the overall supply and "more extra water" to compensate for the reduction in over-extraction rates down to the natural yield of the Coastal Aquifer. Desalination (which adds new water volumes to the system) and wastewater treatment (which enables the recycling of existing water volumes) are not only combinable but must be complementary.

Desalination produces high quality water, yet at a high energy (and environmental) cost; as will be discussed in section 5 of this chapter, the technology is therefore only "justifiable" for serving the "basic need" category of water consumption (i.e. mainly drinking and cooking). Especially in poor regions such as Gaza, it is unthinkable to use desalinated water for agricultural purposes; the latter can, nevertheless, be handily covered by treated wastewater. Treated wastewater can also be used to recharge the Coastal Aquifer, with natural soil filtration through percolation or injection (as done by *Veolia Water* in Berlin for instance). Treated wastewater can also be conveniently used to form a so-called "water wall" against seawater intrusion (as done by *Suez* in South Bay, California, for instance). To summarize, treating wastewater and recycling it has five advantages: 1) it reduces risks of diffused contamination, 2) it virtually reduces demands for extra water and it can be used 3) for the agriculture sector, 4) for aquifer recharge, 5) as well as a barrier against seawater intrusion. A sound strategy to alleviate Gaza's water crisis should be based on a combined solution relying on both desalinated seawater and treated wastewater.

The following figures schematize the above explanation for the need to shift from the current situation of "insufficient water" with a risk of water "insecurity", depleting aquifer, lacking wastewater treatment and desalination (Figure 14) to a situation of "sufficient water" strongly based on a combination of both technologies (Figure 15). Note that the "water autonomy" in the water production is fundamental in this suggested alternative.

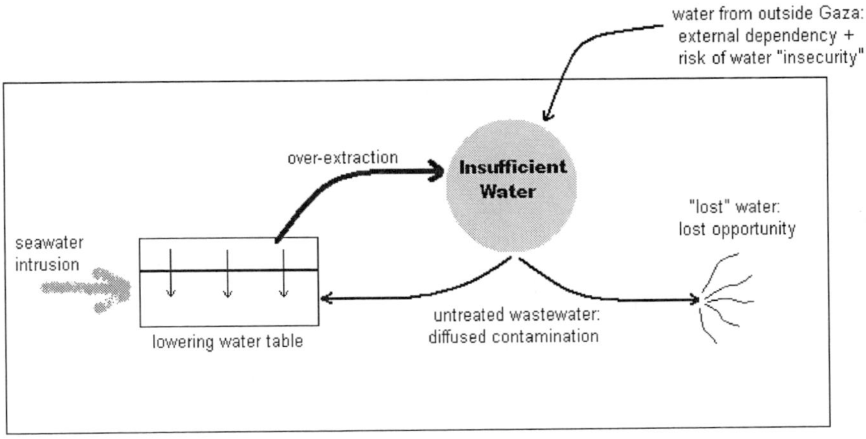

Source: Own elaboration.

Figure 14. Schematic representation of Gaza's current water situation: relying mainly on over-extraction of the Coastal Aquifer

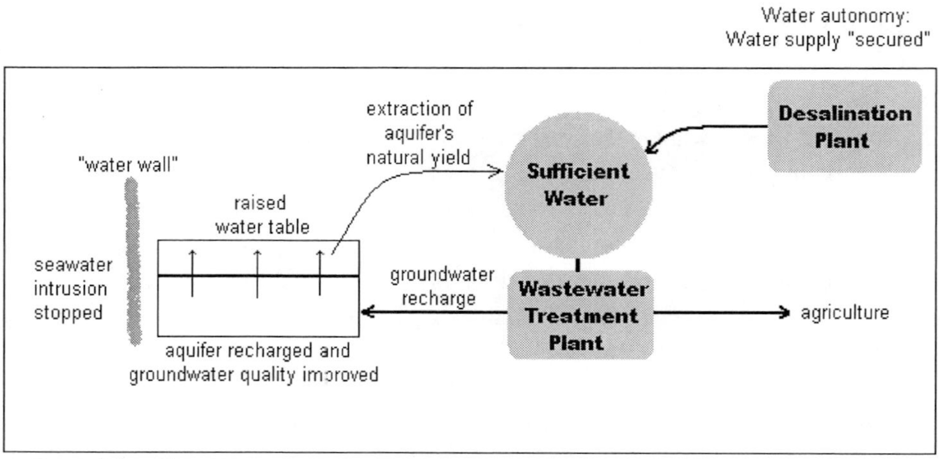

Source: Own elaboration.

Figure 15. Schematic representation of Gaza's ideal water situation: relying on autonomous wastewater treatment and seawater desalination

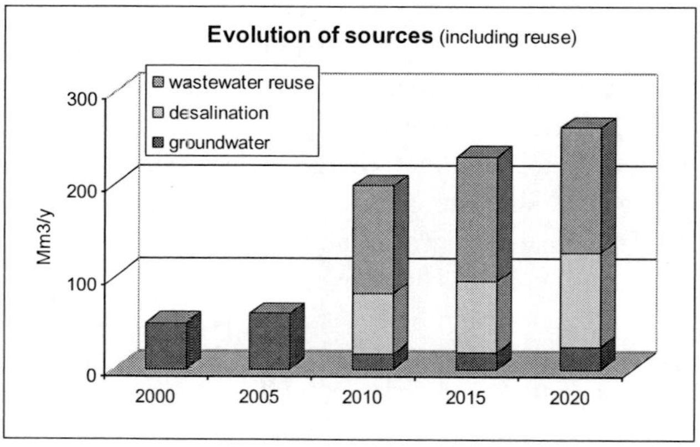

Source: Own elaboration.

Figure 16. Evolution of the groundwater extraction and desalination production in the Gaza Strip

Figure 16 shows the gradual shift from groundwater dependency to desalination and recycled treated wastewater.

Designing a desalination-based solution

Weinthal et al. [44], have integrated the results of geochemical studies and numerical modeling of the Gaza Strip's groundwater system. They come to the ground-breaking conclusion that an important part of the salinity of the Gaza Strip are to be accounted to the natural South-East / North-West flow of saline groundwater from Israel towards the Gaza Strip (Figure 17).

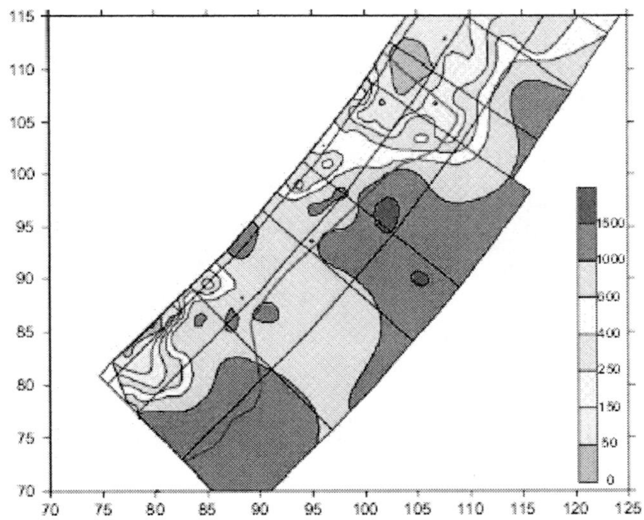

Source: [44].

Figure 17. Salinity gradients in the Coastal Aquifer of the Gaza Strip showing high salinity concentrations along the Israeli-Gaza border

In other words, this means that the Gaza Strip's groundwater calamity is not the result of a classic "upstream-downstream" dispute, where Israel's water-related actions would be the cause of all flaws. Quite on the contrary, what Weinthal et al. [44] suggest is that the pumping of the saline groundwater undertaken by Israel along the Israeli-Gaza border line actually reduces the salinization process of Gaza's aquifer system. As Weinthal et al. argue:

> "[i]ronically, while it could be assumed that Israel's pumping of the ground water might have an adverse effect on water quality in the Gaza Strip, we argue instead that Israel's pumping of the ground water before it reaches Gaza is actually preventing further deterioration in the quality of the aquifer within the Gaza Strip." [44: p.656]

Nevertheless, although the salinization process described by Weinthal et al. [44] can be considered to originate from the natural underground flow, it is important to emphasis that the problem is evidently further exacerbated by the over-pumping rates carried out by the wells scattered around the Gaza Strip. Their hydrodynamic model shows how an increased pumping rate along the border, would result in a net improvement of Gaza's groundwater quality but only when combined with easing-off the excessive extraction rates within the Strip. From these results, one may suggest a water management scheme to the Israeli-Palestinian Joint Water Commission, by which saline groundwater would be extracted from a series of new wells to be drilled along the East border of the Gaza Strip; this would both, reduce the natural saline underground inflow into the Coastal Aquifer system and serve as brackish feed-water to a desalination scheme.

Figure 18 shows what could be called a "double salinization barrier" scheme, which combines both sea- and brackish water desalination as well as wastewater treatment for irrigation and groundwater recharge. The series a newly drilled wells along the East border intercept the saline underground flow and turn it into freshwater through brackish water desalination units. From Weinthal et al. [44], one could suggest a pumping rate of around 12

Mm^3/y from those wells. The pumping of the wells around the Gaza Strip should be phased off. The remaining freshwater production would then come from large-scale seawater desalination facilities. After consumption, the wastewater is treated and the resulting effluent used for the injection of a "water wall" acting against seawater intrusion. Note that the seawater intrusion phenomenon will be already minimized thanks to the reduction in pumping rates of the existing wells. Treated wastewater is also used for irrigation, part of which will percolate back into the subground. Injection of treated wastewater into the underground system has three objectives: 1) the erection of a "water wall" to block seawater from intruding into the freshwater aquifer, 2) replenishing stored reserves and raising the water table level and 3) taking advantage of the natural filtration function of the subsoil to achieve higher groundwater quality (hence, only secondary treatment is necessary for the injected wastewater).

The following fundamental point must be emphasized: a scheme supposed to alleviate the water crisis from which the Palestinian population suffers and grant them with higher water security, requires that any pumping undertaken at the border to feed brackish desalination plants is located within the Gaza Strip and fall under Palestinian control.

Desalination (within Gaza): A political tool towards increased autonomy

According to the Oslo Accords, Israel is required to provide 10 Mm^3 annually to the Strip. It has been years that Israel fails to fulfill its pledge and only 5 Mm^3 are sold. With the inauguration in August 2005, of the 100 Mm^3/y desalination plant in Ashkelon, Israel offered to provide the Strip with those additional 5 Mm^3. Besides the current Hamas turmoil, until now the PA rejected the deal, partly for technical motives and the much higher costs, but the underlying reason is that acceptance would definitively annihilate the chance for Gaza to have its own desalination plant. Indeed, the PWA rightly fears that once the deal is closed, the international community would cancel their financial and political support for the Strip to obtain their own desalination plant on the grounds that such facility is not any more necessary since the Oslo Accord are satisfied. The justification for this desalination plant to be located within the Strip is the independency and security such facility would guarantee to the Palestinian people. The most important point is the location of the plant. In the present context, "location" has to be taken both in its classical (geographical) sense, but also in its political sense of "areas/spheres of control", which in the case of the OPT are always very contentious.

The main issue of the Israeli-Palestinian desalination debate is indeed related to the water security issue. Palestinians need their own production facilities to ensure water rights, self-dependency and autonomy. The underlying discussion lies in the territorial control Israel holds and their ability to regulate water transfers to the OPT and to "turn off the tap" at any moment, for any reason. Under this perspective, water becomes a powerful military blackmailing and extortion tool, through which an entire region can be dehydrated. For Palestinians, water supply controlled by Israel remains, therefore, one of the greatest dreads. Hence in the case of Gaza, the justification for the technology is primarily of a political order, which also leads to the postulate serving as foundation for this case study: a scarcity situation where the dominant component is political, and therefore labeled "politically-constructed scarcity".

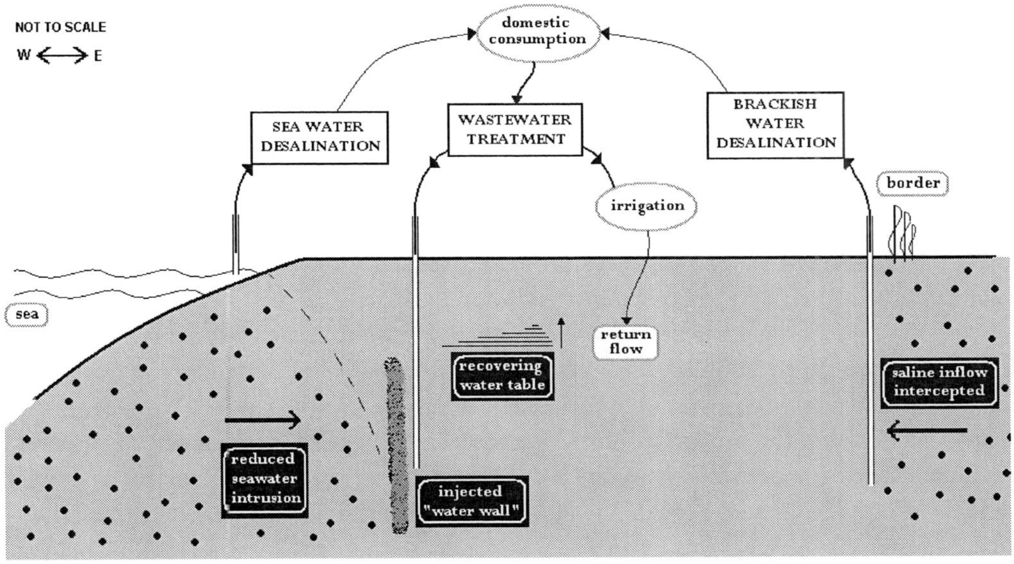

Source: Own elaboration.

Figure 18. "Double salinization barrier" scheme suggested for the Gaza Strip

Desalination as a peace technology

The desalination technology, especially when driven by autochthonous renewable energy sources, truly offers the means to avoid those loudly acclaimed "water wars". In the Middle Eastern context, desalination for the livelihood of the Palestinian people of the Gaza Strip could become a peace tool. In this sense, the technology should be directed towards the poor (guaranteeing basic needs), the environment (reducing the hydro-stress on the Coastal Aquifer) and peace (by by-passing the conflict on water resources shared between the OPT and Israel). Similarly, if desalination would be used by Israel to *substitute* some of the water resources diverted from the West Bank's Mountain Aquifer and, in this way allow the Palestinian population to use their rightful share located within their territory, desalination plants in Israel would also work towards peace. Currently, Israeli desalination plants produce *additional* water for the growing water demands of Israel's coastal cities. In the meantime, diversion from the Mountain Aquifer did not lessen.

The dispute over shared water resources is to be considered as an important deadlock of any "roadmap" suggestions to peace. Desalination has the potential to bypass theses political impasses. With the introduction of the technology, the Israeli-Palestinian dispute over water resources is no longer a zero sum game but truly changes the problem frame; from a physical to a political essence. Desalination potentially contributes to an enhanced equity in water allocation between Palestinians and Israelis.

Desalination for Gaza probably remains more costly than an intra-OPT water diversion scheme that would transfer water from the West Bank to the Strip. However, under current political conditions, the latter does not seem feasible. Nevertheless, one can only wonder to what price peace is valued. If a secured, autonomous and PWA controlled water system is considered as a key component to develop a peaceful outcome to the Israeli-Palestinian conflict, how does the cost of desalination compare to the billions of dollars spent by the international community for the past decades. How does the cost of desalination compare to the cost of occupation? In this respect, Shuval [45] suggests Israel to provide the OPT with an extra 200 Mm^3/y. At around US$ 0.5/m^3 of desalinated water, the resulting annual cost of US$ 100 million to purchase this water is insignificant to achieve peace in the region.

Under the Geneva Convention, it is the occupier's duty to ensure that the basic needs of the occupied population are covered. This is of course far from the reality observed on the ground. Needless to say, the above suggestion of Israel funding such a desalination scheme is quite simply utopia at the present stage. The day the diplomatic and development aid freeze around the current Hamas turmoil is resolved, the necessary funding will probably have to be organized by the international community.

The present case study intended to show, firstly, that the roots of the water crisis in the Middle East are more political than geophysical; secondly that feasible solutions can be found, but that these solutions ultimately require strong political good-will. Desalination, when located within the borders of the Gaza Strip and integrated to a management strategy combining treated wastewater recycling and aquifer recharge proves to offer a readily available solution. Not only would it alleviate the physical freshwater water scarcity and palliate the environmental disaster of Gaza's seriously depleted Coastal Aquifer, but it also has the potential to tackle the politically-constructed water scarcity the Palestinian population is additionally subjected to. Until then, with its 1.4 million detainees, the Gaza Strip remains the biggest open-air prison on earth.

4. Structural Scarcity: Chennai

Water conflicts are numerous in India: water access issues linked to caste division, interstate water sharing disputes, massive dam construction leading to displacement calamities (e.g. Tehri and Narmada), inter river basin transfer schemes, controversial water privatization and groundwater overexploitation. Conflicts such as the rural-urban disputes or the corporation vs. individual/community disputes have to be understood as "redistribution issues", where development itself is questioned and development styles argued over.

This fourth and last case study looks at a different type of water conflict; the conflict of the Poor. This case study depicts the daily battle for vital water in slums of the southern Indian metropolis of Chennai. A water conflict with little visible violence and seemingly very different than the Gaza Strip, yet quite similar in essence. Submerged by another kind of violence, more diffused, more subtle, but probably just as harmful in terms of casualties caused by water-borne diseases. Yet another primarily anthropogenic scarcity: an illustration of water-related violence of men against men. Contrarily to the case of the OPT, it is more of an intra- than an inter-societal battle; a battle of socio-economic status, of caste division, of urban space hierarchies, of selective water flows, of inequitable application of the desalination technology.

This case study focuses on the newly proposed desalination project for Chennai city. The Tamil Nadu government called for the construction of a 100 million liters per day (MLD) seawater desalination plant to tackle Chennai's mounting water crisis. The tender for a 25 year deal was signed in September 2005. The present case study analyses to what extent this additional water improves (or not) water accessibility of those currently suffering most from Chennai's mounting water crisis. Around one third of Chennai's population lives in slums, generally unconnected to the city's main water network [46]. The intention is to evaluate the necessity for this technology in the light of other cheaper and environmental-friendlier options: with an average of 1200 mm of annual precipitation, Chennai's rainwater harvesting potential is huge – it is about 10 times more than the precipitation rates found in most places that use desalination. Also, most of the city's wastewater remains untreated and its re-use potential ignored. To understand the drivers behind the desalination project, one has to dissect the motivations by looking at the broader water management context: the mismanagement (unmonitored and unregulated overexploitation) of urban groundwater, the urban - peri-urban dispute (farm water snatched away for the urban uses), as well as Chennai's Krishna and Cauvery river transfers.

Empirical evidence from this case study shows how water politics in Chennai has very little to do with freshwater scarcity, but more with urban politics ; water governance can be ultimately analyzed through a political urban ecology approach, using a yardstick of the equitable allocation of vital resources among different sections of society as a point of reference. This will serve as a springboard for introducing the concept of "technological equity" that will be developed in section 5 of this chapter.

4.1. Chennai's Water Resources

Chennai is the capital city of the state of the South Indian state of Tamil Nadu. After Delhi, Mombay and Hyderabad, it is the country's fourth largest city and amongst the 40 largest cities in the world. Chennai City counts over 4.3 million inhabitants; the Metropolitan Area around 7 million. Chennai's two major rivers, Cooum and Adyar, run through the city but without natural flow for several months during the dry season. Chennai faces an important water (and sanitation) crisis and inequitable water access between rich and poor is very visible. Chennai has indeed a very large proportion of poor inhabitants; the most evident manifestation of this poverty are the city's slums, hosting around one-third of the population.

Chennai has a tropical climate with a relatively high rainfall, between 1100 mm/y and 1300 mm/y. However, rainfall occurs in spells, downpours are too intense to be properly harvested with current facilities and most of this water, therefore, flows into sea without contributing to the city's water supply. Historically, Chennai's water supply was guaranteed by a complex ramification of networks, leading the collected rainwater into storage facilities: *eris* (tanks), temple tanks, ponds and dugwells that enabled groundwater storage and also brought water into town from the Cholavaram, Redhills and Poondi reservoirs.

Only half the city's households have a direct grid connection, and the average water availability is often as little as 30 lpd. In former days, however, Chennai was a water-rich town, with every household having access to water from personal wells in their backyard. Formerly, the Cooum and Adayar rivers together with the rainwater harvesting network were sufficient to meet Chennai's water needs. Now, with 50 million liters of untreated wastewater being daily discharged in Chennai's waterways, the Cooum and Adayar turned into pestilent sewage conduits. Also, the city's temple tanks became unoperational due to lack of appropriate maintenance. Groundwater then increasingly became Chennai's main water source. But due to over-exploitation and ensuing falling water tables, groundwater had to be tapped further and deeper, resulting in ever higher monetary and environmental costs. Water in Chennai is now highly rationed, distributed intermittently and, during dry periods, often interrupted for several days in a row.

Urban or rural water? Inter- and intra-sectorial resource allocation conflicts

Despite its ever increasing water requirements, Chennai somehow managed to survive its water crisis. How? Simply by extending its ecological footprint far beyond the city's borders, tapping water resources ever further into its peripheral regions; in a way, by diverting part of the urban water crisis onto the peri-urban areas. While Chennai manages to survive it dehydrates the populations and agricultural sector of its surroundings.

The Madras Metropolitan Area Groundwater Regulation (Act 27), which was initially designed for the purpose of water conservation, rapidly became a way to legalize commercial groundwater exploitation. In practical terms, what this regulation truly did was to allow Metro Water, Chennai's public water service operator, to carry out free groundwater pumping in 285 villages (augmented to 305 villages by the 2002 amendment) surrounding Chennai. Although the Act was elaborated precisely to restrict extraction only for domestic purposes, in the context of Chennai's mounting water crisis, commercial water exploitation by private service operators became unstoppable. This was especially stringent during the 2004 drought and through the influence this lobby progressively acquired.

In other words, besides the inter-state conflict (e.g. between Chennai and the farmers of Andhra Pradesh of the Krishna river dispute), in Chennai Metropolitan Area, both the inter- and intra-sectorial water resource allocation are pressing issues. The competition over water is notable between urban and peri-urban/rural (i.e. between domestic, industrial and agricultural uses) as well as between users from the same sector within the urban context (between rich and poor domestic consumers). The latter will be further discussed.

Aquifer tapping: Suicidal water management

Pumping rates exceeding by far natural groundwater recharge has led to serious aquifer exhaustion. In 10 years, the water table has fallen by 7 to 10 meters throughout the city; the North of Chennai and its coastal areas being the most affected. As urban community-shared dugwells ran dry, overexploitation was accelerated by replacing them with ever deeper borewells. The ever more frequent failures to reach the water table or to encounter freshwater lenses is a clear indicator of the increasing groundwater depletion. This has resulted in serious seawater intrusion, causing irreparable damage to the coastal aquifers as far as 15 km inland. Many borewells have been reported to yield brackish water with very high TDS rates, unfit for human consumption.

Coveted rivers: A hydro-structuralist failure

Other attempts to increase water supplies were to harness and divert the Krishna and Cauvery rivers to quench Chennai's growing thirst. These inter-basin river transfers taking place in Tamil Nadu are a representative phenomenon of the all-India river interlinking project. The underpinning approach is to consider rivers as a modifiable piped grid. As a summary of the history of Chennai's increasing water scarcity: water sources were constantly tapped further and deeper, at ever rising costs.

Table 4. Summary of Chennai's main water sources and their water yields as projected by Metro Water under normal monsoon conditions

Source	Location	Water Supply (MLD)
Surface water	Chennai's peri-urban reservoirs (Chembarambakkam, Poondi, Cholavaram, Redhills, Porur Lake and Erattai Eri)	200
	Kandaleru dam (175 km away – water from Krishna river)	130
	Veeranam lake (235 km away – water from Cauvery river)	180
Groundwater	Well fields in Araniyar-Koratallaiyar basin, southern coastal aquifer and municipal sources	233
	Borewells in Neyveli-Panruti belt (when the Veeranam dries up)	90
Total		833

Source: Based on [47], Metro Water[5] and personal communications with Metro Water.

Widening water gap

Chennai obtains its water from two main sources: surface and groundwater. The summary of Chennai's supply capacity is presented in Table 4.

Despite the efforts outlined above and the projected water supply capacity, Chennai has experienced cyclic water shortages, which resulted in a mounting water scarcity crisis. The rapidly growing urban population, combined with the inability of Metro Water to meet its projected supply rates and the irregular character of monsoonal rainfalls further intensified the crisis (Figure 19).

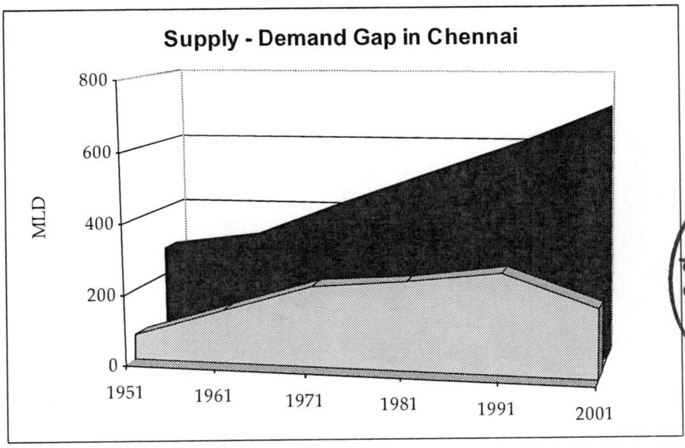

Source: Own elaboration based on Metro Water[4].

Note: The theoretical demand for domestic requirements were calculated on the basis of 100 lpd and an additional 120 MLD was accounted for the industrial sector.

Figure 19. Evolution of the water gap existing between Metro Water's supplies and theoretical demand of Chennai's domestic and industrial sector

[5] Metro Water (2006): http://www.chennaidesal.info

Under normal rainfall conditions, Chennai has an availability of 75 lpd; actual consumption is, however, less than 60 lpd and drops to 30 lpd during drought years. As seen from Figure 19, Chennai's theoretical water demand is around 700 MLD; the resulting gap is important but widens even further in drought years, when Metro Water's supply ability drops to around 180 MLD or during extreme periods, as in 2004, when Metro Water was unable to deliver any piped water to its customers for many consecutive weeks. With Chennai's population expected to rise to 9 million by 2011, this would increase the city's daily water needs to 1450 MLD.

Growing rich-poor water divide

A strong correlation exists between income and water consumption: an extra 10 liters for every additional US$ 150 in monthly income. Evidently, the amount of water used on "basic needs" (drinking and cooking) remains approximately the same in every household; this is the inelastic water demand. The greatest fluctuations are to be attributed to secondary domestic uses such as bathing, toilet flushing, washing, cleaning and (possibly) gardening; i.e. the income elastic portion of water demand. Figure 20 presents a graphical representation of household consumption categories with the lowest income block further decomposed to account for Chennai's slum population living with around US$ 0.45 [46].

Also, according to Vaidyanathan and Saravanan [48], there seems to be an inverse relation between the dependency on strictly municipal water and income. Indeed, over half the water needs in poor households are covered by Metro Water; this dependency gradually decreases to less than 10% in the highest income category. Wealthy households mostly rely on own borewells (with small RO units operating within the house to treat the pumped water) and private tanker operators for bulk water consumption. Evidently, as investment costs of obtaining freshwater constantly increases, the gap between the "water poor" and the "water rich" ever widens.

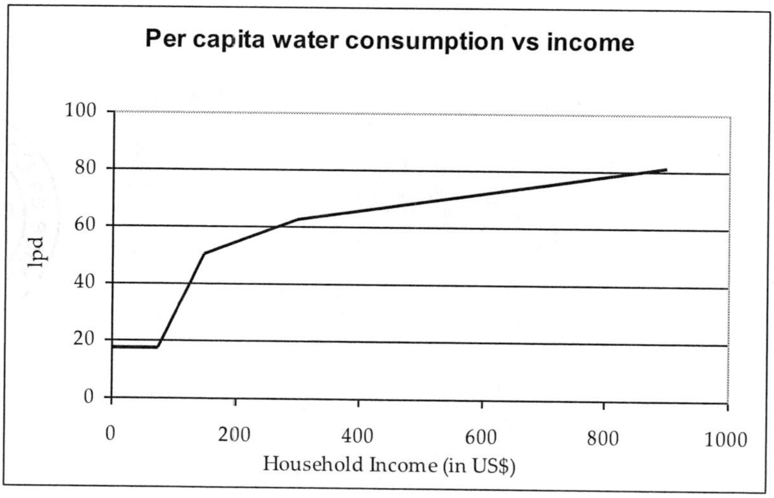

Source: Own elaboration based on [48].

Figure 20. Relating water consumption to income

Now, following the logic of the greater dependency on Metro Water's public water supply as income decreases, one might think that the forthcoming desalination plant will benefit poorer households in relative terms, as it falls under the public water category. This assumption omits, however, that this additional desalinated resource will a) be pumped and distributed through the piped network and b) it assumes that this water will then be sold at the same price. The water poor section of the population is, however, the one remaining unconnected to the urban network – and forced to rely entirely on un-piped public water: mostly communal dugwells and Metro Water's public tanks.

4.2. Water in Chennai's Slums

Slums are urban settlements inhabited by the poorest population strata (mostly from backward castes) under very challenging socio-economic conditions, with highly limited access to water and sanitation. As a reminder, one third of the population of Chennai city lives in so-called slums, that is approximately 1.5 million people and less than 1% of Chennai City's slum population have municipal connections [46]. Slum populations therefore rely primarily on public fountains, open wells and water tanks. Long waiting queues are usually observed at any supply points. The coverage level of public water points is low and 620 persons share on average the same source. Also, the frequency and reliability of supplies are generally poor as 35% of slum dwellers do not have a daily access to water but only on alternate days or less and half the urban slum dwellers are over 50 meters away from the nearest water point. Since slums are annexed rather than integrated to the city, they are often located at the tail end of pipe systems, which suffer from lower head pressures and consequently less water. 85% of slum dwellers have to survive with less than 20 lpd; 30% with less than 2 lpd [46].

The potential of small-scale desalination applications with an autonomous grid has been verified in two coastal slums (Nochikupam and Nochinagar) where the desalination technology could be putatively said to be primarily serving basic needs. Such small-scale, decentralized desalination units supplying water through an autonomous, leak-free, easily maintained PVC piped network. This proves indeed a viable alternative to water tanker, especially if an organized fetching system is set up in cooperation with local water and sanitation coordinators. It should be noted that although desalination is currently more expensive than other available options, privately operated tankered water is sometimes 10 times more expensive. Here again, emphasis needs to be laid on the purpose served by the technology.

4.3. Chennai's Minjur Desalination Project: The Technological Fix Solution

The two small-scale brackish RO desalination units operating in the slums of Nochikupam and Nochinagar have a total production of 0.5 MLD, covering the basic water needs of around 40,000 poor people. The newly proposed project of building a 100 MLD desalination plant at the Minjur site is, however, very different in nature.

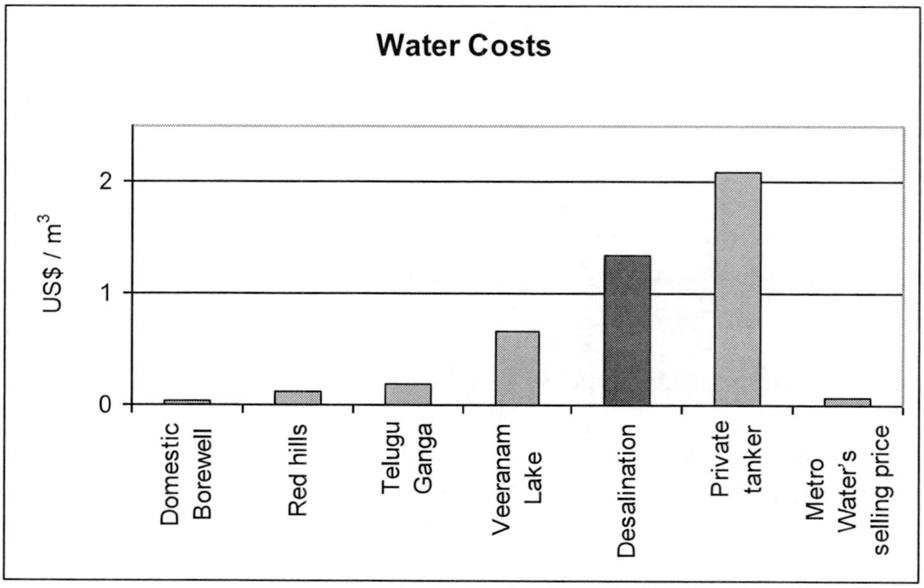

Source: Own elaboration based on [47] and personal communication with Metro Water.

Figure 21. Comparative water production costs

The Chennai Desalination Company Limited was created in June 2004 to coordinate the project. The contract was signed for 25 years on a Design, Build, Own, Operate and Transfer basis, and the promoters are IVRCL Infrastructure & Projects Ltd, (Hyderabad, India) and BEFESA CTA (Spain). The plant will be located in Minjur district (about 35 km North of the centre of Chennai). The desalinated water will be pumped to the underground sump of Redhills reservoir and distributed from there, through the existing urban grid. Besides its investment cost of US$ 150 million, operational costs of the RO process are estimated to be around US$ 1.45/m^3. This is about 10 times the cost of water produced from the Red Hills plant or the Telugu Ganga project, and double that of the New Veeranam project (Figure 21). The investment costs are significantly higher than the necessary funds to rehabilitate all of Chennai's water bodies and rainwater harvesting facilities. Also, reusing wastewater from an effective treatment process for non-drinking purposes and/or groundwater recharge would only cost a fraction of this, around US$ 0.09/m^3. Under this perspective, the desalination option appears as a five-star solution to Chennai's water needs.

Some pending questions

Considering what has been said above, some major questions remain unanswered:

How will the desalination option improve Chennai's groundwater overexploitation problem?

Desalination is often praised as an alternative to fossil groundwater mining or over-exploitation of coastal aquifers. In the case of Chennai this argument, however, does not hold as current over-pumping rates are not likely to stop: the desalination plant is designed to provide *additional* water not *alternative* water.

What will be the long-term effect of brine pollution on Chennai's coast?

The plant will discharge 137 MLD of brine at the Minjur site. This is likely to have adverse effects on local marine ecosystems and consequently Chennai's fishermen -mostly slum dwellers.[6]

Where will the additionally required energy come from and what will be its associated environmental impact?

As for the rest of India, coal is the main energy source for electricity generation in Chennai. The desalination process will, therefore, be responsible for the emission of 30,000 additional tones of CO_2 and, with the mounting nuclear trend, the generation of half a ton of nuclear waste each year.

Who will bear the cost of the water produced?

Cost recovery will result extremely difficult, not to say impossible, with the desalination option. How are these costs going to be met? If this desalinated water is provided through the city's network to all consumers, what will be the repercussions on domestic water bills? How will the most needy sections of the population benefit from it as they are mostly unconnected to the network?

Who will this desalinated water (not) benefit?

The desalinated water will be distributed through the existing piped network. This water will, therefore, not reach the city's slum as they remain unconnected to the urban supply network. Desalination will, therefore, benefit, in the first place, households having direct water access and is unlikely to benefit slum settlements. The new-to-come desalination plant was never thought to be geared towards those that most need it. Excluding the Poor from the urban water services has been a historic constant in Chennai's water management. Notwithstanding, the desalinated water from the projected plant, may well contribute towards improving livelihood conditions in slums, if part of this additional water supply is transported by tanker to the unpiped areas of town. However, this is not Metro Water's intention and no plan has been envisaged to ensure this to happen. Unless the link is not made between "desalination" and "the poor", Chennai's new plant can impossibly be advocated for. In this respect, it is interesting to underline that the water needs of two slums in Chennai are partly covered by small desalination plants, supplying water through an autonomous grid. These systems offer a good illustration of "desalination serving basic needs".

4.4. Missed Alternatives

Considering Metro Water's suggested desalination project, it becomes clear that Chennai needs to re-consider and assess its options. To solve Chennai's water crisis various solutions exist. They are not only combinable but complementary. Firstly, Chennai's important precipitation should be harvested while public water bodies, reservoirs and temple tanks are re-operationalized. Since monsoonal precipitation occur over short time periods, it is

[6] These long-term ecological impacts were not properly addressed by the Rapid Environmental Assessment (see http://www.chennaidesal.info).

important that the runoff flow is thoughtfully divided into its various components to optimize its use: surface water storage, groundwater recharge, contingency and immediate use. Secondly, treated wastewater should be re-cycled and used for aquifer recharge. Thirdly, the bulk water transfer schemes from Krishna and Cauvery should be optimized since the infrastructure is already in place. Fourth, water conservation and demand management should be taken as a standard measures across all technologies and supply schemes. Only once these four solutions have been fully implemented and their potential purposefully exploited, can the desalination option be considered.

The first option mentioned above deserves further comments. Chennai has always been a city dependent on surface water, and around 3600 tanks and lakes are reported to have existed around the city in former times. The functionality of around 2000 of them could be easily recovered. These lakes and tanks are connected to each other through inflow channels, which are, however, often clogged or in disuse hampering their proper functioning. It is estimated that two years' supply to the city was released into the waterways during 2005 monsoon due to a lack of storage capacity. If properly maintained, this network of water bodies and interconnecting channels can store more water than Chennai needs while still recharging sufficiently groundwater for drier seasons. With over 1,200 mm of rain falling over the city's 174 km^2, rainwater harvesting offers an annual potential of 223 MCM (610 MLD), that is more than Chennai's domestic needs.

Overall, this shows that the desalination project -although it is touted as *the* solution to Chennai's water crises- is an inappropriate option. Moreover, under the current design, it is extremely unclear how it will serve the needs of the city's most water-starved communities, namely the urban poor. It will divert enormous amounts of public funds away from necessary investments that would allow to conserve and produce water in a much more sustainable and cost-effective way.

4.5. Discriminatory Water Allocation: Inequity and Structural Scarcity

Homer-Dixon defines "structural scarcity" as arising "from inequitable distribution of resources" [28: p.5]. As has been shown, a large divide exists in terms of domestic water consumption between Chennai's slums and the other neighborhoods. The same differences apply in terms of water quality as well as supply frequency and reliability. The terminology of "structural scarcity" also refers to the fact that daily available, good quality water strongly lacks in slums due to an infrastructural deficit. Slums are unconnected to the piped grid, and remain, therefore, at the margins of the urban water network. "Structural scarcity" also because of the evident lack of interest the city's public service operator shows towards slums. Indeed, as explained above, the new desalination plant is not intended to alleviate the water lack of those that most suffer from it. Finally, "structural scarcity" because the wrong options have been chosen to tackle Chennai's water problem and the water crisis of slums in particular. The structural deficit is essentially caused by ill-suited and above all inequitable water governance schemes.

Hierarchies of urban space: When "natural" water becomes "social"

It is striking to see how Metro Water elaborated –in its early days- a methodology of "social classification". Indeed, an official system to categorize the population based on the annual property rental value of the residence (i.e. correlated to wealth) distinguished between different types of consumers and service connections. Under these regulations, households living in residences which were valued to be under a certain threshold were not eligible for a private connection. The neo-liberal reforms undertaken within Metro Water, have put this "social classification" system back on the agenda in a reinforced manner [49].

Mosse notes that "the connections and interdependencies of hydrology interweave with those of caste-class and kinship, business and politics" [50: p.4]. Although this assertion arises from the analyses of tank irrigation systems in South India, the way water flows around a city like Chennai is still very much categorized according to the same social segregation. This can be best observed by comparing the socio-economic status of Chennai's neighborhoods and the expansion of the water supply network (including tanker deliveries). For instance, some northern districts of Chennai -although these neighborhoods feature high population densities and are among the oldest areas of the city- are poorly supplied whereas relatively new developed localities such as Annanagar enjoy top round-the-clock services. Those areas feature much lower population density rates, but not surprisingly, greater incomes and host a lot of influential people. Hence, what could be described as "hierarchies of urban space" are produced or at least exacerbated through discriminatory water allocation patterns; in other words, artificially maintained scarcity. This is precisely the definition of "structural scarcity". According to these geographical hierarchies, water allocation in the distribution network strongly differs from one zone to another within the city. The "method of allocation" involves a constant process of calibration, modifying the water flows throughout the city. Allocation is less depended on day-to-day fluctuation in availability, than on strictly political motives. As argued by de Wit [51], particular policy implementations in Chennai are very much influenced by political considerations. In this respect, the "machine-style politics" progressively became predominant, where "machine-style politics" refers to:

> "a specific type of relationship between people (voters) on the one hand and political parties and governments on the other. The machine-politics model postulates that voting behaviors is not (so much) determined by ideology, but rather by the expectation or actual receipt to some material inducements or rewards" [51: p.33).

These rewards may be promises for more secured employment or housing, but in the case of Chennai, certainly also (free) access to minimum water quanta. Such policy promises prove to have a tremendous influence on voting behavior and become indeed very powerful electoral weapons.

Since the early days of the Corporation of Madras, the needs of impoverished households lacking direct access to water within their dwellings were inventively met through the 4000 public fountains scattered across the city. Notwithstanding, it is striking to note how in a brochure published by the Corporation back in 1957, these public fountains were said to be "mostly situated in unimportant streets and by-lanes" [52: p.40]. These "unimportant streets and by-lanes" were in fact the city's numerous slums. As can be seen from the following quotation –taken from an early Administrative Report of the Corporation- slums in Chennai have been deeply despised:

> "A large number of the floating population in the City of Madras, many of whom are beggars, have no houses and live in the streets and other public places. These people, without

contributing anything to the public revenues take advantage of the municipal water supply and other public conveniences provided by the Corporation." [53: p.1]

Under these considerations, the countless water buckets lining up in front of the water tankers can be seen as materialized structural scarcity; yet an artificially maintained scarcity or better said a geographically determined scarcity producing hierarchies of urban space. As Mathur [54] showed for the whole of India, there is a direct correlation between (low) income and the dependency on alternative water sources, that is alternative to the mainstream piped network. According to him, over 60% urban dwellers who fetch their daily water supply directly from rivers, canals, public fountains and water tanks live with less than US$1.2 per day. The case of Chennai's slums seems thus to be fairly representative.

The aim of the four case studies presented in this chapter was to discuss whether desalination could improve scarcity situations in a cost-effective way (including environmental and social externalities), and in a manner suitable for reaching those who most need it. Focusing on these issues has cast light on the different purposes served by the desalination technology. The energy-intensiveness and other drawbacks of the desalination technology are particularly problematic from an ethical standpoint if they do not contribute to maintaining livelihood. Indeed, when the application of the desalination technology fails to meet this objective, then traditional water heritage and social integrity are dismissed while environmental degradation, economic volatility and technological "lock-in" are engendered for the sake of sustaining luxury (as in the case of water for tourist services in Lanzarote). Nevertheless, when these considerations are mirrored at other cases where desalination purposefully alleviates "structural scarcity" (in the two coastal slums of Nochikupam and Nochinagar) or "politically-constructed scarcity" (as could be done in the case of the Gaza Strip), then the lofty function of the technology to satisfy basic water needs can be verified. The analysis of these case studies points at a normative gap between the desalination technology and equity. This gap will be filled in section 5 of the present chapter.

5. Equity and the Purpose Served by the Desalination Technology

"One reason the environment in the twentieth century changed so much is because prevailing ideas and politics –from an ecological perspective- changed so little."

(John Mc Neill [55: p.325])

International water law[7] stipulates "equitable utilization and allocation of water resources" as one of its founding principles. However, the notion of equity still lacks solid theoretical foundation in the current water management literature. From the case studies, the postulate emerges that an ethical analysis is required prior to the implementation of the desalination technology. This leads to the conceptual proposition of "technological equity" that opens the debate on the rationality of the objectives, of the use and of the purpose served by technology. From the analysis of sustainability and scarcity, this section elaborates on the

[7] as elaborated under the 1997 UN "Law of Non-Navigational Uses of International Water Courses" and the Helsinki Rules

links existing between technology and equity. Science, Technology and Society (STS) studies suffer from a normative deficit when it comes to assess technology from an equity point of view. This deficit can be overcome by calling upon the "principle of responsibility" and the postulate is that putting equity first opens up the range of possibilities; through reflections on technological equity, more sustainable and effective solutions can potentially be put on the table.

In Chennai, as opposed to the other cases, there is a wide palette of technologies available to tackle the water crisis; as modern and traditional technologies confront, the discussion also involves the issue of technological *choice* and technological *change*. The concept of technological *change* entails confrontations between constructivist and deterministic views of the evolution of technology in society where experts, scientists and explorers physically invent and develop new technologies out of determination and pure deterministic research. The debate on technological change implies that this purely deterministic view is not accurate, pointing out that a technology is at least partly a social-construct and very dependent upon societal dynamics. There should be a reflected incentive to be given by society to determine on which technological path it wants to embark since the choice of technological trends renders "technological change" a consciously chosen process. In this sense, should not a reflection on the equitable application of a given technology form part of this technological change process? If the aim is to achieve a more equitable and just society, the equity of a given technological implementation should be questioned and the changes of technology made along the lines of securing "technological equity".

5.1. Technological Design and Technological Choice

What drives the decision in favoring one particular technology over another, when it comes to making a choice? Choosing one technology over another is, in fact, designing a pathway. When one chooses a particular technology one chooses a (technological) design, i.e. primarily a concept, an idea, a construct in the realm of imagination that precedes the material artifact itself. The concept of *technological design*, does not only refer to the conventional definition of *design*, that is the non-textual representation of an artifact (through a drawing or a mathematical equation, for instance) but also to the essence of that particular technology: as Papanek [56:p.4] puts it, its "conscious and intuitive effort to impose meaningful order". For instance in the case of tank irrigation in South India, Shah [57] describes how a certain type of technological design establishes (or maintains) power relations between different groups of farmers. She shows how tail enders in the irrigation process (i.e. the most disfavored) usually correspond to the section of Indian society that are historically the most challenged (i.e. lower caste farmers):

> "The powerful farmers of the head reach remain in a powerful position by removing main canals that can possibly rotate water between head and tail end and converting water distribution completely to field to field irrigation. That means powerful farmers reproduce the design of field to field irrigation and consequently their power as well. Designs thus are not only socially arranged but they reproduce social arrangements." [57:p.270]

Indeed, in the light of Papanek's definition, picking one technological design from a range of available choices hails a whole lot of purposive actions. "Imposing meaningful order" is also the act of striving towards a particular purpose and often relates to a –or rather- *the* solution. A technology and its design, is not conceptualized in its inert material form but in terms of its utility. Hence, technological choice is very strongly value-laden in the sense that it is not chiefly the technology *per se* that is selected, but rather the outcome that is believed it may achieve, the function it is meant to serve. This means, firstly, that taking a decision and coming to a particular technological choice involves a whole range of assumptions, expectations and subjective judgments based on a specific interpretation of the world. Secondly, that this decision process has, nevertheless, some intentional attributes and that it is fully deliberate in nature (even though there might also be non-intentional secondary effects). Thirdly, that there is not necessarily one single right answer, but there could be many different pathways to the very same problem; consequently, the myth "one size fits all" should be dismissed. Finally, it raises the following questions: What are the vested interests and hidden agenda lying beneath "*the* solution"? How is "*the* solution" codified, wrapped in a rhetorical commercial packaging? What are its narratives and which discourse does it use? Simply put, it opens the debate on power – or what is here coined as "technological power".

5.2. Power, Codes and Narratives

The codes associated with a given technology include the narratives used to argue in favor (or against) the selection/choice of its design. The different case studies presented in this work intended to decrypt the power interests that lie beneath the codes of technological choice. Power at work will often attempt to obscure its real agenda, by normalizing, naturalizing, politicizing the choice made as being objective and disinterested. However, Winner [58] captures well this discussion with his concept of *technological code* by demonstrating how technology -by means of its artifacts and design- can be very political. He illustrates this by 'decoding' how the specific design of a bridge can, for instance, deny certain population groups (mainly poor and ethnic minorities) from accessing certain parts of a city. This concept helps to decipher the links between the technological design, its power dynamics and ethical ground. In the case of Israel, offering desalinated water to Gaza but with the technology under Israel's control codifies its role as a political instrument of control. In the case of Chennai, the fact that the desalinated water will be distributed through the existing water network, and therefore bypassing the slum areas, can thus be seen as "coded" with inequity.

Another key aspect is to explore how technological change is influenced by power relations. This helps explaining how a given technology is shaped – under existing power dynamics. In absence of democratic checks, a given technology (and its "attached practices") can be easily monopolized by certain actors, granting them with substantial control over others. Also, the decision to choose for a particular technology (desalination instead of rain water harvesting, for instance) can be inherently related to the dominance of a particular group of social actors. In this sense, power relations are articulated by producing precisely intended technological choices, or, put differently, technological choices are crystallized as the result of power relations.

Power is therefore both an aspect of the desalination technology as well as the shaper of the setting in which it is used. In other words, the introduction of the technology delivers power –or better said- has the ability to alter the existent power arrangements and, at the same time, those power dynamics intrinsically determine how desalination will be used, and what for. Indeed, Sangren [59] explains how power reproduces social processes while reinforcing itself by organizing the social settings in which it acts. This self-feeding loop means that the actors of those power dynamics introducing the desalination technology, ultimately do this with the view to maintain or increase their control -or simply put- their power. Nevertheless, it is important to keep in mind that technological choices are shaped in inherently political arenas and that these choices cannot become vehicles for democratization if lacking an ethical compass. Without such compass, technological change leads to situations of anomy.

5.3. Technology, Ethics and Responsibility

As Rabelais said: "*Science sans conscience n'est que ruine de l'âme*" ("science without consciousness is the ruin of the soul"). Humans have the moral duty to remain the masters of technologies they conceive; technology has to remain a "human-delegate". Society has to be considered "responsible" of its technological choices and the introduction of these technologies into the world. As Jonas [60] does, we must embark onto a "search of an ethics for the technological age" by asking:

> "Is there moral profit in this intrinsic progressivism of science and technology? […] Do science and technology through their progress contribute to the general moral betterment?" [60:p.168-169]

The forward-pushing motion of science and technology can be called "progress"; but which direction does this motion take? Only ethics can answer these questions, since technology does not merely perform tasks but what ought to be seen as value and ethics laden duties. This implies a new reflection on ethical principles, which puts forward a "principle of responsibility":

> "The lengthened reach of our deeds moves *responsibility*, with no less than man's fate for its object, into the centre of the ethical stage. Accordingly, a theory of responsibility, lacking so far, is set forth for both the private and the public sphere. Its axiom is that responsibility is a correlate of power and must be commensurate with the latter's scope and that of its exercise." [60 – emphasis as in original text]

Under this perspective, Metro Water –as a State agency- is to be judged as highly *ir*responsible for not ensuring that a significant share of the additional water provided to Chennai through the construction of the new desalination plant is dedicated to the section of the urban population that most needs it, namely the city's slum dwellers.

5.4. Technology and Equity

Although the concept of "equity" implies some ascription to equality, as Gandhi wrote: "[m]y ideal is equal distribution, but so far as I can see it is not to be realized. I therefore work for equitable distribution" [61]. The difference between "equality" and "equity" could be that, while the former strives towards the ideal of egalitarian shares of a finite resource, the latter sets as minimal condition that every member of a given community is at least guaranteed to have its basic needs fulfilled while sharing the finite resource. In other words, equity is the suboptimum where, if inequalities in the allocated shares do exist, theses discrepancies don't jeopardize the livelihood of others.

Acknowledging the fairness, freedom and even-handed attributes of equity, when looking at the link between equity and technology one should question *who* benefits (and *who* doesn't). The overarching question of (water-related) technological equity implies an analysis of the allocation and/or appropriation of the produced resource across social class, economic status, and gender. In India for instance, the fact that water rights are typically assigned to landowners entails substantial effects on gender equity issues, as control over land is generally vested to men, and consequentially so is the control over water access. In turn, the burden of meeting the household's water needs generally falls on women (and children), at the opportunity cost of education or other empowering and productive activities.

In this chapter, it is argued that equity is a *sine qua none* condition for meeting some fundamental aspects of "sustainable development". Equity should, therefore, be the binding element of its economic, social and environmental pillars. Water encompasses those three dimensions, by sustaining basic and economic needs, through its linkage with social and political power, and by maintaining the ecological balance. Thus, the main interest lies in the strive towards equitable water allocation -both in process and outcome, where, the access to natural resources are understood as possible means to reduce poverty by improving equity [62]. Besides the initial conditions, the concern of equity in water allocation should not only be limited to equity in outcomes but also equity in rule content and equity in processes, where, as explained by Sangameswaran,

> "[e]quity in rule refers to the extent to which equity concerns are incorporated in the design of the water project itself, an issue related to the vision of development. Equity in processes refers to the extent to which there is equitable participation in different processes [...] over a period of time." [63:p.16]

So, equity should always be looked at in the four temporalities : initial conditions, content, process and final result. This raises the key question of how to assess the level of equity of a project, and more specifically, of a given technology.

5.5. Technological Equity: Towards an Equitable Use of Desalination

The core of the debate between moderate and radical constructivism in STS studies, is that the latter challenges the main assumption of the moderate school that scientific and technological development can be explained by social factors and accuses it of replacing "technological determinism" by another form of determinism: "social determinism". Instead, radical constructivists consider both technology and society as (constructed) fabrications. This means that one cannot determine what the specific social context looks like but only give a certain interpretation to the role played by relevant actors. Although the radical approach might be pertinent, it nonetheless results in a serious normative deficit, especially for the analysis of the present chapter that intends to clarify the links between equity and technology. As Keulartz et al. argue:

> "radical constructivists are committed to a "methodological agnosticism", not only about what kinds of societal influences shape technology but also about what kinds of social, political, and moral impacts technology has". [64:p.13]

The present chapter intends to overcome the normative deficit of this "methodological agnosticism" by refocusing the debate around profoundly human and value-laden questions of livelihood. Thus, it adopts an asymmetrical social approach. In other words, it overtly recognizes its biased, "crypto-normative" view by favoring the weaker side (like the occupied, the slum dwellers, the lower castes, the women) in the controversies emanating from the analysis of the desalination technology. In this sense, the present chapter rebels against the analytical coldness of the radical social constructivist approach that (because of its self-imposed normative restrictions) is obliged to limit itself to map conflicts in terms of allies and opponents, strategic negotiations, and tactical maneuvers. Such approach leaves no room for ethical deliberation and position taking. As Hamlett claims,

> "It may be time for constructivist analyses to move beyond the descriptive examination of the social dynamics of technology to a more proactive approach on the larger issues critics identify". [65:p.114]

By introducing the below outlined concept of "technological equity" a "should" discourse is indeed engaged. Kant's old dictum "thou canst because thou ought" should -according to Jonas [60]- be reversed to: "thou ought because thou canst". Furthermore, under the conception of nonreciprocal contractual responsibility and the positive duty of power, it is argued here that this dictum could be turned into "thou ought because thou rule".

In other words, the "new ethics of the technological age" [60] implies a responsibility imperative that inexorably shifts from *willing* to *obligation*. While Kant primarily focused on the individual, this chapter looks first and foremost at the broader social context. Since those who rule are organized in institutions, the aforementioned shift from *willing* to *obligation* can only be canalized through human agencies. Sharing limited resources (such as desalinated water for instance) requires social arrangements. These arrangements are affected by technological choice, by its design and by the way technology mediates the allocation of resources: the so-called "attached practices" of technology. The latter can be either inclusive or exclusive. In Chennai, for instance, under the current plan the attached practices of the desalination technology are exclusive but could be inclusive if the institutional and structural frame is altered. When those resources are insufficient to meet everyone's demand, it is not only the institutional rules that determine whose demands are met and whose not, but also the way resources are inherently allocated and distributed by that given technological choice and design. This is what could determine if a given technological design and technological choice resulting in a particular application of the desalination technology is equitable or not.

However, traditional ethical vocabulary falls short when dealing with technological culture. This chapter thus suggests the term "technological equity". It establishes a semiotic link between equity and technology, helps to overcome the aforementioned normative deficit in STS studies and contributes to the debate on the democratization of resource utilization and allocation.

"Technological equity" focuses on the way technology may be shaped by concerns of enhancing equity; i.e. the way a technological choice and design needs to be made or adjusted in order to increase equity. In other words, it comes down to the outcome of the given technology: does it lead to enhanced equity or does it reinforce well-established inequitable power arrangements? Referring back to Figure 2 in section 1, the concept of technological equity links the purpose served by the desalination technology (in terms of alleviating *what* dominant component of scarcity: physical, social, political, structural) and *how* it is done: by enhancing equity (or not). So, the concept of "technological equity" implies both an analysis of the situation and a process that could potentially alter the situation. For instance, to what extent does the use of desalination for the alleviation of socially-constructed scarcity (like Lanzarote) enhances equity? Not very much, since in this particular case, the technology primarily serves luxury behaviors while adversely affecting the global environment and harming ecosystems. Comparatively, to what extent would desalination enhance equity if directed towards alleviating the politically-constructed scarcity in the Gaza Strip or the structural scarcity slum dwellers suffer from in Chennai's neglected neighborhoods? The same logic applies not only for the adjustment of technological design, but also of technological choice. In the case of Chennai, for instance, it could be agued that in absolute terms rainwater harvesting is likely to have a higher technological equity than the desalination option, partly because the former can be operated autonomously and in a decentralized manner (i.e. directly benefiting the urban slums) whereas the latter is linked to institutional (centralized) power arrangements. Figure 22 presents a graphical representation of the "technological equity" concept.

On the left-hand side of Figure 22, "equity" is only secondary to "technology". Although the equity criteria remains a part of technology, technology is exogenous to equity and the decisional choice (i.e. referring back to the concept of "technological choice") and design (i.e. referring back to the concept of "technological design") of the given technology are mainly dictated by the existing power arrangements. Technology is, therefore, represented in the "realm of power".

On the right-hand side, equity is considered first. Technological choice and technological design shape, adjust, develop and modify the given technology but in view of enhancing the overall equity of the system. Technology is endogenous to equity, and becomes one of its inherent dimensions. Although technology is still influenced by decisional choice and design, they lie in the "domain of ethics", with the "imperative of responsibility" [60] guiding the way these decisions are taken.

In the first instance, the technology arrangement is trapped and lacks a "human face", as Schumacher [66] would say. The only way to break away from these self-determining power dynamics is by re-locating it in the domain ethics (and democracy) where the technology arrangement can then relate to equity. Under this perspective, technological equity can be understood as a *process* leading a given situation from one system (left) to another system (right). Technological equity starts by reversing the priorities, putting equity first.

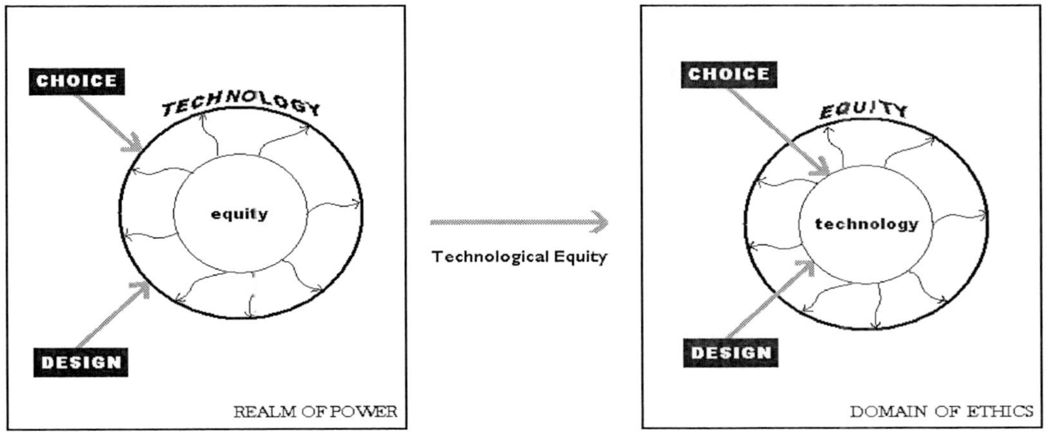

Source: Own elaboration.

Figure 22. The "Technological Equity" concept

The starting point for the present reflection was the question of how to assess technology from an equity point of view and more precisely, what needs to be looked at, if equity is to be enhanced in the development, implementation and use of a given technology. The intention is indeed to use the concept of "technological equity" as a process to enhance the overall equity of a situation. This would result in the need to develop a tool to assess technology: Technological Equity Assessment (TEA). Recalling what has been said above [62,63], technological equity refers to 1) the initial conditions, 2) the content, 3) the process and 4) the outcome, and TEA could serve as a basis to evaluate equity considerations along those four temporalities. Equity considerations must be upstreamed in the decision making process in order to become its *raison d'être*. Ultimately, through a TEA, technological options would be selected or adapted according to the given situation in view of improving the overall equity of the system.

To give an illustration, in the case of Chennai for instance, it is the fact that the desalinated water will be distributed through the network that shapes the social arrangements emerging from the technology; that is a social arrangement where, the poor -who are not connected to the city's main water grid- are excluded from the official purpose of the plant, namely to provide additional water to the urban population. Thus, it is that particular aspect of the technological design (i.e. the desalinated water distributed through the network) that makes the whole technological choice questionable (i.e. inequitable). Having said this, the same technology could also operate very differently and ensure that the poor benefit from it (as observed in the slums of Nochikupam and Nochinagar). Modifying its design and gearing the desalinated water towards the urban slums would enhance equity. An enhanced "technological equity" is the enhancement of equity among users through the use of technology (by adjusting allocation and access provisions). In the case of Gaza, the technological design criteria "location of the plant" ultimately determines its purpose. Therefore, the concept of "technological equity" and the TEA tool suggested here has nothing to do with the desalination technology *per se*, but with its intentional design, with the way it is consciously used, the purpose it is meant to serve.

The case studies presented in this chapter mostly deal with relatively short-term, local and primarily anthropocentric considerations of the desalination technology. Having said this, a longer time frame was adopted for the comparison of the pre- and post-desalination era, the anthropocentric scale expanded as environmental pollution associated with the use of the technology was considered and the local character of the case study approach intended to illustrate global trends. The notion of technological equity can indeed be applied to broaden the analytical scale of purely anthropocentric, localized and short-term perspectives. As Tàbara [67] posits, S*ustainable Equity* could be seen as overcoming risks of "immediatism" (concerned with short-term vested interests), anthropocentrism and excessive localism. The concept of technological equity can be applied to an expanded reference frame where reflection encompasses a global world outlook, on a long-term inter-generational scale with a co-evolutionary view transcending the rigid boundaries of human and non-human. Figure 23 shows three vectors which scales inflate as the overall equity of the system increases.

In this view, it is not a matter of opposing equity considerations to efficiency or sustainability, but to consider equity as part of efficiency and sustainability. Through the multi-scalar perspective presented in Figure 23, the postulate is that enhanced equity leads to enhanced sustainability (in it's *strong* sense [69]).

As a means of concluding, Table 5 offers the first guidelines of what a TEA frame could take into account. It first recapitulates the main elements of the analysis presented in this chapter by approaching the case studies through the "technology equity" concept as presented in Figure 22 and 23 and then suggests some guidelines according to the four temporalities of equity as presented above.

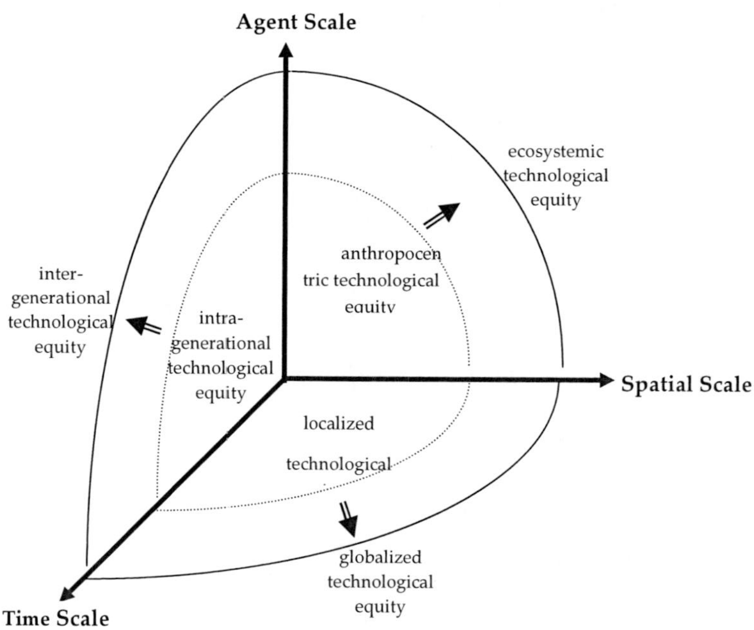

Source: Adapted from [68].

Figure 23. Extended scalar reference of Technological Equity Assessment

Table 5. Recapitulative table of the four case studies and first guidelines for a TEA of desalination.

	Laâyoune	Lanzarote	The Gaza Strip	Chennai
Present water consumption	The poorest 60% of the population in the city of Laâyoune consumes 29 lpd.	Lanzarote's residents consume an average 120 lpd; tourists in standard hotels consume 460 lpd and 1000 lpd in luxury hotels.	Since 90% of the water in the Gaza Strip is non-potable, consumption is around 15 lpd. Israeli settlers consume nearly ten times more (of fresh, clean water) per capita then the (brackish and polluted) water consumed by Palestinians.	Average water consumption is about 60 lpd; but in slums, typically less than 20 lpd. A 100 MLD desalination plant is to be constructed.
Main agent/power conflict	The development policy from "Northern Morocco" to control the Polisario movement is partly reflected in the Sahraouy vs. Northern Moroccan water consumption habits.	The 'increased supply' management approach is indirectly dictated by the tourist industry. Although no declared conflict exist, there is a serious questioning of the bolted tourist development and a tourist moratorium has been implemented in the 1990's.	Ongoing confrontations have been taking place ever since the creation of the State of Israel in 1948. A strong control of Gaza's "water tap" has been put in place by the Occupier.	Strong state authority over slum policies and control of water provisions. The conflict can also be seen as a slum vs. non-slum neighborhood polarity.
Technological design, choice, change and use	Two water distribution systems run in parallel: one traditional (brook water), one technological (desalinated water). The former used essentially by the local Sahraouy population; the latter by the Northern immigrated population.	Any sustainable alternative to desalination would necessarily involve a reduction in consumption quanta of the tourist industry and network losses. There is also the question of the energy source used and the ways to limit greenhouse gas emissions triggered by desalination.	Current groundwater over-exploitation leads to serious resource degradation. A large-scale desalination scheme is to be under-taken. In order to enable desalina-tion to alleviate the water dispute, production has to fall under Palestinian control. Desalination should go hand in hand with treated wastewater recycling.	The city's groundwater system suffers from over-exploitation. The desalination option is pushed forward by State authority, while the traditional water culture of temple tanks (rainwater harvesting) is being dismissed and wastewater treatment not sufficiently used.
Purpose served by the desalination technology	The relatively modest desalination production mainly serves to alleviate a scarcity situation, which dominant component is geophysical and climatic (labeled *physical scarcity*).	The large per capita desalination production is primarily dedicated to sustain the water demand of (luxury) hotels. In a naturally arid geophysical environment, this entails a *socially-constructed scarcity*.	The naturally arid conditions of the Gaza Strip are strongly exacerbated by military occupation (very high population density, aquifer over-exploitation, water rights scorned) leading to a situation of scarcity where the dominant component is political (labeled *politically-constructed scarcity*).	The public piped water system by-passes the poor population living in the city's slums. This *structural scarcity* is essentially caused by ill-suited and inequitable water governance schemes. It also refers to the infrastructural deficit and to the evident lack of interest the city's public service operator shows towards slums.

Table 5. (Continued)

		Laâyoune	Lanzarote	The Gaza Strip	Chennai
Multi-scalar considerations		In the long-term, brine pollution will increasingly become a threat to the local marine ecosystem and to the fishermen community.	Besides brine pollution, the energy-intensiveness of desalination becomes very questionable ethically; especially, when future generation, other species and population groups are taken into account.	Aquifer over-exploitation appears even more serious when considering their long-term effects. Also, the energy-intensiveness of the technology and brine pollution are a major concern; renewable energies offer great potential for desalination, although their scope remains somewhat limited compared to (cheap) fossil fuels.	
	Laâyoune	**Lanzarote**	**The Gaza Strip**	**Chennai**	
Guidelines according to the four temporalities of Technolo-gical Equity	Initial conditions	The limited water demands are satisfied by local provision systems; needs are adapted to natural conditions.	Prior to the introduction of the technology, the situation was radically different with no tourist industry and very low water consumption.	Since the beginning of the conflict, water shares are highly inequitable.	Slums suffer from inequitable water provisions.
	Content	The use of desalination somewhat contradicts the local water culture.	Although there is no fierce inequity felt on the island, the tourist industry nonetheless dictates the insular desalination production.	Inequity lies at the heart of military occupation.	Inequitable water provisions are correlated to socio-economic conditions also reflected in the Indian caste system.
	Process	The introduction of the technology reflects the technocratic imposition of the Northern Moroccan authority.	Civil society can exert some political pressure on the tourist industry.	Inequitable decision-making process, notably through the veto right Israel exerts in the Joint Water Committee.	Although slums can exert political pressure by acting as "vote banks", the current situation is characterized by a lack of transparency, public participation.
	Outcome	The unacceptability of desalinated water in Sahraouy living habits illustrates that technological change is hardly	Equity in outcome is partly dependant on the technological design (energy consumption, energy share and	The outcome of any desalination scheme greatly depends on its design: location of the plant, control over production,	Under current conditions, the new desalination plant is unlikely to benefit the slums and the peri-urban areas.
		imposable	brine discharge), yet strongly linked to the evolution of the tourist industry and its consumption patterns.	security	This could however be different if slums were connected and/or if small-scale desalination plants with autonomous grids were introduced in slums.

6. From Physical to Social Scarcity: An Ecological Economics Approach

"To believe that exponential growth may last eternally in a limited world, you must be a crazy or an economist"

(Kenneth Boulding)

This last section puts the desalination technology back into its broader socio-economic context, making use of the lessons learnt in the case studies. The following discussion provides clues to contrast situations in which desalination alleviates socially-constructed water scarcities with cases where desalination serves to meet "basic human needs" and sustaining livelihood. Also, the incessant necessity to increase water supply capacities is analyzed from a "need satisfactor" [12] point of view.

6.1. Demand versus Supply

The global freshwater scarcity crisis we are facing is the consequence of both population and consumption growth combined with declining natural water resource stocks mainly due to pollution and unsustainable resource exploitation. This results in a generalized deflection towards what can be seen as a social scarcity pathology. Several responses then emerge.

On the one hand, there is an urge to increase water supply capacities. In this respect, the desalination technology offers part of the solution, as was verified in the cases of the Gaza Strip and Chennai's slums. On the other hand, there is growing recognition for the need of demand-based water management to provide sounder options. "Water demand management" as suggested by the European Water Framework Directive (EWFD) [70] simultaneously rests on the three pillars of Sustainability: economic (''full cost recovery'' principle), social (proactive ''public participation'') and environmental (aiming at restoring ''good ecological status'' of water bodies). The EWFD strives towards re-establishing the balance between supply and demand by providing more goods and services using less water. In Spain, the *Nueva Cultura del Agua* (*New Water Culture*) emerged as a civil society movement, which uttered its rising mistrust for paternalistic hydraulic policies correcting primarily socially-constructed water scarcities and serving oligarchic interests. A movement representing social actors from different fronts cried for a new "water culture" that would acknowledge the multiple dimensions of environmental, social, economic, political, ethical and emotional values of this vital resource [71]. This New Water Culture argues for a fundamental shift from "hydraulic works promotion" towards an "economic management" approach and strives to integrate both socio-economic and environmental aspects.

This can also be understood as the turning away from management strategies founded in "hard" technologies towards "softer" approaches [72]. The twentieth-century was marked by centralized water policies extensively relying on the construction of massive infrastructure such as dams and inter-basin river transfers. While the latter brought considerable benefits, they are also associated with dramatic social, economical, and environmental drawbacks. As Gleick posits, carefully planned and managed centralized infrastructures must be complemented with softer and small-scale decentralized facilities:

> "The soft path for water strives to improve the productivity of water use rather than seek endless sources of new supply. It delivers water services and qualities matched to users' needs, rather than just delivering quantities of water. It applies economic tools such as markets and pricing, but with the goal of encouraging efficient use, equitable distribution of the resource, and sustainable system operation over time. And it includes local communities in decisions about water management, allocation, and use." [72:p.1529]

The rainwater harvesting option presented for the case of Chennai is a clear illustration of a "soft" technology alternative and reminds of Schumacher's notion of "intermediate" or "appropriate" technology [66]. From there also arises the suggestion that a "soft path", entailing more decentralized and participative technologies would be associated with higher levels of equity. Ultimately, such demand-side approaches aim at minimizing the need for additional supplies (through hard-technological choices) and try to avoid "supply-creates-demand" type vicious circles. And finally, on the global scale, the new (and softer) Water Culture strives towards enhanced water access equity.

6.2. Ever Growing Demands

The principle of "domination of nature" that led at the end of the 19th and especially during the 20th century to a productivist water management approach results in a well-established hydraulic structuralism, strongly rooted in engineering and technical sciences. This can be verified particularly in Spain that traditionally focuses on the supply-side of water management; the issue of water quality and pollution has only truly appeared on the agenda in the late 1980s. Water in Spain has been a key issue for development and economic growth, especially in the southern and interior regions. Water for tourist services and irrigated agriculture have continuously increased demands. Part of this demand has been met by a plethora of large-scale hydraulic works that ''vertebrated'' and ''re-equilibrated'' the territorial hydrology system, especially since the 1960s [73].

Say's Law states that 'supply creates its own demand, which will exhaust supply'. Although Say verified this principle by looking at the economy on the whole, it also seems to hold within the water sector itself. Indeed, increasing supplies will be insufficient because they increase demands even faster and ultimately, supplies will always be fully used up. In such context, it is interesting to see to what extent -what Martinez-Alier [2] calls- the Jevons' paradox[8] also applies to desalination: do we automatically use more water as desalination gets cheaper? Decreasing production energy costs seem indeed to subsequently trigger a rebound effect on water consumption [75].

Furthermore, the long unquestioned success of the hydraulic structuralism approach produced the sensation that water scarcity problems could be entirely solved by increasing supplies. As this conception spread amongst population and tourist industry, the traditional water culture prudence progressively eroded, generating rising consumption patterns. Ironically, additional supplies seem to create a serious contradiction in which a ''water squander'' culture -subsequently triggering socially constructed water calamities-

[8] Jevons [74], in his analysis of the production and consumption of coal in Great Britain, pointed out back in 1865, that the higher efficiency of steam engines paradoxically led to an increasing use of coal by making it cheaper relative to oautput.

increasingly emerges in a natural contexts of absolute scarcity. Considering the irrigation quotas of private gardens, Domene et al. [76] shed light on the intrinsic relation existing in the Metropolitan Region of Barcelona between level of income, household typology and water consumption. Under such a perspective, the "North Atlantic-type lush green lawn" and the "private swimming pool" are to be considered as "positional goods" [77] in Spain's Mediterranean context. As verified in the case of Lanzarote, similar arguments hold when considering the Spanish tourist industry or when analyzing -through a "Virtual Water" approach [78]- intensive irrigated agriculture in the arid Almeria and Murcia regions. In these desert conditions, tomatoes grown under highly irrigated quotas, embodying 90% water, are exported to the humid North. The massive tourist industry and agro-business (with over 27,000 ha of greenhouses) to be found in the Almeria region result in an average consumption of more than 3000 lpd in one of the areas of Europe with lowest rainfall. Unsustainable local aquifer exploitation and mass-scale subsidies -incentive for inefficient irrigation- are not sufficient any more. Desalination must now fill the ''deficit''.

6.3. Desalination: Meeting Need or Greed?

In this view, the new Spanish National Hydrological Plan (PHN), based on extensive desalination, still maintains water supplies well above real necessities, strongly contradicting the natural aridity context. How does supplying ever more water to a desert-type region in order to satisfy imported recreation habits clearly not adapted to the local environment, solve drought problems? The main concerns lie in 1) the indirect growth and consumption stimulation effect that the additional desalination facilities hold (triggering "socially-induced environmental impacts" [79]) and 2) the purpose served by those facilities.

When production scales are excessive and uses inadequate, do desalination technology lock-ins not turn physical scarcities into social ones? In such case, how different really is the new PHN from the former much contested ideology underpinning the Ebro river transfer [80]? Under current plans of massively implementing the desalination alternative without consideration of its questionable uses in the tourist and agricultural sectors, does not the myth ''water for all'' still prevail; or is it shifting towards ''water for anything''? Although energy consumption, monetary costs and environmental impacts of desalination may present some clear overall advantages to the former Ebro-transfer option, to what extent do the management trends underpinning the desalination alternative differ from the former PHN? How does it deal with demand and consumption patterns? Just as in the former Plan, the additional water will quench the thirst not only of the intensive plastic-tented irrigation agriculture of Almeria but also of the numerous new tourist developments and golf courses. Although such facilities are to be charged 10 times more for their water than farmers, in this view, would not a widespread implementation of desalination plans still remain in the realm of "Increased Supply" management approach, clashing with the *Nueva Cultura del Agua* and the EWFD that insists for desalination to be the solution of last resort? In fact, similarly to gargantuan water-diversion projects, this kind of application of the desalination technology holds the unrealistic hope of a supply-side technological fix, which delays the onset of demand-side based alternatives.

The divide between different water uses becomes clearer as does the differentiation between ''technological change'' and ''technological equity''. Where mismatches exist between desalination serving “basic needs” and water services for tourism (maintaining “luxury”), the environmental impacts exposed earlier on in this chapter (brine and greenhouse gases) become particularly concerning and ethically questionable.

6.4. Desalination in an Oil-based Society

An intrinsic relation exists between the desalination solution in meeting water demands and the energy dependency of its socio-economic context. Despite unavoidable exhaustion and adverse environmental impacts, fossil fuels in general and oil in particular, can be understood as the main driver of modern society. This fact increasingly causes serious problems as Hubbert curves show how global oil production will peak around 2010, calling the end of cheap energy [81,82]. It took us less than 200 years to use up half the 300 million years old carbon stock stored through photosynthesis in plants dating from the carboniferous era; we are currently witnessing the “last hours of the ancient sunlight” [83]. As Illich would say, no wonder why our oil-based society has become “hysterically concerned with machine fodder” [84:p.20].

From such a perspective, desalination can be considered as a process transforming oil (or other fossil fuels) into freshwater: i.e. shifting problems from one scarce resource (water) to another (non-renewable energies). The volatility of a socio-economic system entirely reliant on cheap, readily available and non-renewable energy is likely to trigger further environmental degradation and geopolitical conflicts. This reality becomes even more stringent in places totally depending on desalination, where the “oil addiction” reaches as far as vital water needs: that is what can be understood by “technological lock-in”.

There is a trend towards an increasing energy-cost of obtaining energy [85]. Renewable energies certainly hold great potential to satisfy part of current demands at lower environmental costs. Although omnipresent and inexhaustible, renewable sources are, however, much more diluted and require vast amounts of energy to concentrate highly site-specific flows in order to carry out useful work. With smaller Energy Return on Energy Input (EROI), they will not be able to fully substitute the non-renewable resources that currently quench our thirst for growth and consumption.

The forthcoming oil peak sets a definite turning point in modern history. In view of gradually easing off our addiction to oil, phasing out and switching towards other (less polluting) energy sources deems necessary. However we have to bear in mind that the latter are unlikely to provide us with the same luxurious living standards to which we have become accustomed in the West thanks to cheap oil [86]. In this perspective, the energy issue of the desalination option becomes ever more concerning. Generalizing the implementation of sound demand-side based strategies seems therefore paramount in view, not only to increase the efficiency of water uses but also to ensure equity of access. The ever worsening gap between consumption levels of Rich and Poor are intrinsically related to the ideology of “salvation through growth”, leitmotiv of modern western culture. The rush for economic growth ultimately leads to ever increasing consumption patterns.

6.5. Water Scarcity in a Consumption Society

The so-called "developed countries" consume about 80% of the world's total resources while accounting only for around 20% of the world population. At the same time, water demands of the "developing countries" are the fastest growing in the world. Furthermore, out of the billions of liters produced each year by desalination plants worldwide, only a deriding fraction is consumed as drinking water. This means that a considerable amount of the energy-intensive desalinated water is actually used for toilet flushing, cleaning or wasted in leakage. Not only is the water itself being wasted, but also the energy, the material resources and the human effort necessary to satisfy these high demands. Additionally, the environmental impacts entailed by brine discharge and greenhouse gas emissions are not accounted for in monetary terms. The environment continues to bear the cost as new desalination plants are built to keep up with rapid demographic and economic growth, which is in fact becoming *uneconomical*. This situation arises from the strongly anchored belief that human needs (such as water demand) will necessarily be satisfied by economic growth, which will automatically enhance quality of life.

The evolutionary economist Schumpeter warned about the confusion existing between *growth* and *progress*, where (vertical) economic growth, as described in this context, is an increase in quantity and cannot therefore be sustainable indefinitely on a finite planet, whereas (horizontal) economic progress is an improvement in life quality and may well be sustainable as it does not necessarily rely on greater energy and material throughputs. So, the term "sustainable growth" is an obvious oxymoron [87].

Economic growth as measured conventionally with GDP at constant prices reveals problematic in several ways. GDP has long been criticized for excluding non monetary transactions such as the very meaning of the word "economy" implies[9]. Rising profits and income per capita are considered -by neo-classical economics- as an adequate indicator to assess *standard of living*, often associated with *quality of life* and ultimately confused with *societal happiness*. However, the costs of wars, toxic farming, climate change, oil spills, depression or cancer while profitable for the few and added to GDP, should really be subtracted to reflect societal progress[10]. Hedonist psychology shows that the correlation between GDP and happiness in the western world no longer holds.

In addition to the distribution conflicts that arise from an indicator as biased as GDP, how can economic growth be sustained on limited stocks of natural capital? In a finite system such as the Earth, an expanding economy cannot last infinitely. Ecological economists like Daly [90] argue that technology and resource substitution (through ingenuity) cannot continuously outrun depletion and pollution and that basic relation of man-made and natural capital is one of complementarity, not substitutability. Moreover, as the "ecological footprint" points out, despite our increasing technological sophistication, mankind remains in a state of "obligate dependence" on the productivity and life support services of the ecosphere [91].

---

[9] Originally, the economy (*oikonomia*) has no chrematistic intention *per se* and simply refers to the "management of the house".

[10] Daly and Cobb [88] developed the 'Index of Sustainable Economic Welfare', which accounts for eighteen aspects of our economic lives that GDP ignores, making (positive) contributions from unpaid household labour, and (negative) contributions from resource depletion, income inequality and environmental damage. Similarly, *Redefining Progress* elaborated the 'Genuine Progress Indicator' and El Serafy [89] the 'Green GDP'.

As Daly shows [90], there is a maximum scale of the economic subsystem, a point beyond which further physical growth, while possible, costs more than it is worth. This allows a clear distinction between economic and *uneconomic* growth. Once economic growth increases ecological costs faster than production benefits, it becomes a true uneconomic growth, impoverishing rather than enriching, and its measure, GDP, then becomes “a gilded index of far-reaching ruin” [90]. Indeed, Max-Neef's “threshold hypothesis” shows that:

> “for every society there seems to be a period in which economic growth (as conventionally measured) brings about an improvement in the quality of life, but only up to a point -the threshold point- beyond which, if there is more economic growth, quality of life may begin to deteriorate.” [92:p.117]

This led Max-Neef [12] to differentiate between *needs* and *wants*, where the latter are shaped by culture and individual personality. Hirsch's “positional goods” [77] explore this point by distinguishing between *absolute* and *relative* deprivation in the societal context. Hirsch argues about the tragedy of positional goods where material artifacts ultimately do not provide satisfaction –as prestige from distinction rapidly fades away when generalization takes place- and create congestion.

A fundamental question for ecological economics (arguing that the limiting factor in development is no longer manmade capital but remaining natural capital) could, therefore, be expressed in the following terms: considering the anticipated load of the human economy, are supply systems such as desalination, adequate to meet water demands in a sustainable way, while simultaneously maintaining the general life support functions of the ecosphere? In this “environmental limits” perspective, concerns that the desalination technology will be primarily directed at what could be loosely called “water luxuries” rather than supplying poor communities with “livelihood water”, seem well founded.

By the neo-classical definition, economics is the science of efficient allocation of scarce resources among competing ends through the price system, but ignoring the distribution (equity) issue. In this sense, the added-value of water services for tourism for instance, can be easily recognized. Similarly, the subsistence counterpart can be discarded since not noticeably affecting GDP. Ecological economics questions the allocation of life-supporting resources by market laws and argues for the incommensurability of values [93] as an alternative to value monism[11]. Multi-criteria theory [95] can be suggested as an alternative to (current) utility theory. This is also why, from an ecological economics perspective [96], the continued growth of macroeconomic systems is considered to be both ecologically unsustainable and existentially undesirable. Rich countries should question their wasteful consumerism and ever growing appetite for exhaustible resources and embark on Sustainable Degrowth [97] (meaning less use of materials and energy while sustaining and maintaining happiness), or the “prosperous way down” –as Odum called it. Because in the long term, “either we adapt with deliberate process or have these changes forced on us with damaging repercussions” [86].

[11] Implying “that all objects of utility have some common characteristic that allows them to be compared” [94:p.212].

CONCLUSION

"The world has enough for everyone's need, not for everyone's greed."

Mahatma Gandhi

The *problématique* of this chapter was to discuss the (un)sustainability of desalination as a long-term solution, especially with regard to socially-constructed freshwater scarcities. The global freshwater scarcity crisis we are facing is the consequence of both population and consumption growth combined with declining natural water resource stocks. The present analysis distinguished between two broad categories of freshwater scarcities: one is to be seen as the consequence of primarily climatic and geophysical factors, whilst the other results from fundamentally anthropogenic factors. The first category was labeled *physical* scarcity, as illustrated by the case of Laâyoune. The second one was further divided into *socially-constructed* (illustrated by the case of Lanzarote), *politically-constructed* (the Gaza Strip) and *structural* scarcity (Chennai's slums). In each case, the desalination technology played a significant role. From the comparative approach of the first two case studies, it has been shown that there is a generalized deviation towards a *social* scarcity pathology. In the following two case studies, the political/governance aspects acted as a predominant trigger. A fair reallocation of resources –both in terms of water quantities and access rights- seems only too legitimate in cases such as the Israeli-Palestinian nexus. Equally, this also applies for intra-urban structural arrangements, as seen in the case of Chennai's slums.

In accordance with the paradigm of *strong* sustainability, this chapter suggested the concept of "technological equity". Indeed, mismatches exist between desalination serving basic human needs and sustaining luxury (e.g. water services for tourism or water intensive irrigation), or maintaining well established power structures (through military occupation or socio-economic segregation). The energy-intensiveness of the desalination technology becomes of particular concern and ethically questionable for the latter cases. In these situations, traditional water heritage and cultural water memory are neglected and social integrity as well as human rights are scorned. In the meantime, environmental degradation, economic volatility and technological "lock-in" are engendered for the sake of sustaining luxury and human greed for power and money. The assessment of technology should be inscribed in an ethical frame that would allow pondering the rationality of objectives, uses and purposes served by desalination. The energy, monetary, social and environmental costs are then to be considered in a case-specific manner, i.e. in relation to the use of the desalted water. The underlying postulate is that, by "putting equity first" more sustainable and effective solutions can be obtained. "Technological equity" was defined as a process through which the overall equity of a given situation would be enhanced. Some guidelines were suggested for the development of a Technological Equity Assessment (TEA), an analytical frame that would evaluate the equity level of a given situation.

In the case of Lanzarote, with the sudden eruption of large quantities of easily available freshwater, a centuries-old water saving culture was dismissed and humid climate kind consumption patterns adopted. This paradoxically led to a situation of water squandering in a natural context of absolute scarcity, which subsequently triggered socially-constructed water calamities. In this particular case, the desalination technology gives the illusion that the limits naturally set by the hydrological cycle do not exist. However, unlimited abundance is a hoax and remains one of the most virulent economic myths.

Having said this, it is important to acknowledge the great potentials the desalination technology offers when put to proper use, particularly with regard to water conflicts, as discussed in the case of the OPT. Moreover, desalination could become a real solution for meeting basic human needs in many poor regions of the world. This was illustrated by the case studies of a couple of Chennai's slums, where the population was provided with a so-called "water life-line" by small-scale desalination plants at a cost of approximately 3 kWh/m^3.

In 2005, the *European Declaration for a New Water Culture* based on the ethical foundations laid by the *Nueva Cultura del Agua*, recognized the different functions and values of water. The purposes of the desalination technology should be differentiated along the same lines as the uses of water are categorized in this Declaration, meaning:

- as a top priority, "Water for Life": guaranteeing universal access to drinking water and sanitation for basic human needs,
- followed by "Water for general interest purposes": preserving health and social cohesion,
- and finally, "Water for economic growth" as a third level of priority: managed efficiently, under economic rationality.

Acknowledging that socially-constructed scarcity is an inherent characteristic of *economic goods* and not of *human rights*, saving-incentive policy interventions should be implemented; for instance, properly designed water pricing and tariff structures, which would be an instrument to moderate luxurious demands. Socially-constructed scarcity must be managed in order to regulate the demand/supply nexus. State-of-the-art seawater desalination (involving costs around US$ $0.6/m^3$) truly offers "Water for Life" perspectives if it is not high-jacked by "Water for economic growth" or "Water for domination and control" purposes. So, regardless of the management model (public, private or PPP), desalination should primarily serve its lofty function of meeting human *need* rather than human *greed*, in other words sustaining *livelihood* rather than satisfying *luxury* or maintaining *power*.

The issue of "energy for desalination" was also addressed. Although, decreasing marginal energy costs have enabled rapid growth of worldwide desalination production, the energy source plays a fundamental role when determining the sustainability level of the technology. Turning fossil fuels into water might be questionable in the long term. The "oil for water" pathologies shed light on how desalination displaces costs and environmental externalities spatially, temporally and onto other species. Bearing in mind the forthcoming oil peak, water is then produced by using up limited and polluting energy sources as well as scarce greenhouse gas sinks: the desalination technology involves an uncomfortable trade-off between water and energy requirements.

In the last section of this chapter, the ideology underpinning the increased supply frenzy was analyzed. The intention was to shed light upon the intrinsic parallelism existing between the economic growth paradigm and the management trend of ever increasing water supplies. As seen from the case of Spain, socially-constructed scarcity generally emerges from an "increased supply" approach. However, these production patterns and sustained growth trends put at risk the freshwater resource basis. This highlights, once again, the limits of the economic process as a subsystem of the ecosphere. The socially-constructed urgent call to massively implement ever greater desalination facilities correlates with the "salvation through growth" dogma, speeding up the production-consumption carousel. This growth ideology inherently leads to "consumerism as a need satisfactor". This is, however, far removed from meeting basic human needs for the Poor in the South, living in acute physical scarcity situations.

Desalination should remain a solution of last resort, chosen only after all appropriate water demand management measures have been implemented. This applies especially for countries featuring inefficient water utilities, high water losses and weak sector policies. Ultimately, saving water rather than developing new supplies is often the 'next' best source of water, both from an environmental and economic perspective. As explained in this chapter, rationalization and equitable allocation of available water is the most sustainable and efficient solution to the problem of freshwater scarcity.

From such a perspective, achieving a *World Water Contract* [98] should be supported, especially for populations suffering from environmental injustice and conflicts over resource access. Indeed, "water wars", as Shiva [99] understands them, also include the "war" between conflicting water management paradigms. These include bellicose divergences on how water is distributed and which values are attributed to it. In addition, reviewing the history of water conflicts, nations have cooperated much more than they have fought over the sharing of lakes, rivers and other water bodies. In this view, the real water war is far more bloody and silent: 6000 daily deaths caused by diarrhea due to lack of sanitation. Seen in these terms, the lack of water which the Gazian population and Chennai's slum dwellers suffer from, can truly be considered as war.

ACKNOWLEDGMENT

I would like to express my deepest recognition to my mentors Dr. Joan Martinez-Alier, Dr. David Tàbara and Dr. Sybille van den Hove d'Ertsenryck. I would like also to thank Dr. David Saurí, Dr. Erik Swyngedouw, Dr. Luis Serra, Dr. Ferran Izquierdo Brichs, Dr. Estaban Castro, Dr. Giorgos Kallis and Dr. David Barkin for their comments and advices; to my friends at the ICTA, especially Vincent Moreau, Thomas Koetz, Claudio Cattaneo, Eneko Garmendia, Xaví Cazorla, Gonzalo Gamboa, Sigrid Muñiz, Elena Domene, Rosa Binimelis, Eli Roca as well as Anna Borfo, Anna Doroteo and Loli Garcia. In Lanzarote, I would like to thank Quino Miguélez, Idoya Cabrera, Manuel Plasencia and the Fundación César Manrique; in Morocco, Moha and Hassou; in India, Dr. Sharad Lele, Dr. Priya P. Sangameswaran, Dr. Ramana, Dr. Esha Shah, Dr. Ajit Menon and his family, Dr. Karen Coelho, Ms. Shobha Iyer, Mr. Arivalaganto and the CISED at ISEC; in the Occupied Palestinian Territories, Ms. Rima Abu Middain, Mr. Khaled Shahwan, Ms. Rebekah Kosinski, Dr. Rabi and the PAPP/UNDP. A special thanks to family and friends for support and love, especially to Renate von Medeazza and Nadj Feyder.

This research was funded by the IQUC and BE scholarship.

References

[1] Kranzberg, M. (1997). Technology and History: 'Kanzberg's Laws'. In T.S. Reynolds & S.H. Cutcliffe (Eds.) *Technology and the West. A Historical Anthology from Technology and Culture. Chicago*, USA: The University of Chicago Press, 5-20.

[2] Martinez-Alier, J. (1987). *Ecological Economics*. Oxford, UK: Blackwell.

[3] Lotka, A. (1945). The law of evolution as a maximal principle. *Human Biology, vol. 17*, 167-194.

[4] Sterrer, W. (1993). Human economics: A non-human perspective. *Ecological Economics, vol. 7*, 183-202.

[5] Shiklomanov, I. A., Shiklomanov, A. I., Lammers, R. B., Peterson, B. J. & Vorosmarty, C. (2000). The dynamics of river water inflow to the Arctic Ocean. In: T*he Freshwater Budget of the Arctic Ocean. Dordrecht*, The Netherlands: Kluwer Academic Publishers.

[6] Al-Weshah, R. (2002). The role of UNESCO in sustainable water resource management in the Arab World, *Desalination, vol. 152*, 1-13.

[7] Catton, W. (1986). Carrying capacity and the limits to freedom. *Social Ecology Session* 1, XI World Congress of Sociology, New Delhi, India.

[8] UNDP, (2006). Beyond scarcity: Power, poverty and the global water crisis. *Human Development Report*. NY, USA: UNDP.

[9] World Water Council, (2000). *World water vision; Making water everybody's business*. London, UK: Earthscan.

[10] IPCC, (2001). *Climate Change 2001: Impacts, Adaptation and Vulnerability*. Geneva, Switzerland: UNEP.

[11] Naredo, J. M. (2003). *La encruijada de la gestión del agua en España*. Archipiélago, *vol. 57*, 1-17.

[12] Max-Neef, M., Elizalde, A. & Hopenhayn, M. (1986). *Desarrollo a escala humana : una opción para el futuro*. Uppsala, Sweden: Fundación Dag Hammarskjöld.

[13] Wangnick, K. (2002). The region leads in seawater desalination. *Paper presented at the IDA World Congress on Desalination and Water Reuse,* Manama, Bahrain.

[14] MacHarg (2001). The evolution of SWRO energy recovery systems. *Desalination Water Reuse, vol. 11*, 49-53.

[15] Raluy, R. G., Serra, L. & Uche J. (2005). Life cycle assessment of desalination technologies integrated with renewable energies. *Desalination, vol. 183*, 81-93.

[16] UNEP / MED POL (2003). Sea Water Desalination in the Mediterranean: Assessment and Guidelines. *MAP Technical Reports Series No. 139*. Athens, Greece: UNEP / MAP.

[17] ONEP, (2003). Expérience et perspective de dessalement à l'ONEP. *Paper presented at the Journée nationale sur le dessalement,* Rabat, Morocco.

[18] INALSA, (2004). Datos Estadísticos 2003. *Oficina Técnica*. Lanzarote, Spain: Inalsa.

[19] Centro de Datos, (2003). *Anuario estadístico 2002*. Lanzarote, Spain: Cabildo de Lanzarote.

[20] UNEP / MCSD / Blue Plan (1999). Report of the Workshop on Tourism and Sustainable Development in the Mediterranean (Antalya, Turkey, 17-19 September 1998). *MAP Technical Reports Series No. 126*. Athens, Greece: UNEP.

[21] INALSA, (2001). *Informe para Consejo de la Biosfera*. Lanzarote, Spain: Inalsa.

[22] INALSA, (2003). *Inalsa Informa (2)*. Lanzarote, Spain: Inalsa.

[23] Fundación César Manrique, (2001). *Lanzarote: el Papel de la Crisis.* Lanzarote, Spain: Fundación César Manrique.

[24] Roca, J., Puig, I., Hercowitz, M. & Hernández, O. (2003). *Fiscalidad y medioambiente en la isla de Lanzarote, LIFE Lanzarote 2001-2004: Exploración de nuevas líneas de Actuación.* Lanzarote, Spain: Observatorio de la Reserva de Biosfera - Cabildo de Lanzarote.

[25] INALSA, (2003). *Datos Estadísticos 2002.* Oficina Técnica. Lanzarote, Spain: Inalsa.

[26] Palestinian Hydrology Group, (2004). *Water for Life. Israeli assault on Palestinian water, sanitation and hygiene during the Intifada.* WaSH Monitoring Report. Jerusalem, Israel: PHG.

[27] UN, (1992). *Water Resources of the Occupied Palestinian Territory.* Committee on the Exercise of the Inalienable Rights of the Palestinian People. NY, USA: UN.

[28] Homer-Dixon, T. & Kelly, K. (1995). *Environmental scarcity and violent conflict: The case of Gaza. Occasional Paper, Project on Environment, Population and Security.* Washington D.C., USA: American Association for the Advancement of Science and the University of Toronto.

[29] UNEP, (2003). *Desk Study on the Environment in the Occupied Palestinian Territories.* Nairobi, Kenya: UNEP.

[30] Water Resources Action Program, (1994). *Palestinian Water Resources – A Rapid Interdisciplinary Sector Review and Issues Paper.* Ramallah, OPT: WRAP Task Force.

[31] PWA / USAID, (2000). *Coastal Aquifer Management Program, Integrated Management Plan.* prepared by Metcalf & Eddy. Gaza, OPT: PWA.

[32] PWA, (2000). *Prospects for Water Desalination in Gaza.* Ramallah, OPT: PWA.

[33] *Palestinian Economic Council for Development and Reconstruction* (2001). Ramallah, OPT: Palestinian Water Strategic Planning Study, PA.

[34] Swyngedouw, E. (2004). *Social Power and the Urbanisation of Water – Flows of Power.* Oxford, UK: Oxford University Press.

[35] Roy, S. (1995). *The Gaza Strip: the political economy of de-development.* Washington D.C., USA: Institute of Palestine Studies.

[36] Roberts R. S. & Emel J. (1992). Uneven Development and the Tragedy of the Commons: Competing Images for Nature-Society Analysis. *Economic Geography, vol. 68(3)*, 249-271.

[37] Gleick, P. H. (1994). Water, war, and peace in the Middle East. *Environment, vol. 36 (3)*, 6-15, 35-42.

[38] Weizman, E. (2002). *The Politics of Verticality. series of articles and photo-essays* obtainable from: http://www.opendemocracy.net/debates/article-2-45-806.jsp

[39] Halper, J. (1999). *Dismantling the Matrix of Control.* News from Within, October 1999.

[40] PA (2005). *Medium Term Development Plan 2005-2007.* Ramallah, OPT: Ministry of Panning, PA.

[41] Palestinian Hydrology Group & PA, (2005). *Evaluation of Small Scale Brackish water Desalination Plants in Gaza Strip.* Ramallah, OPT: PA.

[42] Mason, B. (2003). Geologists call for desalination of Gaza Strip's water. *Nature, vol. 426*, 110.

[43] PWA, (2000). *Water Desalination Plan.* Ramallah, OPT: PWA.

[44] Weinthal, E., Vengosh, A., Marei, A., Gutierrez, A. & Kloppmann, W. (2005). The water crisis in the Gaza Strip: Prospects for Resolution. *Ground Water, vol. 43(5)*, 653-660.

[45] Shuval, H. (2005). An Equitable Reallocation of the Shared Water Resources Between Israelis and Palestinians. *Paper presented at FAME 2005, Alternative World Water Forum,* March 17-20th 2005, Geneva, Switzerland.

[46] Tamil Nadu Slum Clearance Board, (2005). *Pre-Feasibility Study for Identification of Environmental Infrastructure Requirements in Slums of Chennai Metropolitan Area.* Chennai, India: TNSCB.

[47] Srinivasan, R. K. & Kozhisseri, D. (2004). Emergency. *Down To Earth, vol. 13(15)*, 23-38.

[48] Vaidyanathan, A. & Saravanan, J. (2004). *Household water consumption in Chennai City. A sample survey.* New Delhi, India: Centre for Science and Environment.

[49] Coelho, K. (2005). Unstating 'the public': an ethnography of reform in an urban public sector utility in south India. In: D. Mosse & D. Lewis (Eds.), *The aid effect: Giving and governing in International Development.* London, UK: Pluto Press.

[50] Mosse, D. (2003). *The Rule of Water: Statecraft, Ecology and Collective Action in South India.* New Delhi, India: Oxford University Press.

[51] de Wit, J. W. (1996). *Poverty, policy and politics on Madras slums : dynamics of survival gender, and leadership.* New Delhi, India: Sage.

[52] Corporation of Madras, (1957). *Madras City Water Supply. Brochure: Inauguration of the Rapid Mechanical Filters.* Chennai, India: Corporation of Madras.

[53] Corporation of Madras, (1932). *Administration Reports 1931-32.* Chennai, India: Corporation of Madras.

[54] Mathur, O. P. (1994). The state of India's urban poverty. *Asian Development Review, Studies of Asian and Pacific Economic Issues, vol. 12(1)*, 32-67.

[55] Mc Neill, J. (2000). *Something new under the sun. An environmental history of the 20th century.* New York, USA: The Penguin Press.

[56] Papanek, V. (1997). *Design for the Real World: Human Ecology and Social Change.* London, UK: Thames and Hudson.

[57] Shah, A. (1998). Developing Rainfed Agriculture: Implications for Women. In: C. Datar (Ed.), *Nurturing Nature: Women at the Centre of Natural and Societal Regeneration.* Bombay, India: Earthcare Books.

[58] Winner, (1985). quoted in Shah, E. (2003). *Social Designs. Tank Irrigation Technology and Agrarian Transformation in Karnataka, South India. Hyderabad, India: Wageningen University Water Resources Series,* Orient Longman.

[59] Sangren, S. (1995). Power against ideology: a critique of Foucaultian usage. *Cultural Anthropology, vol. 10(1)*, 3-40.

[60] Jonas, H. (1979[1984]). *The imperative of responsibility. In search of an ethics for the technological age.* Chicago, USA: The University of Chicago Press.

[61] Gandhi, M. (1927). *Young India,* 17 March 1927.

[62] Boyce, J. & Pastor, M. (2001). *Building Natural Assets: New Strategies for Poverty Reduction and Environmental Protection.* Amherst, MA, USA: Political Economy Research Institute.

[63] Sangameswaran, P. (2005). Equity in Community-based sustainable development: a case study of watershed development in western India. *PhD thesis,* University of Massachusetts Amherst.

[64] Keulartz, J., Schermer, M., Korthals, M. & Swierstra, T. (2004). Ethics in technological culture: a programmatic proposal for a pragmatic approach. *Science,* Technology and Human Values, *vol. 29(1)*, 3-29.

[65] Hamlett, P. W. (2003). Technology theory and deliberative democracy. *Science,* Technology and Human Values, *vol. 28 (1)*, 112-140.

[66] Schumacher, E. F. (1973). *Small is beautiful: Economics as if people mattered.* NY, USA: Perennial Library.

[67] Tàbara, D. (2006). Power, responsibility and situational ISA. Exploring some false dichotomies in Integrated Sustainability Assessment. *Working paper of WPI of EU-funded MATISSE research project.*

[68] Tàbara, D. (2002). Sustainability Culture. In: Governance for Sustainable Development, Barcelona: Advisory Council for Sustainable Development, International Institute on Governance & Government of Catalonia. *Papers de Sostenibilitat, vol. 2*, 53-85.

[69] Neumayer, E. (1999). *Weak Versus Strong Sustainability: Exploring the Limits of Two Opposing Paradigms.* Cheltenham, UK: Edward Elgar.

[70] European Parliament and Council of the European Union, (2000). Directive 2000/60/EC of the European Parliament and of the Council establishing a framework for the Community action in the field of water policy. *Official Journal of the European Communities*, OJ L 327, Brussels, Belgium.

[71] Arrojo Agudo, P. & Martínez Gil, F. J. (1999). *El Agua a Debate desde la Universidad, Hacia una nueva cultura del agua*. Zaragoza, Spain: C.S.I.C.

[72] Gleick, P. H. (2003). Global Freshwater Resources: Soft-Path Solutions for the 21st Century. *Science*, *vol. 302*, 1524-1528.

[73] Esteban, A. & Naredo, J. M. (2004). *Ideas y propuestas para una nueva política del agua en España.* Bilbao, Spain: Bakeaz.

[74] Jevons, W. S. (1965). *The Coal Question: an Inquiry Concerning the Progress of the Nation, and the Probable exhaustion of our Coal Mines.* NY, USA (1st edition by Macmillan, London, 1865).

[75] Meerganz von Medeazza, G. (2004). Water desalination as a long-term sustainable solution to alleviate global freshwater scarcity? *A North-South approach. Desalination, vol. 169*, 287-301.

[76] Domene, E., Saurí, D. & Parés, M. (2005). Urbanization and sustainable resource use: the case of garden watering in the Metropolitan Region of Barcelona. *Urban Geography*, *vol. 26(6)*, 520-535.

[77] Hirsch, F. (1976). *Social Limits to Growth,* Cambridge Mass, USA: Harvard University Press.

[78] Allan, T. (2003). Virtual Water: The water, food and trade nexus. Useful concept or misleading metaphor? *Water International* (IWRA), *vol. 28(1).*

[79] Meerganz von Medeazza, G. (2005). Direct and socially-induced environmental impacts of desalination. *Desalination*, *vol. 185*, 57-70.

[80] Meerganz von Medeazza, G. (2004). Desalination as a sustainable alternative to inter-basin water transfers? The case of Catalonia. *Paper presented at IV Congreso Ibérico del Agua,* Tortosa, Spain.

[81] Deffeyes, K. S. (2001). *Hubbert's Peak: The Impending World Oil Shortage.* New Jersey, USA: Princeton University Press.

[82] Campbell, C. J. & Laherrère, J. H. (1998). *The End of Cheap Oil: Global production of conventional oil will begin to decline sooner than most people think, probably within 10 years.* Scientific American.

[83] Hartmann, T. (1999). The Last Hours of Ancient Sunlight. NY, USA: Harmony Books.

[84] Illich, I. (1974). *Energy and Equity.* London, UK: Calder & Boyars.

[85] Cleveland, C. J., Kaufmann, R. K. & Stern, D. I. (1998). Analysis: Aggregation and the role of energy in the economy. *Ecological Economics, vol. 32*, 301-317.

[86] Odum, H. T. (2000). *The Prosperous Way Down.* Boulder, Colorado, USA: The University Press of Colorado.

[87] Latouche, S. (2004). *Survivre au développement.* Paris, France: Mille et une nuits.

[88] Daly, H. & Cobb, J. B. (1989). For the Common Good. *Redirecting the Economy toward Community, the Environment and a Sustainable Future.* Boston, USA: Mass Beacon Press.

[89] El Serafy, S., Yusuf, J. & Lutz, E. (1989). *Environmental accounting for sustainable development.* Washington D.C., USA: The World Bank.

[90] Daly, H. (1999). Steady-state economics: avoiding uneconomic growth. In: J.C.J.M. van den Bergh (Ed.), *Handbook of Environmental and Resource Economics.* Cheltenham, UK: Edward Elgar.

[91] Wackernagel, M. & Rees, W. (1997). Perceptual and structural barriers to investing in natural capital: economics from an ecological footprint perspective. *Ecological Economics, vol. 20(1)*, 3-24.

[92] Max-Neef, M. (1995). Economic growth and quality of life: a threshold hypothesis. *Ecological Economics, vol. 15*, 115-118.

[93] Martinez-Alier, J., Munda, G. & O'Neill, J. (1998). Weak comparability of values as a foundation for ecological economics. *Ecological Economics, vol. 26*, 277-286.

[94] Gowdy, J. & Erickson, J. D. (2005). The approach of ecological economics. *Cambridge Journal of Economics, vol. 29*, 207-222.

[95] Munda, G. (2004). Social multi-criteria evaluation: methodological foundations and operational consequences. *European Journal of Operational Research,* vol. 158, pp. 662-677.

[96] Georgescu-Roegen, N. (1971). *The Entropy Law and the Economic Process.* Cambridge, Mass, USA: Harvard University Press.

[97] Georgescu-Roegen, N. (1979 [2006]). *La décroissance : Entropie, écologie, économie* (third edition). Paris, France: Sang de la terre. (first edition: Demain la décroissance. Lausanne, Switzerland: Pierre-Marcel Favre).

[98] Petrella, R. (2001). *The Water manifesto: arguments for a world water contract.* London, UK: Zed Books.

[99] Shiva, V. (2002). *Water Wars: Privatization, Pollution, and Profit. Boston,* USA: South End Press.

In: Desalination: Methods, Cost and Technology
Editor: Irena A. Urbonienė pp. 325-339

ISBN: 978-1-61668-909-4

Chapter 10

Renewable Energy Sources for Seawater Desalination -Present Status and Future Prospects*

***Eftihia Tzen*†**
Center for Renewable Energy Sources, 19th klm
Marathonos Ave, Pikermi 19009, Greece

Abstract

The need for desalting seawater is becoming more and more pressing in many parts of the world. In the Mediterranean, water needs are definitely increasing. An increase of around 30% of the water demand by the year of 2010 has been estimated for southern and eastern countries.

Seawater is practically an unlimited resource of water and seawater desalination has a long time been a major source of water in many regions of the Mediterranean.

The matching of Renewable Energy Sources (RES) with seawater desalination is a slowly growing market and thus mainly due to the high cost of the produced water.

Parameters like the size of the unit, the design, the energy consumption and availability and other factors that should be considered for the final decision of technologies selection will be examined in this work.

Several RES desalination units have been installed around the world. Several combinations such as photovoltaic units with reverse osmosis plants, wind energy with reverse osmosis and solar thermal distillation plants become in operation, most of them within research programmes. Several scientific publications on the matching of the two technologies exist and a lot of work has been done in particular from research centers. However it seems that we are still at the beginning.

There is no doubt that world's water needs can be covered and satisfied if also non-conventional resources are utilized. There is no doubt that world's energy needs can be

* A version of this chapter was also published in Desalination Research Progress, edited by Daniel J. Delgado and Pablo Moreno, Nova Science Publishers. It was submitted for appropriate modifications in an effort to encourage wider dissemination of research.

† Email: etzen@cres.gr

covered and satisfied if also Renewable Energy sources are utilized. The matching of the two technologies under certain circumstances is an important prospect to the present and future potable water needs.

INTRODUCTION

In the Mediterranean, the water needs are continuously increasing. Desalination has for a long time been a major source of water in many parts of the world. Countries like Spain, Malta, and UAE cover a large part of their water needs by the use of desalination and particularly seawater desalination.

Seawater desalination is a steadily growing industry. The market for seawater desalination has a significant upturn during the last years. Driven by the increasing world population and the diminishing fresh water sources, a result of global warming, environment destruction many countries have constructed or are constructing desalination plants to mainly cover potable water needs.

Desalination plants exist in regions that have hot climates, relatively low and unpredictable rainfall and in general in regions where conventional water resources are not able to cover the water demand. Seawater desalination accounts for a worldwide water production more than 35 million m^3/day with major players the Middle East and specifically Saudi Arabia, the UAE and Kuwait [5]. Other areas in which desalination is of growing importance and become more prominent is the Mediterranean Sea, the Red Sea, the coastal areas of California, China and Australia.

Several desalination processes have been developed but not all of them are reliable and in commercial use. The most important processes are split into two main categories, thermal (or distillation) and membrane processes. The most widely used thermal processes for sea water desalination are: Multistage-Flash Distillation (MSF), the Multi Effect Distillation (MED or ME) and the Vapour Compression (VC) process. Membrane processes consist of Reverse Osmosis (RO) and Electrodialysis (ED) processes. Electrodialysis is confined to the desalination of brackish water while Reverse Osmosis is used for both, brackish and sea water desalination. The traditional dominating MSF process is continuing to lose its market share to RO and MED, due to improvement of membrane technologies and the cost advantage.

Desalination is an energy intensive technology. The energy input to the plant may be thermal, (defined in terms of units of water produced per unit of steam or per 2,500 kJ used) mechanical or electrical, (expressed in kWh/m^3). In recent years energy consumption by desalination has been reduced significantly, especially in RO process, however, still is a major limitation for the desalination technology.

Furthermore, over the years, the cost of water produced in seawater desalination plants has dropped considerably, while the cost of water produced by conventional treatment plants has risen, due to the over-exploitation of aquifers and the intrusion of saline water, and also to the generally increasing contamination of water. It is obviously that desalination becomes day by day a more competitive water technique.

Combining Desalination and Renewable Energy Sources (RES) would generate a sustainable source of fresh water as well as energy. This combination is of major importance as it limits and reduces air pollutant emissions and green house gases generated by combustion of fossil fuels. Moreover, the cost of the conventional energy sources is expected to continue increase. This will make the cost of renewable energy sources highly competitive against fossil fuel.

Additionally, it often happens that the geographical areas where water is needed are well gifted with renewable energy sources (RES), be it solar, wind, and in specific areas geothermal energy.

A renewable energy driven desalination plant can be designed to operate coupled to the grid and off-grid (stand-alone or autonomous). The matching of the desalination process to a renewable energy source is fairly complex, especially regarding autonomous operation.

Desalination processes are best suited to continuous operation. The majority of the renewable energy sources are distinctly non-continuous and are in fact intermittent often on a diurnal basis. Unpredictable and non steady power inputs, such as in the case of renewable energy sources, force the desalination plant to operate in non optimal conditions and may cause operational problems. An energy storage system could achieve the required power stability, although it will increase the total system cost.

In general, the matching of renewable energy sources to desalination processes is a technical challenge with major problems associated with their intermittent character and total system cost.

The selection of the most suitable RES desalination combination is an important factor on the success of the installation. Selection of the technologies is based on several parameters, such as site conditions, local circumstances, energy availability, etc. The "best matching" for a particular application will be the one that reliably produces water of significant quality and quantity at reasonable cost.

SELECTION OF TECHNOLOGIES

The choice of a desalination process over the other is based on several parameters such as feed water type, product water quality and quantity, energy source and availability, etc. In order to select a desalination process a number of factors need to be taken into consideration. The main points are discussed below:

1. Feed water type - Thermal distillation plants are mainly used for the desalination of seawater, RO process is used for the desalination of brackish and seawater while ED is preferred for brackish water desalination.
2. Water quality- Thermal distillation plants produce distillate water of low salinity, around 10 to 20 ppm TDS (parts per million Total Dissolved Solids). By contrast the TDS of product water from RO and ED is usually around 250-500 ppm TDS.
3. Plant capacity - Thermal distillation plants are typically available and good performed in large unit capacities. The exception to this is the VC plants which are used in small and medium scale applications (up to 2500m^3/day). MED plants are used in medium and large scale applications. Small scale MED plants have also been developed. Membrane plants can be easily adapted to any plant size and can be found in all capacity ranges.
4. Construction time - Thermal distillation plants require more time than a membrane process. The construction time of a large MSF plants can take more than 2 years. Large RO plants can be constructed and installed in half time period. Small RO

plants can be produced and installed in almost one month. Due to the modularity of the process an RO unit could be fitted in container/s reducing also the construction time due to the civil works limitations.

5. Economics – The total cost of the water produced by a desalination plant includes the investment cost as well as operating and maintenance cost. The energy cost is of major concern. With upgraded designs and improved energy efficiencies, desalination units can currently deliver fresh water from the sea at costs less than 0.8€/m^3.

Another parameter that should be taken into consideration is the remoteness of an installation. If the plant is to be remotely sited, the technology of choice should be robust and as independent of support from outside of the area as possible, both in respect of spare parts and chemicals as well as operator expertise.

Selection of site regarding seawater intake and brine disposal is another important factor that should be taken into consideration. The potential environmental impacts of desalination projects need to be evaluated intently.

The choise of a renewable energy source for electricity production is influenced on factors such as the RES potential at the specific site, land availability and accesibility, type of connection (grid connected or stand alone), budget constrain and financial support. These factors will affect the final decision and unit electricity cost. Also in this technology selection, the study and concern of the environmental impacts (civil engineering works, roads accesibility, etc) and in several cases the corrective measures are crucial and should be taken in consideration (Figure 1).

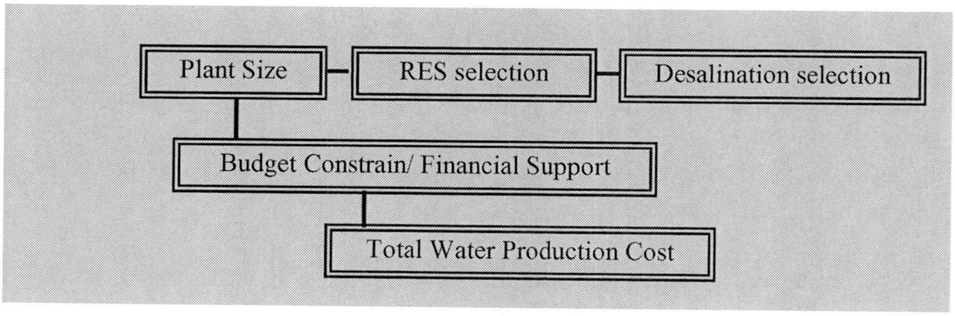

Figure 1. Parameters for the decision a RES Desalination plant installation.

Once a decision has been made to investigate the possibility of building a RES driven desalination plant the true cost of water in the locality should be established. The final unit water cost of a RES desalination system is "site and system specific". Each system is a different case; different site data, different design, impacts and cost.

When several alternative RES - desalination schemes are applicable for a specific case, the final decision concerning the most prominent combination should be based, once again, on criteria such as:

- Commercial maturity of technology. An appropriate way to validate this is by examining the performance of similar existing applications
- Local support availability (installers, technicians, etc.)

- Simplicity of operation and maintenance of the system

The above factors, in conjunction with available technical information (feed water quality, output water requirements, RES potential) provide a starting point for the engineer and the decision maker.

RES DESALINATION STATE -OF-THE-ART

Worldwide, several RES desalination pilot plants have been installed and the majority has been successful in operation. Virtually all of them are custom designed for specific locations and utilize solar, wind or geothermal energy to produce fresh water.

Energy driven desalination technologies mainly fall into two categories. The first category includes distillation desalination technologies driven by heat produced by RES, while the second includes membrane and distillation desalination technologies driven by electricity or mechanical energy produced by RES (see Figure 2). Many of these combinations may not be viable under certain circumstances, but they are included in the researcher's assessment.

Between the several technical feasible combinations of RE technologies and Desalination, some seem to be more promising in terms of technical, economic feasibility and performance than others. However their applicability strongly depends on the factors discussed before, such as desalination plant size, local availability of renewable energy resources, and quality of the feed and produced water.

A renewable desalination plant can be designed to operate coupled to the grid or off-grid (autonomous system). Where the system is grid connected, the desalination plant can operate continuously as a conventional plant and the renewable energy source merely acts as a fuel substitute. Where no electricity grid is available, autonomous systems have to be developed which allow for the intermittent nature of the renewable energy source. The latter case poses the problem of renewable energy variability because most energy systems lack an inherent energy storage mechanism. Desalination systems have traditionally been designed to operate with a constant power input and unpredictable and non-steady power input impose the desalination plant to operate in non-optimal conditions [2]. An energy storage device for the stability of the power input from RES and mainly from wind and solar energy sources is recommended. RES Desalination systems operate under variable power conditions they have not been proven yet their performance and economic advantage.

Autonomous systems are mainly of small-scale. The majority of small-scale installations have been developed so far are within research projects. The size standards of RES desalination plants are different than in conventional systems. For instance, as large desalination plant is characterized a plant with a daily production of more than 10.000m^3. For RES desalination matching, a small system usually is characterised a desalination system having a capacity of few liters up to 50m^3 daily production. A medium RES desalination system could be characterized a system of 50 m^3/day to approximately 150 m^3/day. From more than 150 m^3 daily production the systems characterized as large and usually are grid connected or a combination of both, stand alone and grid connected. In Table 1 several seawater RES desalination installations are described.

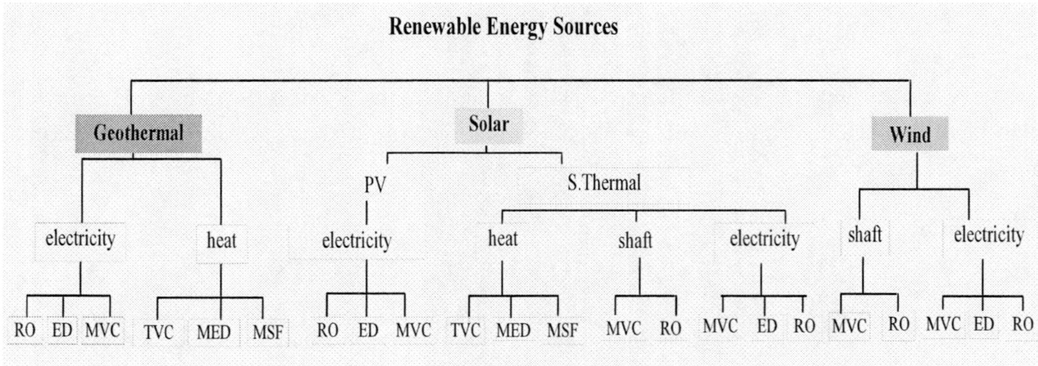

Figure 2. RES Desalination Combinations [1]

Over the last two decades in particular, numerous desalination systems utilizing renewable energy have been constructed. Most of these plants have been installed as research or demonstration projects. Several of these plants do not exist. Most likely the majority are pilot units and after the end of the project there is no interest or budget to continue their operation [2]. However, the lessons learned have hopefully been passed on and are reflected in the plants currently being built and tested. Table 1 provides some of the most known RES desalination plants.

Table 1. RES Desalination applications

Plant Location	Des. process	Des. capacity m^3/h	RES	Autonomous/ Grid connected	RES installed power	Commission Year
La Paz, Mexico	MSF	0.41	solar	A	352 m^2 collectors	1980
Isl. of Suderoog, Germany	RO	0.25-0.37	wind	A	6 kW W/T	1980
Saudi Arabia	RO	0.135	solar	A	8 kWp PV	1981
Lampedusa, Italy	MSF	0.3	solar	A	408 m^2 collectors area	1983
Safat, Kuwait	MSF	0.41	solar	A	220 m^2 collectors area	1983
Abu Dhabi, UAE	MED	3.3	solar	A	1862 m^2 collectors area	1984
Vancouver, Canada	RO	0.41	solar	A	4.8 kWp PV	1984
Ohsima island, Nagasaki	ED	0.41	solar	A	25 kWp PV	1986
Almeria, Spain,[12] CIEMAT, DLR	MED	3	solar	A	2.672 m^2 collectors area	1987
Lampedusa, Italy[13]	RO	3+2	solar	A, G	100 kWp PV	1990
Lipari island, Italy	RO	2	solar	A	63 kWp PV	1991

[12] The Solar thermal MED plant at PSA (Almeria, Spain) was developed by CIEMAT (Spain) and DLR (Germany) in 1987, initiating the Solar Thermal Desalination Project (STD), which was completed in 1993.

[13] In 1995 and after 5 years of successfully operation was decided the modification of the system and its incorporation into the island grid. The PV system then was used as backup.

Table 1. (Continued)

Plant Location	Des. process	Des. capacity m^3/h	RES	Autonomous/ Grid connected	RES installed power	Commission Year
Borkum Island, Germany	MVC	2	wind	AS	45 kW W/T	1993
Fuerteventura, Spain	RO	2.3	wind diesel	A	225 kW W/T	1995
Therasia Island	RO	0.2	wind	A	15 kW W/T	1997
Pozo Izquierdo, SDAWES project	RO EDR MVC	8 x 1.0 7.9 2.08	wind	A	2x230 kW W/T	1995
Syros island, Greece	RO	2.5- 37.5	wind	A+G	500 kW W/T	1998
Pozo Izquierdo, DESSOL project	RO	0.4	solar	A	4.8 kWp PV	1998/2000
Kimolos island, Greece	MED	3.3	geoth	A	Geothermal, 61oC	2000
Pozo Izquierdo, Spain, SODESA project	Distillati on	0.025	solar	A	$50m^2$ collectors area	2000
Keratea, Greece PAVET Project	RO	0.13	solar wind	A	4 kWp PV, 900 W W/T	2002
Oman, MEDRC project	MED	1.0 (in 9hr)	solar	A	$5.34m^2$ collectors area	2002
Pozo Izquierdo, Spain, AEROGEDESA project	RO	0.80	wind	A	15 kW W/T	2003
Loughborough Univ,UK	RO	0.5	solar	A	1.54 kWp PV	2003
Loughborough Univ,UK	RO	0.5	wind	A	2.5 kW W/T	2003
Beni Khiar, Tunisia	MED	0.15-0.20	solar	A	108 m^2 collectors area	2003
ADIRA project	RO Solar still	0.004-0.01	solar	A	PV Solar distillation	2003-2007
Milos island, Greece[14]	RO	2x1000	wind	G	850 kW W/T	2007
Heraklia island, Greece[3]	RO	3.3	wind off shore	A	30 kW W/T	2007

The ADIRA project, as mentioned in Table 1, is one of the most recent demonstration projects, supported by the European Commission under the contract number ME8/AIDCO/2001/0515/59610. The aim of this project is the provision of fresh water by the use of small RES desalination units at several remote areas of Morocco, Cyprus, Jordan, Turkey and Egypt. The water capacity of the autonomous RE desalination units will range between 100 l/day to $10m^3$/day. Around 10 systems for brackish and seawater desalination will be developed within this project, most of them of PV RO combination. Only in Sinai in Egypt a solar still unit for seawater desalination will be developed. The project will not only cover the technical issues but also legal, social, economical and organizational matters [13].

[14] Both plants are developed within the Operational Programme "Competitiveness" of the Greek Ministry of Development. (Measure 6.3, Measure 4.5, respectively)

Several other projects have been currently elaborated regarding the development of large scale solar thermal plants and other RES desalination combinations. AQUASOL (2002-2006) project aims on the development of a seawater desalination technology based on multi-effect distillation that is energy efficient, low-cost and has a zero discharge [14]. MEMDIS (2003-2006) and SMADES (2003-2006) projects concerns with the development of innovative autonomous solar membrane distillation plants.[15]

From the literature is arising that a large percentage of the RES desalination installations for seawater and brackish water desalination owns to the combination of photovoltaic systems with RO. As it is shown from Figure 3, an also famous combination is the solar thermal distillation followed by Wind RO.

The unit water cost from the already installed RES desalination plants ranges from around 1€/m^3 up to 10€/m^3. For example for Lampedusa PV RO plant, during its autonomous operation, a unit water cost of around 6.5 ECU/m^3 (1996) has been reported while for the autonomous PV RO plant in Pozo Izquierdo, Gran Canaria (DESSOL project) a unit water cost of 9€/m^3 (2000) has been reported. Regarding wind RO applications, for wind RO plant in Pozo Izquierdo (AEROGEDESA Project) a unit water cost of 3-5€/m^3 has been considered [14]. For the geothermal MED plant in Kimolos island, Greece a unit water cost of around 0.8 €/m^3 has been estimated [10].

By comparison, the costs to produce fresh water from ground and/or surface water in western countries range from \$0.40 to \$0.75/m^3 before distribution (Global Water Report, 2001). In Western Australia the water cost (including distribution) varies from \$0.45-0.61/m^3 in the Metropolitan area \$2.42/m^3 along the main water scheme pipeline and to more than \$6.05 in remote areas. Water price in Europe is on average \$0.70 with highest price in Germany (\$1.41), France (\$1.04) and UK (\$0.93), and lower price in Italy is \$0.35 [12].

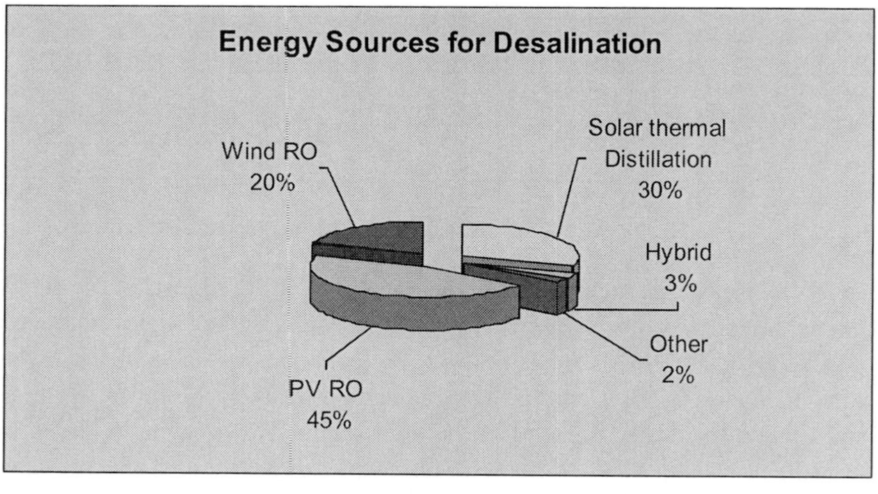

Figure 3. Energy Sources for Desalination

[15] AQUASOL project: Ref. no EVK1-CT-2001-00102; MEMDIS project: Ref no. NNE5/819/2001; SMADES project: Ref no. ICA3-CT-2002-10025.

CURRENT RES DESALINATION MARKET

The RES market has been steadily increasing over the last 20 years. Solar energy, by means of photovoltaic as well as solar heating, and wind technology, has had years of commercial acceptance all over the world and in a wide variety of applications. Geothermal technology market has a strong background and a continuously growing market, however is more or less site specific. Additionally, seawater desalination market is well established with a large potential of applications worldwide.

The market of the RES desalination systems is fairly new. In recent years several suppliers provide in the market modular RES desalination solutions, most of them of standard and large sizes. Two German companies work with Wind RO combination and provide autonomous and/or grid connected Wind RO systems in the range of 175 m^3/day to 2000 m^3/day. Several other companies seem to be ready for their entering in the market with large scale solar distillation systems and Wind Mechanical Vapor Compression seawater desalination units. According to the market suppliers several large pilot projects are planning to be installed in the near term in Tan Tan city of Morocco, in the United Arab Emirates and in Aqaba, Jordan. The expected increase on the RES desalination market is expected to raise the competitiveness and reduce the cost of such systems [15].

CONCLUSIONS

Keeping in mind the climate protection targets and strong environmental concerns, future water desalination around the world should be increasingly powered by solar, wind and other clean natural resources. Although, significant experience on the RES Desalination and successful applications exist, there is much room for improvement by a technical and economical point of view.

Picture 1. 900kW W/T, CRES, Keratea

A lot of work has been done on a theoretical basis but not on a real basis. Real installations of medium and large size, operating under hard climate conditions, by non experienced staff, local market limits, etc. will improve the efficiency and reliability of these systems. The next step is the increase of the RES desalination market potential and the commercial acceptance. The achievement of the above issues will reduce the cost and will increase the accessibility of RES desalination plants.

Picture 2. 4kWp PV plant, CRES, Keratea

Picture 3. 130 lt/hr RO unit, CRES, Keratea [11]

Picture 4. Off-shore Wind RO unit, Greece

Picture 5. The well and geothermal pump of the plant in Kimolos island, Greece

Picture 6. 80 m^3/day MED in Kimolos island, Greece [10]

Picture 7. The Wind Park and the desalination domes in Gran Canaria

Picture 8. VC plant - Plate Heat exchangers

Picture 9. 500kW W/T in Syros Island, Greece

Picture 10. RO plant in Syros island, Greece [10]

Picture 11. MED desalination unit in PSA, Almeria

Picture 12. Solar collectors filed layout, PSA Almeria [3]

REFERENCES

[1] Baltas, P., Tzen, E. & Perrakis, K. et al, (1998). *Desalination Guide Using Renewable Energy Sources,* THERMIE Programme, E.U, *CRES, Greece*, ISBN 960-90557-5-3.

[2] Morris, R. (1999). Renewable Energy Powered Desalination Systems in the Mediterranean Region" *United Nations Educational Scientific and Cultural Organization*, July.

[3] Report on the Status of Autonomous Desalination Units Based on Renewable Energy Systems, Co-ordination Action for Autonomous Desalination Units based on Renewable Energy Systems, *ADU RES Project*, INCO-CT-2004-509093, October 2005.

[4] *Study: Water Desalination Worldwide for seawater and brackish water 2006-2020-2015*, Helmut Kaiser Consultancy, 2007.

[5] Wangnick K. (2004). *IDA Worldwide Desalting Plants Inventory, Report No 18*, June 2004.

[6] Lattemann, S. & Hopner Th. (2008). Environmental Impact and Impact Assessment of Seawater Desalination, *Desalination*, *220*, 1-15.

[7] Papapetrou, M., Epp, Chr. & Tzen, E. (2007). Autonomous Desalination Units Based on Renewable Energy Systems-A Review of Representative Installation Worldwide, Rizzuti L. et al, *Solar Desalination for the 21st Century,* Springer, ISBN 978-1-4020-5506-5, 343-353.

[8] Tzen, E., Theofilloyianakos, D. & Kologgios, Z. (2008). Autonomous Reverse Osmosis Units Driven by RES, Experiences and Lessons Learned, *Desalination*, *221*, 29-36.

[9] Augsten, E. (2007). *Desalination by Means of Wind and Sun, Sun and Wind Energy, Vol. 2*, ISSN 1861-2741 74714, 32-36.

[10] Tzen, E. & Morris, R. (2003). Renewable Energy Sources for Desalination, *Solar Energy*, *75*, 375-379.

[11] Barron, O. (2006). *Desalination Options and their Possible Implementation in Western Australia: Potential Role for CSIRO Land and Water,* CSIRO National Research Flagship Initiative.

[12] www.adira.info

[13] www.adu-res.org

[14] www.enercon.de, www.synliftsystems.de, www.kerenergien.de

Evaluators Name & Title: Dr Konstantinos Perrakis, Electrical Engineer

INDEX

A

D

F

G

H

I

J

K

L

M

N

O

P

Q

R

T

U

V

W

Y

Z